S0-BSV-981

VNR Metric Handbook

Foreword

The United Kingdom is presently going through the trials of conversion to the metric system, particularly in the areas of monetary exchange and dimensional control in the design of buildings. Advantages to both are obvious—especially in international exchange and other relationships with continental Europe. To help simplify the design/dimensioning process in the architectural field, the following 'graphic standards' in the metric system have been compiled and authored by Leslie Fairweather, ARIBA, and Jan A. Sliwa, DIPIng, DipArch, ARIBA.

Since the United States has long been on the metric system in its currency, implementation of this method in office practice should not be as difficult for the stateside architect. Far more demanding, however, will be the problems of the manufacturing community to adjust to the new system in terms of output. For an interim period, manufacturers will have to alter existing machinery settings and eventually, in many instances, new production equipment may be required. Once the nation does embark on the metric system such familiar references as 2×4's and 4×8 panels will evaporate from the professional's vocabulary. For an indefinite period preceding the changeover, professional journals and technical literature should dimension details in feet and inches as well as in metric data. Complete acceptance could come within the next decade.

As many American architects are already preparing drawings and specifications for projects to be erected in countries where metrication has long been the practice, this handbook will have immediate application. For all other practitioners it should also have great interest and value.

Burton H. Holmes
Executive Editor
Progressive Architecture

Contents

The change to the metric system

Basic metric data
A General

B Building types

C Environment and structure

References

Appendix

Van Nostrand Reinhold Company

Authors

This handbook has been written, compiled and edited by:

JAN A. SLIWA DipIng DipArch RIBA; head of Research and Development Unit, Oxford Regional Hospital Board; member of the RIBA Professional Services Board; RIBA representative on the Building Divisional Council of BSI; and Architects' Journal guest editor for metric and related topics.

LESLIE FAIRWEATHER RIBA; Technical editor of the Architects' Journal.

Jan Sliwa   *Leslie Fairweather*

Preface to second edition

The handbook was first published as three special issues of the Architects' Journal in March 1968.
At that time there were many areas of confusion and doubt among those responsible for the change to metric, including the notorious controversy over the use of comma or point as the decimal marker. The AJ, in common with a number of official bodies and Ministries, followed the current BSI ruling and adopted the comma. This decision was later rescinded and the decimal point was then officially approved. This new edition of the handbook incorporates all decisions and official recommendations and standards, including use of the decimal point, issued since the first edition.

Five new sections have been added and the remainder have been enlarged and brought up to date. The conversion factors and tables are now considerably expanded and are included at the end of this handbook for convenience of reference.

Author: Fairweather/Sliwa
Title: THE VNR METRIC HANDBOOK
Code Number: G 234–000–8

Contributors

Considerable help was given in the preparation of particular sections of the handbook, which is gratefully acknowledged.

Section 6 Decimal currency
ANN RATHBONE MA BArch RIBA; formerly Assistant technical editor, Architects' Journal

Section 8 Anthropometric data
SELWYN GOLDSMITH MA RIBA; formerly Assistant technical editor, Architects' Journal

Section 16 Farm buildings
JOHN B WELLER DipArch RIBA; Architect in private practice

Section 20 Theatres
RODERICK HAM RIBA AADip; Architect in private practice and member of the Association of British Theatre Technicians

Section 22 Sports and swimming
G. A. PERRIN RIBA DipTP AMPTI; Consultant architect to the National Playing Fields Association and head of research group, The Polytechnic, Regent Street, London
REX SAVIDGE DipArch, DipTR MSIA RIBA; Architect in private practice
WEST GERMAN TECHNICAL ADVISORY BUREAU, Cologne

Section 23 Educational buildings
ROGER FITZSIMMONS; Assistant partner with Peter Falconer and Partners, responsible for schools

Section 27 Heating, thermal insulation and condensation
STEVEN V. SZOKOLAY MArch (L'pool) RIBA; Senior lecturer in environmental sciences, The Polytechnic, Regent Street, London

Section 28 Lighting
PETER JAY MA FIES; Electrical consultant in private practice

Section 30 Structural design
DAVID J. ADLER BSc DIC CEng AMICE; Civil engineer, Building Design Partnership

Section 31 Selective bibliography
PATRICIA CALDERHEAD; formerly librarian, Architects' Journal

The editors acknowledge help given by:

The Concrete Society
Institution of Structural Engineers
British Pre-cast Concrete Federation
Cement Makers' Federation
Brick Development Association
Timber Research and Development Association
Timber Trade Federation
Aluminium Federation
Copper Development Association
Lead Development Association
Zinc Development Association
Gypsum Plasterboard Development Association
British Gypsum Ltd
Fibre Building Board Development Organisation
Wood Wool Slab Manufacturers' Association
Pilkington Bros Ltd
Metal Windows Federation of Great Britain
British Woodwork Manufacturers' Association

1 Introduction to metric

Contents

1 The handbook

The purpose of this handbook is to offer those engaged in the planning, design and construction of buildings a grounding in the principles of the metric system*, with a reasonably comprehensive spread of basic data presented in metric terms. It forms part of a larger metric service offered to architects by the Architects' Journal in which advice and information on all matters relating to the metric change are regularly featured.

The first edition of this handbook (March 1969) was published in response to the British Standards Institution document *Change to the metric system in the construction industry—questionnaire* (PD 5767:1966), which includes AJ information sheets and design guides in the proposed list of essential reference publications of an 'industrial' nature which would require metrication under item 3(b) of the BSI programme for the metric change (see table 1 p4). Because of uncertain knowledge about the dimensional co-ordination principle of controlling dimensions at that time it was difficult in the first edition to do more than offer generalised data and advice. However, the emerging picture is now clear enough to enable publication of this second edition of the handbook in which firmer guidance and more consistent data have been possible.

1.1 Layout

The first part of the book, under the general title *The change to the metric system*, discusses some of the implications and difficulties of the metric change before giving a description and explanation of the system itself in section 2. Two further sections in this part deal respectively with decimal currency and the work of public sector and other official bodies involved in the change.

The bulk of the book is concerned with presenting *basic metric data* for a wide range of building types and for environmental and structural design. The book is completed by a selective bibliography and a selection of common conversion factors and tables.

1.2 For and against the change

It is not intended to argue for or against a changeover to the metric system but to attempt a balanced picture of the issues involved. The decision has been made: it is time to start acquiring metric knowledge. Preparation of the first sets of metric plans started in January 1969 and what is urgently required now is basic design information to familiarise architects with the new system.

1.3 The new dimensional environment

The use of metric scales and measuring instruments should offer no problems, but the process of acquiring metric data in order to build up a mental picture of the new dimensional environment will take time and can be achieved only by a gradual assimilation of the new sets of values.

The breakaway from the foot/inch convention will perhaps not come easily to those with a deep-rooted attachment to the duodecimal system. All that is going to happen, however, is the change from one convention of measurement to another, to gain simplicity of operation and standardisation in line with 85 per cent of the world's population. A practical and emotionally neutral atmosphere will help the architect to approach this new system with a more responsive mind, thus making the learning process easier.

2 Historical background

The scientific evolution of the eighteenth and nineteenth centuries, and the industrial revolution of the nineteenth century, accelerated the movement towards unification of existing systems of weights and measures—all by name apparently interchangeable but in practice unrelated.

Academic efforts to introduce the decimal system of measurement began in the sixteenth century sponsored by astronomers and physicists. The search for definable and stable standards of weights and measures culminated in 1790 in France in the proposal to establish a metre as the new decimal unit of length expressed in geodetic terms and defined as the length of a simple pendulum with a swing of one second at sea level on latitude 45deg.

Although adopted by the French in the same year its final official recognition did not take place in France until 1840. In the next stage of evolution the metre, while still relating to geodetic factors, became associated with a more constant and physically measurable standard. The metre was defined as the distance between two points or lines and marked accordingly on a metal bar of platinum and iridium (90 per cent Pt and 10 per cent Ir)—the length of the marked unit representing one ten-millionth part of the earth's meridional quadrant.

2.1 Final definition of a metre

The growth of science and technology necessitated further refinements in the basic formula of a metre and the final stage of evolution brought about replacement of the standard metal bar as an international prototype unit by a new definition of a metre in which the physical determination of the wavelength of light plays a dominant part. The most recent pronouncement in 1960 states that the metre is equal to 1 650 763·73 wavelengths in vacuum of the orange radiation $(2p_{10} - 5d_5)$ of a krypton atom of mass 86.

The importance of creating alternative means of determining the metre by the formula of light wavelengths lies in the fact that the natural standard of length can be reproduced independently anywhere in the world without having to refer to the international prototype metre in Sèvres in France, where the platinum-iridium metal bar is kept.

---

* Although the change to the use of metric is, strictly, a change to the use of SI units (Système International d'Unités), the term 'metric' has been used in this handbook as its meaning is commonly understood and most official and other publications employ the term

2.2 Architecture not radically affected

The mathematical ratios and formulae of the metric system form a background quite different from that of the anthropometric relationship of the foot/inch system, but in spite of the differences it is unlikely that architecture will be greatly affected by the changeover. The judicious use of proportions, ratios and dimensionally co-ordinated sizes available within the metric system will finally help people to understand the simplicity of operations it offers.

3 Implications and opportunities

The cost of the undertaking is directly proportional to the smoothness of the changeover, and depends on a balanced understanding by all of the issues involved.

An exaggerated view of the difficulties that may occur, shrouded by a veil of mystery of the unknown, will undoubtedly put a brake on a rational approach to metric education. Equally, a superficial attitude of hurried activity will breed contempt through overfamiliarity. In the first case the architect will leave education until too late, expecting others to solve all the problems for him: each will wait for the other to start and the result will be a last-minute rush. In the second case spectacular results may be mistaken for the quality and depth of knowledge that will be required in the process of learning.

The importance of the metric change lies not so much in the physical aspect of having to acquire and assimilate a new system of measurements, as in the *implications* of the changes that it may bring about.

3.1 Importance of establishing user requirements

Unlike its development on the Continent, the metric system is coming to Britain in an age of high technological development and in an era of design influenced by the collective demands of consumers, and not solely by a crafts tradition or the manufacturer's idea of standardisation.

The hitherto manufacturer-orientated standards will have to be revised during the metric changeover to take into account first and foremost users' requirements and then, and only then, the industrial problems of production.

Questions to be asked will include not only those relating to the previously accepted specifications which were required to satisfy industry, statutory undertakings and the distributor, but also the searching questions which must be put by the designer in order to design a product fit for the purpose. The designer anxious to link the future of the production industry with the metric changeover will now put questions of a different order: that of a consumer. They will deal, in addition to the aesthetics of the design, with environmental planning, anthropometrics, economy in terms of initial and maintenance cost, obsolescence and replacement.

3.2 User requirements check list

The questions to be asked by every architect involved in the metric change will follow the sequence of a check list. He must establish:

1 What changes there have been in the pattern of human behaviour since the original basis for the design standard was laid down;

2 What changes have taken place since then in the social, economic and physical condition of the human being;

3 What the effect will be of the introduction of controlled environmental conditions on the new metric design;

4 What is the best functional metric dimension (size) for the new design;

5 Will the design need to be dimensionally co-ordinated with other units;

6 What the overall dimensional pattern of the future is to be from the user's point of view;

7 To what range of controlled dimensions the product should conform;

8 What range of modular sizes is required;

9 What the best material will be to achieve the desired function honestly and economically;

10 What labour will be required in production, and for which market it is intended;

11 How much it will cost;

12 What is to be the expected life of the product;

13 Will it satisfy the appropriate tests.

Failure to establish these factors during the course of the metric change may result in a straightforward translation of dimensions and standards—a process which could have results disastrous for designer, producer and consumer alike.

3.3 What the architect can do

A great deal of the work in asking these questions will be outside the province of an individual architect and no doubt they will be considered by the functional panels of the BSI committees representing the manufacturers and the public and private sectors of the industry. This is not to say however that an architect can remain inactive—he must quickly begin to assimilate the new metric values in the light of the check list above, constantly keeping up to date on the co-ordinating data published by authoritative sources.

4 Programme of metric change

The official programme for the metric change is contained in two BSI documents:

PD 6030: 1967 *Programme for the change to the metric system in the construction industry* (see chart I p4)

PD 6249: 1967 *Dimensional co-ordination in building: Estimate of timing for BSI work* (see chart II p5,6)

The chart of the programme (table I) and the official announcement of the Ministry of Housing and Local Government (Circular 1/68 *Metrication of house building*, January 1968) confirm that the public sector of the building industry and anybody engaged in design for the public sector will between 1st January 1969 and 31st December 1971 switch over to drawings and documents in metric terms for new contracts. This means that local authorities will be able to submit metric plans at layout stage to the ministry for approval after 1st January 1969, and at tender stage after 1st January 1970. The contractors will change to construction based on metric drawings for new contracts between 1st January 1970 and 31st December 1972.

In Scotland, the Scottish Development Department circular 27/68 is the corresponding document to the Ministry of Housing and Welsh Office circular 1/68. The Scottish circular differs from the Ministry of Housing circular in reference to housing standards.

The time factor is therefore critical if the stages of the changeover are to be fulfilled within the prescribed period. The private sector while not duty bound officially to start metric drawings on 1st January, 1969 would be well advised to follow suit and synchronise its changeover with the public sector.

4.1 Manufacture of metric components

The most difficult time of the transition period will be caused by the manufacturers. Early decisions on sizing of metric components is absolutely vital to the manufacturer whose production line is affected by delays and debates. To help him with the main dimensional framework the ministries, through their Sub-committee for Components Co-ordination, set up a separate interdepartmental Component Co-ordination Group (CCG) under R. Grunberg to unify the dimensional requirements of the public sector. In a comparatively short time, the CCG collated data supplied by the Department of Education and Science, Ministry of Housing and Local Government, Ministry of Public Building and Works and Scottish Development Department into a set of documents called DC (Dimensional Coordination) Statements issued through the Ministry of Public Building and Works. These documents form the basis of the key British Standard on dimensional co-ordination, BS 4330: 1968 (metric units) *Recommendations for the co-ordination of dimensions in building: Controlling dimensions.*

It is important to note that while the public sector has issued statements in which it gives definite guidance as to the controlling dimensions, it does not concern itself with the establishment of actual sizes of components. It must be firmly understood that establishment of dimensional ranges of components, work sizes, standards of thicknesses and sizing will belong to various technical committees of the BSI. The BSI role as a link between public and private sector is underlined by its nomination by the Government to act also as the major organisation responsible for co-ordination of the metric changeover programme.

4.2 Building regulations
The first stage in the metrication of the Building Regulations, as indicated on the Programme Chart under item 3(a), has been complied with by the Ministry of Housing and Local Government by publishing a straightforward translation of present imperial weights and measures values into their exact metric equivalents: *Public Health Act 1961 The Building Regulations 1965: Metric Equivalents of Dimensions.* The figures are given purely for information and as such did not have to go through any parliamentary procedure.

A full explanation of the procedures to be adopted in the metrication of Building Regulations and other statutory documents, and the implications for architects, is given on p8.

4.3 Programme of change for architect's offices
A network analysis of the programme to guide offices in their own application of PD 6030 is available as PD 6421 *Network illustrating requirements for commencement of construction using dimensionally co-ordinated products.*

Each office should use these two documents in planning its own programme of change. Note especially that architects are *not necessarily* expected to start all new projects in metric from now on unless conditions are propitious. The size, type and suitability of project, availability of metric sized products and components, preparedness of the architect, his staff and other members of the design and construction team, and the financial effect on the client are some of the factors to be considered before deciding to go metric.

4.4 Guidance from BSI
Current guidance from BSI helps to clarify this question of timing and choosing the most suitable projects for metric working. In discussing metric contracts the BSI statement reads:

'Whereas PD 6030 clearly envisaged a preponderance of early metric projects designed to incorporate mainly foot/inch components but described in metric equivalent terms, it is now felt that unless these projects can be related to BS 4330 *Controlling dimensions*, they should be actively discouraged. This is not to say that foot/inch components should not be used if they can be made to fit within the metric dimensional framework set out in BS 4330. This advice, if widely followed, should have the effect of producing a fairly gradual start to metrication of projects from which feedback of experiences might be made to have worthwhile value. Time should allow costs yardsticks to be reviewed more meaningfully and product manufacturers would get a chance to judge more accurately the potential markets for their newly-sized components. The bulk of the change to metric project design would therefore be concentrated into the latter half of the period 1969-1971 when more newly-sized components should be available.'

'In offering this advice BSI is aware that it puts even greater responsibility onto those with large scale procurement programmes since it is expected that it is from these programmes that the production of new products should be initially financed. Given effective feedback to industry, the incorporation of newly-sized products in, say, a large Government contract should allow consequential specification of the same products in smaller private contracts, thus providing the manufacturer with continuity of production and the specifier with assurances of suitability and availability.'

'It is essential that the designer of the smaller project takes steps to verify the availability of the components he is specifying and to help him in this, MOPBW and RIBA are to collaborate in the provision of a feedback service on just what new components are being produced for large procurement programmes and in what quantities'.

5 Suggested office procedure

The influx of programmed and unprogrammed metric data, and the necessity to acquire a metric 'dictionary', will put a strain on individual practices of architects, QSS, engineers and general contractors. It is suggested that in order to avoid multiplication of effort, every office delegates one group of professional men within the office or, in the case of a small office, an individual, to deal exclusively with the metric changeover. Collation of relevant material, statutory or otherwise, should be confined to one section of the office. The same group would then be responsible for preparing the first metric drawings, tenders and contracts, and for guiding others within the office on metric procedure.

The important issues to be studied by the group will include phasing of long term schemes according to the BSI programme, as well as checking the rate of availability of metric components. Finally the group will ensure that no dual dimensioning is entered on drawings and that all metric activities are proceeding with the minimum amount of translation. Uniformity of approach, even within one office, is of paramount importance to the success of the metric changeover.

Table 1 BSI programme chart for the change to metric (from PD 6030)

Date for the change to decimal currency, 15 February 1971

Key: ||||| Period for preliminary preparations ▮ Period during which the bulk of the change will be taking place ⋮⋮⋮ Period during which residual changes will probably continue

#		1966	1967	1968	1969	1970	1971	1972	1973	1974
1	Time taken to produce the programme	▮	Published February 1967							
2	Preparatory studies									
	(a) Time taken for BSI to produce its construction industry guide to the use of the metric system	▮	Published February 1967							
	(b) Time required for BSI to produce key dimensional recommendations based on user studies	|||||	▮	⋮⋮⋮						
3	Essential reference publications									
	(a) Time required to make available in metric terms essential reference publications of an *official* nature	|||||	▮	▮	⋮⋮⋮					
	(b) Time required to make available in metric terms essential reference publications of an *industrial* nature	|||||	▮	▮	⋮⋮⋮					
4	Products for which dimensional co-ordination is essential									
	(a) Time required for manufacturers to provide technical information in metric terms for their products *as they are now produced*		▮	▮						
	(b) Time required for BSI to produce metric dimensional recommendations and British Standards for these products	|||||	|||||	▮	▮	▮	▮	⋮⋮⋮		
	(c) Time required for manufacturers to change to full production of new metric dimensionally co-ordinated products				|||||	▮	▮	▮	⋮⋮⋮	
5	Products which are dimensionally related to those in item 4									
	(a) Time required for manufacturers to provide technical information in metric terms for their products *as they are now produced*		▮	▮						
	(b) Time required for BSI to produce metric dimensional recommendations and British Standards for these products	|||||	|||||	▮	▮	▮	▮	▮		
	(c) Time required for manufacturers to change to full production of new metric dimensionally co-ordinated products				|||||	▮	▮	▮	⋮⋮⋮	
6	Products which are not dimensionally related to those in item 4									
	(a) Time required for manufacturers to provide technical information in metric terms for their products *as they are now produced*		▮	▮						
	(b) Time required for BSI to produce metric dimensional recommendations and British Standards for these products		|||||	▮	▮	▮	▮	▮		
	(c) Time required for manufacturers to change to full production of new metric dimensionally co-ordinated products				|||||	▮	▮	▮	⋮⋮⋮	
7	Products which are required to have only sensible metric sizes and values									
	(a) Time required for manufacturers to provide technical information in metric terms for their products *as they are now produced*		▮	▮						
	(b) Time required for BSI to produce metric standards for these products		|||||	|||||	▮	▮	▮	▮		
	(c) Time required for manufacturers to change to full production of their products to the new metric standards			|||||	|||||	▮	▮	▮	⋮⋮⋮	
8	Time required for manufacturers to produce all measuring instruments for the construction industry calibrated in metric terms		|||||	▮	⋮⋮⋮					
9	Time required for designers and quantity surveyors to change to production of drawings and documents in metric terms *for all new contracts*					▮	▮	▮	⋮⋮⋮	
10	Time required for main contractors and subcontractors to change to construction based on metric drawings and documents produced under item 9						▮	▮	⋮⋮⋮	

This vertical bar shows the date by which the change to metric should be effectively complete ⟶

Table II Dimensional co-ordination in building: estimate of timing for BSI work (from PD 6249)

Functional group	Subgroup	Example items	1967	1968	1969	1970	1971	1972
1 Structure	Vertical element		▒▒▒▒▒					
		Columns; walls		███	▏▏▏▏▏▏▏▏			
		Sills; lintels			███	▏▏▏▏▏▏▏		
		Wall ties				███	▏▏▏▏▏▏	
	Horizontal element		▒▒▒▒▒					
		Beams; roofs; floors		███	▏▏▏▏▏▏▏			
		Foundations; toppings			███	▏▏▏▏▏▏▏		
	Combined elements	Portal frames; pitched trusses			███	▏▏▏▏▏▏		
		Staircases; ramps				███ ▏▏▏▏▏		
2 External envelope	Vertical element		▒▒▒▒▒					
		Windows; door sets; cladding		███	▏▏▏▏▏▏▏			
		Insulation; facing slabs			███	▏▏▏▏▏▏▏		
		Subsidiary items			███	▏▏▏▏▏▏		
	Horizontal element	Sheet roof cladding			███	▏▏▏▏▏		
		Roof insulation				███ ▏▏▏▏▏		
		Subsidiary items				███	▏▏▏▏▏▏	
3 Internal sub-division	Vertical element		▒▒▒▒▒					
		Partitions; door sets		███	▏▏▏▏▏▏▏			
		Balustrades; grilles			███	▏▏▏▏▏▏		
		Subsidiary items			███	▏▏▏▏▏▏		
	Horizontal element		▒▒▒▒▒					
		Ceilings		███	▏▏▏▏▏▏▏			
		Floor finishes				███ ▏▏▏▏		
		Subsidiary items				███	▏▏▏▏▏▏	
4 Services and drainage	Supply systems (heating and ventilation, electrical, water, gas)	Stacks; trunking; tanks; heaters; generating units; radiators; boilers; lighting fittings				███ ▏▏▏▏▏		
		Subsidiary items				███	▏▏▏▏▏▏▏	
	Disposal systems (soil, rainwater, refuse)	Manholes; stacks; gulleys; chutes; turntables; ducts; large containers				███ ▏▏▏▏▏		
		Subsidiary items				███	▏▏▏▏▏▏▏	

Table II Dimensional co-ordination in building: estimate of timing for BSI work *continued*

Functional group	Subgroup	Example items	1967	1968	1969	1970	1971	1972
	Communications and protection systems	Distribution boards and boxes; hydrant boxes; hose reel cabinets; sprinkler grids						
		Subsidiary items						
	Mechanical vertical circulation	Passenger and goods lifts; hoists; escalators						
		Subsidiary items						
5 Fixtures, furniture and equipment	Kitchen equipment, storage equipment, furniture	Cooking units; sink units; racks; refrigerators; cupboard spaces; built-in units						
		Subsidiary items						
	Sanitary and laundry equipment	Urinals; baths; cisterns; sinks; wash basins; drying appliances; incinerators						
		Subsidiary items						
	Miscellaneous equipment	Garage equipment; bicycle racks						
		Subsidiary items						
6 External works	(Provisional time-bar pending further studies)							

Key

▦▦▦▦▦ Time for producing controlling dimensions
███████ Time for producing and publishing dimensional requirements in and between functional groups
║║║║║║║║ Time for metrication of relevant British Standards

6 Positive and negative features of the change

The metric changeover should be devoid of any partisan issues setting the imperial system against the metric system if the process of change is to be done efficiently. It is however salutary to evaluate the positive and negative points of both systems.

On the positive side, the metric system gives simplicity in use and ease in learning. It also offers a logical interlock between linear measures, weights, cubic and volume measures—all based on the decimal unit—which dispenses with the fifty-three varieties and combinations of the imperial system and substitutes only three: metre, litre and kilogramme. In joining the 85 per cent of the world's population already using the metric system, the UK will be able to apply its production and export facilities to metric countries: at present it exports 55 per cent of its output to metric using countries. It is important to note that a great many of the export products go to metric but non-Common Market countries, therefore however important the issue of joining the Common Market may be, it should not be quoted as the major reason for going metric.

6.1 Technical publications
The UK's lead in the technological field is acknowledged throughout the world, yet the vast majority of our technical books and publications are set in imperial measures and are consequently of little value to readers in metric countries, even to those who understand English.

The metric changeover, with all new books published with metric data, will open new markets for British technical books and publications. But republication of these books must, in many cases, await the arrival of new standards and controlling dimensions before being reset in metric terms. A bonus for architects will be easier understanding of foreign books and journals—at least of the drawings and tables.

6.2 Foreign markets
In the building field, the change to metric will enable some of the exporting manufacturers in this country to streamline their production runs to just one—in metric. Others, not being in the export field because of dimensional non-compatibility with metric countries, will now begin to look for other markets abroad. Finally, the interchange of technical students and professionals between Britain and other metric countries will bring better understanding of other people's problems and achievements.

6.3 Simplicity of the metric system
The simplicity of the metric decimal system may cause the

user some worry at first as being *too* simple. The habit of going through the various multiplications or divisions of 3, 9, 12 of the imperial system to arrive at the required figure, will leave the user of the metric system puzzled or even cheated by *not* having to perform all those calculations from now on.

There will be an understandable feeling of having missed part of the calculations or of not having done something properly. This may not last long, but it should be noted in order to prepare for a period of insecurity, which only constant practice will overcome. Calculations involving foot unit, yard, square yard, cubic yd, which result in a known progression of values, must now be un-learned. A growing familiarity with the metric decimal progression should however soon dispel anxiety.

6.4 Disadvantages of the metric system

To put the changeover in the right perspective the negative aspects of the metric system must also be realised. Metric values bear no relation to the human stature and as such are impersonal, mathematical expressions of an entirely different concept of values. The foot/inch system incorporates a convenient method of identifying the incremental units of a foot, and one 'feels' this identity in reading the dimensions or values. For example, 2ft 4in as a value has certain known and definite characteristics, while 70 cm or 700 mm is an impersonal unit of a decimal system. The anonymity of the metric system will initially cause some difficulty in remembering certain values.

Compared with a metric decimal system, the ability to divide a foot unit by 1, 2, 3, 4, 6 and 12 can be viewed as either a disadvantage or an advantage. On the one hand the duodecimal system produces an irregular pattern and is difficult to rationalise within a dimensional co-ordination framework; it is also cumbersome and irrational in use. On the other hand it offers great flexibility. And irrespective of whether any move towards multiplicity of components is right or not, the lesser flexibility of the metric system is reckoned by some to be a questionable virtue.

6.5 The next stage

Although not directly the concern of the architect the next difficulty in the application of the metric system in Britain will lie in the sphere of activities connected with heat transfer, refrigeration, air-conditioning, aeronautics and the oil industry—all of which rely heavily on the use of data originating in the US and set out in imperial units.

7 Some practical problems

the primary obstacle facing every architect at the beginning of a metric design will be total lack of 'feel' of new sizes, spaces and values. The lack of confidence in his own dimensional judgment of environment will hit the architect quite forcefully. The only remedy will be to learn the new data as quickly as possible; this must be directly in the new units, without any translation.

There will be, however, a few problems connected with this learning process which must be identified so that the architect can be aware of what is involved.

7.1 The problem of dual data

The BSI programme allows, in its path towards a totally rationalised metric system, a period during which nothing else will be required from official or industrial publications but translation of data into equivalent metric figures. These dual data will thus appear alongside each other in statutory publications as well as in manufacturers' literature. Some publications have already appeared in this form.

It is to be hoped that the extent of this type of information will be kept to the absolute minimum, especially that of a statutory nature. The reason for this is quite simple. An architect learning his first dimensions and sizes of objects or products, his first heights or statutory distances, will no doubt memorise the very first metric figures given to him. Those figures unfortunately will be only temporary—but the human mind, being what it is, will tend to remember the first learned values. To avoid delay and confusion, it is vital that the most important basic data should appear in rationalised metric format at the beginning of the changeover period when everybody will be learning them and not at the end when wrong values would already have been acquired and memorised.

The importance of the right approach at the outset was realised early on by several responsible bodies. The Building Research Station announced that from November 1968 only metric units would be quoted in BRS publications and, where applicable, SI units would be used directly in its technical data. Experience has shown that prolonged use of both sets of units (imperial and metric) has little advantage. Similarly the Ministry of Housing and Local Government in their new Metric Design Bulletins, have given dimensions only in metric units, and this is to be ministry policy for future publications.

7.2 Manufacturers' literature

In fact this decision is a practical and common-sense approach to reality. The BSI programme allowed two years, 1967 and 1968, for manufacturers to provide technical information in metric terms for their products 'as they are now produced' (items 4(a), 5(a) 6(a) and 7(a) of the BSI programme chart—see table I). The economics of producing technical literature would suggest that it does not pay to publish material which would soon be obsolete.

The architect is therefore likely to obtain the bulk of his initial metric data in direct metric units—without the imperial system alongside.

7.3 Availability of products

Further difficulties which the architect will encounter in the first phase of his metric changeover, will lie in assessing what is available and what is not available in rationalised metric sizes. Rationalised metric sizes are sizes which have been metricised in relation to the dimensional co-ordination principle, to controlling dimensions and to technological developments which have taken place since the original standard was prepared.

The industry and BSI will have to indicate to the practising architect, qs and engineer a comprehensive and accurate timetable showing what metric products are to be available on the market and when they are to be available in plentiful supply. The exact metric dimensions of new products will have to be indicated to enable them to be specified when metric drawings are being prepared. Publications showing equivalent or rounded-off metric sizes will not necessarily mean that the material is available in its final rationalised metric form. The superficial rounding-off of the manufacturer's existing dimensions in metric terms without being backed by a concurrent change in production will be extremely dangerous, as it will cause confusion to the architect at the drawing board and to the general contractor on site. The earliest possible establishment of definite standards of overall thickness and sizes of sheet materials, glass, timber and so on will ease the task of the designer in the critical stage of his work.

7.4 Rounded and rationalised dimensions

The following example will show what the architect may be confronted with in the process of the changeover. The thickness of sheet material in its equivalent metric form will be published by manufacturers (according to items 4(a), 5(a) and 6(a) of the BSI programme chart) as $\frac{1}{8}$in = 3·17 mm, $\frac{1}{4}$in = 6·35 mm, $\frac{3}{8}$in = 9·52 mm, $\frac{1}{2}$in = 12·7 mm, $\frac{3}{4}$in = 19·05 mm, 1in = 25·4 mm.

In the process of translation some of these dimensions will be rounded off and will appear as $\frac{1}{8}$in = 3·2 mm, $\frac{1}{4}$in = 6·4 mm, $\frac{3}{8}$in = 9·5 mm, $\frac{1}{2}$in = 12·7 mm, $\frac{3}{4}$in = 19·1 mm or 19 mm, 1in = 25·4 mm. The final rationalised metric thicknesses of materials may take the form of increments of 3 mm, 6 mm, 10 mm, 12 mm and 25 mm. From this process it is obvious that rounding-off may not give the final answer and that a logical rationalisation must be achieved to arrive at the right size for the product. The fractional values involved in the examples quoted will hardly matter in scaled drawings as they are of minute thickness, but cumulatively they cannot be ignored.

7.5 Importance of the millimetre

The value of a millimetre should not be underestimated. Many architects imagine that one millimetre is simply the thickness of a pencil line. In fact 1·5 mm equals $\frac{1}{16}$in, or $\frac{1}{8}$in equals 3 mm–both dimensions important enough not to be ignored especially in assessing the dimensional values of tolerances and gaps for jointing solutions.

8 Conversion of statutory requirements

One of the most important features of the changeover is conversion of the Building Regulations into the metric system.

The basic form of the Building Regulations, as a statutory document, cannot be changed without the approval of Parliament. It is the only document that will indicate, when passed and put into full operation, that the metric system is part of the law of this country. Because, however, a great deal of revision work on British Standards and BS Codes of Practice will, according to the programme chart, go on until 1972 (tables I and II) it will be difficult to finalise the rationalised metric Building Regulations before the end of the changeover period. The work on dimensional co-ordination and controlling dimensions must precede the final issue of the metric Building Regulations as all the dimensional standards contained in this issue will have to coincide with the official governmental standard of controlling dimensions for public buildings and BS 4330. Furthermore, Building Regulations refer to codes and standards of the BSI, the Public Health Act 1936. the Factories Act 1961, weights and measures, pressures, sizes and so on, and all these products and standards must be rationalised (not just translated) in metric terms well before the final version of Building Regulations is metricised.

8.1 Parliamentary procedure

The metric change of the Regulations must therefore be phased to suit the overall progress of the changeover and to conform to parliamentary time and procedure. It is not possible to pass a single all-embracing Act of Parliament which will automatically amend all existing legislation, because the degree of accuracy of conversion of imperial to metric units required will vary between one case and another. The general intention is that there should be an enabling Act which will make it possible for Ministers to introduce the necessary amendments to the legislation which their Departments administer, by means of statutory orders. It is hoped that this enabling Bill will be on the Statute Book by 1971. Much existing legislation is already within the powers of Ministers to amend (Regulations, Orders, Statutory Instruments, etc). Earlier action will be taken in certain cases where this is necessary, since it will be the responsibility of Government Departments to ensure that the change to the metric system in any sector is not impeded by the absence of suitability modified legislation. The new legislation will probably also provide, in many cases for the first time, legal definitions of the metric units which are permitted for all purposes, in addition to providing schedules of legal conversion factors. The first stage of the metrication of the Building Regulations as indicated on the programme chart under item 3(a) has been complied with by the Ministry of Housing and Local Government by publishing straightforward translation of present imperial weights and measures values into their exact metric equivalents: *Public Health Act 1961 The Building Regulations 1965 Metric Equivalents of Dimensions*. These figures are given purely for information and as such did not have to go through the parliamentary procedure.

8.2 Scottish regulations

New space and heating standards for Scotland are to be introduced with the change to metric. They will be the same as those that apply in England and Wales. The new Scottish Development Department bulletin *Metric Space Standards*, outlines the standards of space, fittings and equipment that apply to metric house designs after 1st January 1969. The standards will not apply to housing designed in imperial measure.

New indicative cost tables, which relate to the new metric standards, have also been published.

Housing designed to the new standards will not in all cases comply with the requirements of part xv of the Building Standards (Scotland) Regulations 1963/67. Amended regulations are in course of preparation but it will take time before they become effective. In the meantime, arrangements for granting relaxations will be available under Section 4 of the Building (Scotland) Act 1959 to enable schemes to proceed to the new metric space standards.

8.3 Difficulties in the interim period

The difficulties will arise when, in the interim period, architects working in the metric system will try to relate the heights and spacings given in BS 4330 to those stated in the present Building Regulations. To avoid anomalies in presenting the numerical values it is stated by the MOHLG that in due course the imperial dimensions and measurements in the Building Regulations (and in other relevant Statutes and subordinate legislation) will be changed to metric values, which will normally be rounded values and not exact metric equivalents of the present imperial values. Proposals for such rounded metric values will be issued for comment before incorporation in the legislation.

In the meantime, where the present requirements of the Building Regulations prevent the use of a metric dimension which is otherwise considered desirable (eg to facilitate dimensional co-ordination in buildings) the Ministry would be prepared to consider the possibility of an earlier ad hoc amendment of the regulations.

The AJ *Metric Guide to the Building Regulations* offers the quickest route to operating the present values in metric terms.

9 References and sources

BRITISH STANDARDS INSTITUTION

BS 4330: 1968 (metric units) Recommendations for the co-ordination of dimensions in building: Controlling dimensions CI/sfB (F4j)

PD 5767: 1966 Change to the metric system in the construction industry—questionnaire CI/sfB (F7)

PD 6030: 1967 Programme for the change to the metric system in the construction industry CI/sfB (F7)

PD 6249: 1967 Dimensional co-ordination in building: Estimate of timing for BSI work CI/sfB (F4j)

PD 6421: 1968 Network illustrating requirements for commencement of construction using dimensionally co-ordinated products CI/sfB (F4j)

MINISTRY OF HOUSING AND LOCAL GOVERNMENT AND WELSH OFFICE Metrication of house building Circular 1/68 CI/sfB 81 (F7)

MINISTRY OF HOUSING AND LOCAL GOVERNMENT Public Health Act 1961: The Building Regulations 1965: Metric equivalents of dimensions CI/sfB (A3j) (F7)

SCOTTISH DEVELOPMENT DEPARTMENT Metric space standards Circular 27/68 CI/sfB (F7)

MINISTRY OF PUBLIC BUILDING AND WORKS Dimensional co-ordination for industrialised building:

DC4 Recommended vertical dimensions for education, health, housing, office and single-storey general purpose industrial buildings. 1967 CI/sfB (F4j)

DC5 Recommended horizontal dimensions for educational, health, housing, office and single-storey general purpose industrial buildings. 1967 CI/sfB (F4j)

DC6 Guidance on the application of recommended vertical and horizontal dimensions for educational, health, housing, office and single-storey general purpose industrial buildings 1967 CI/sfB (F4j)

DC7 Recommended intermediate vertical controlling dimensions for educational, health, housing and office buildings and guidance on their applications. 1967 CI/sfB (F4j)

DC8 Recommended dimensions of spaces allocated for selected components and assemblies used in educational, health, housing and office buildings. 1968 CI/sfB (F4j)

All published by HMSO

ARCHITECTURAL PRESS AJ Metric Guide to the Building Regulations. 1968 15s CI/sfB (A3j) (F7)

2 Basic metric system and SI units

Contents

1 Development of original metric system

The original metric system of measures based on a decimal system of units went through several refinements, with sponsors varying between astronomers, geodetic mathematicians, physicists and scientists. The progress of technology necessitated revision of the MKS system (metre, kilogramme, second) defined in 1900, and discoveries in the electro-magnetic field resulted in adding the fourth basic unit of ampère, the unit of electric current, to the existing system of units of mechanics. The MKSA (or Giorgi system) as the new metric system was known, lasted till 1960 when again the growth of thermodynamics, nuclear physics and electronic science caused further refinement of the metric system by adding the degree Kelvin as the unit of temperature and the candela as the unit of luminous intensity. This final set of six basic units became known as 'Système International d'Unités' known in all languages as SI units.

2 Basic SI units

The six basic SI units, which will most likely be the limit of the major units affecting the architect in day-to-day practice, are shown in table I.

Table I The six basic units of the SI system

Quantity	Name of unit	Unit symbol
Length	metre	m
Mass	kilogramme	kg
Time	second	s
Electric current	ampère	A
Thermodynamic temperature	degree Kelvin*	°K
Luminous intensity	candela	cd

* The degree Celcius (°C) will be used in practice, see para 6 p11.

Table II Derived SI units

Physical quantity	SI unit	Unit symbol
Force	newton	N = kg m/s²
Work, energy, quantity of heat	joule	J = Nm
Power	watt	W = J/s
Electric charge	coulomb	C = As
Electrical potential	volt	V = W/A
Electric capacitance	farad	F = A s/V
Electric resistance	ohm	Ω = V/A
Frequency	hertz*	Hz = cycle/s
Magnetic flux	weber	Wb = Vs
Magnetic flux density	tesla	T = Wb/m²
Inductance	henry	H = V s/A
Luminous flux	lumen	lm = cd sr†
Illumination	lux	lx = lm/m²

*The term 'cycle per second' (c/s) has been used in the UK
† sr is the symbol for steradian

Table III A range of multiples and submultiples of SI units

Factor	Multiples and submultiples	Prefix	Symbol
Thousand million	10⁹	giga	G
One million	10⁶	mega	M
One thousand	10³	kilo	k
One hundred	10²	hecto	h
Ten	10¹	deca	da
One tenth	10⁻¹	deci	d
One hundredth	10⁻²	centi	c
One thousandth	10⁻³	milli	m
One millionth	10⁻⁶	micro	u

Table IV Specific uses for selected units

Quantity	Unit symbol	Recommended use	Remarks
Length	km	Transportation	
	m	All uses	
	mm	Large scale detailing, structural calculations and small sections specification	Avoid mixing with values in metres
Mass*	kg	All uses	
	g	Routine analysis and testing	
Area	km² ha (hectare)	Real estate, town and country planning	1 ha = 10000m²
	m²	All uses	
	mm²	Structural calculations and small sections specification	
Volume	m³ mm³	Taking-off quantities and structural calculations	
	l (litre) ml	Volumes of liquids and routine analysis and testing	litre = dm³
Temperature	°C (°K)	Temperature scale	The Celsius scale (C°) will be used in practice
Angle	Rt L	Setting-out geometry	Right-angle
	deg or °	Surveying, etc	Also minutes and seconds
Pressure and stress	MN/m² (N/mm²) kN/m²	Structural analysis	10⁴ kgf/m² = 9·80665 MN/m²

* The metric tonne is not listed above in view of danger of confusion with the British ton during the changeover period but there is no other reason why it should not be used

3 Derived SI units

From the six basic SI units given in table I a number of other units have been derived which are necessary for the calculation of structural forces, electrical and thermal values and other related properties. These are given in table II.

4 Decimal multiples of SI units

A range of multiples and submultiples of SI units is given in table III. Specific uses for selected units, in order to eliminate unnecessary units, are given in table IV.

Table V gives a comprehensive list of SI units and a selection of recommended decimal multiples and sub-multiples of the SI units together with other units or other names of units which may be used (from PD 5686).

5 The unit of force

From the units listed in table II it will be seen that all future work on stress and pressure will contain a new metric unit of force: the newton. A great deal has been written about the newton and the newton/m² but the meaning of these terms is not always explained against the old conventions of pound force and pound force per square inch respectively.

5.1 Effect of gravity

Until the International Organization for Standardization (ISO) accepted the newton as a new unit of force, metric countries were using kilogramme or kilogramme-force units, while Britain was employing pounds or pound-force units. The problem in using these units lies in the fact that the kilogramme as a unit refers to a quantity of matter subject to a gravitational acceleration of 9·81 m/s². The use of such a unit in spheres where gravitational acceleration plays no part at all proved cumbersome and at times inconsistent; the introduction of the SI system cleared away this ambiguity. The newton as a new unit of force has *no* 9·81 m/s² encumbrance of gravitational force and its use will simplify the application of the newton for quantities such as pressure, stress, work and power.

5.2 Objections to the newton

Before SI units were accepted some metric countries introduced the term kilopond for the kilogramme-force unit. The newton supersedes both. There is however quite a body of opinion among the structural and civil engineering profession which does not favour use of the SI system in structural and civil engineering calculations. The objection is not centred on the use of the metric system as a whole but round the use of SI units, with particular hostility towards the newton as a unit of force: structural engineers would prefer the use of kilogramme-force (kgf) or kilopond. The argument, as far as it can be ascertained, revolves around the practicability of using newtons in calculation of stress, pressure and other derived values. It is maintained that introduction of the newton will cause figures in common use to be high, with many digits, and therefore more difficult to memorise and manipulate. In this, the structural and civil engineer would be worse off than the architects, qs or builder who, at the worst, could fall back on existing information from metric countries. The structural engineer cannot do this because the metric countries are still trying to introduce the new SI units among their engineers, and all *their* data is in the old metric units. There is as yet no international standardisation agreed upon multiples and submultiples of newtons required for structural calculations.

5.3 Use of multiple units

The structural engineers' real fear appears to be the physical problems involved in writing compound calculations resulting in large figures with a lack of confidence about which multiple or submultiple of the unit to place at the end. Perhaps their reluctance stems from uncertainty about the values involved, but one can understand their apprehension when looking at the following comparative figures:

1 lbf/in²	=	6895 N/m²
Pressure of 100 lbf/in²	=	7·04 kgf/cm²
	=	689·48 N/m² × 10³

The last figure can be expressed as 689 480 N/m² or in a more manageable way as 689·5 kN/m². The newton and the newton per square (especially the latter) are too small for practical purposes and it is generally understood that multiples of the basic units will be accepted for all practical purposes rather than the basic units themselves.

A slight preference has been expressed for writing multiples of pressure and stress units in the form N/mm² (newtons per square millimetre) rather than MN/m² (meganewtons per square metre). It is recommended that this practice should be adopted but regarded as subject to revision in the light of experience. This recommendation helps a great deal towards future rationalisation but the architect must be aware that it is by no means standardised. Use of kN/m² for load and N/mm² for stress has been endorsed by the Institution of Civil and Structural Engineers.

See also section 30 of this handbook, p160.

6 Units for heating, ventilating, electrical and mechanical engineering

The position of the services engineering professions is one of complete acceptance of the metric system, including the SI units. The IHVE prepared a report which concluded: 'Adoption of the SI system of units would help international relationships. Particularly it would help this industry with its sales to Europe and its association with international professional commercial bodies.' Finally in 'Conclusions and recommendations' it states:

'When a change to the metric system is necessary it is recommended that the heating and ventilating industry should support the adoption of the SI system. No system intermediate between SI and traditional systems should be contemplated'.

Recommendations of this nature, especially coming from professions so closely connected with the use of US technical data, is of great value to a rationalised approach to the changeover.

Acceptance by the services engineers of SI metric units for their future work will affect the architect only marginally.

6.1 Temperature

The degree Kelvin unit belongs to the group of six basic SI units as a quantitative unit of thermodynamic temperature. The units of Kelvin and Celsius (centigrade) temperature interval are identical. A temperature expressed in degrees Celsius is equal to the temperature expressed in degrees Kelvin less 273·15. What this means in simple language is that each grade, degree or interval of the Celsius scale is in magnitude the same as a degree Kelvin, but in relation to the zero degree Celsius, the zero point of the Kelvin scale slides down 273·15 degrees (point of absolute zero), making 0°K = —273·15°C or 0°C = +273·15°K.

For customary temperatures and temperature intervals, the degree Celsius (°C) will be used.

6.2 Pressure units

For calculations of atmospheric pressure, water pressure, steam pressure, force, energy and power the newton will become the basic unit. The significant changes will take place in the calculations of heat values. The familiar Btu unit and even the old metric kilocalorie (kcal) will disappear, giving way to the new unit of quantity of heat, the joule (J). The customary horse-power (hp) unit will change to kilowatts (kW). The thermal transmittance (U value) will become in SI units watt per square metre per degree celsius (W/m² deg c), instead of the familiar U = Btu/ft²h degF.

See also section 27 of this handbook, p147.

6.3 Sound units

Sound insulation values will not change through the introduction of the metric system with the exception of the unit of frequency which will become the hertz (HZ) unit instead of the British cycle per second (c/s) with 1 HZ = 1 cycle per secon (lc/s). The decibel unit (dB) used in acoustic and sound insulation calculations remains unaltered.

See also section 29 of this handbook, p159.

6.4 Lighting units

In lighting, the unit of illumination now called lumen per square foot will give way to the metric unit of lux (1 lux = 1 lm/m²).

See also section 28 of this handbook, p158.

7 References and sources

BRITISH STANDARDS INSTITUTION The use of SI units PD 5686 January 1969 CI/sfB (F7)

BRITISH STANDARDS INSTITUTION The use of the metric system in the construction industry PD 6031 2nd edition December 1968 CI/sfB (F7)

Table V List of SI units and a selection of recommended decimal multiples and sub-multiples of the SI units together with other units or other names of units which may be used (from PD 5686)

An asterisk against a unit means that the unit may be used in the UK but is not yet included in the ISO draft recommendation.

Item No in ISO/R31	Quantity	SI unit	Selection of recommended decimal multiples and sub-multiples of SI unit	Other decimal multiples and sub-multiples of SI unit	Other units or other names of units which may be used	Remarks
Space and time						
1–1.1	plane angle	rad (radian)	mrad μrad		degree (.. °), $1° = \dfrac{\pi}{180}$ rad minute (.. '). $1' = \dfrac{1°}{60}$ second (.. ''), $1'' = \dfrac{1'}{60}$	
1–2.1	solid angle	sr (steradian)				
1–3.1..7	length	m (metre)	km mm μm nm	dm cm	*International nautical mile (1 n mile = 1852 m)	
1–4.1	area	m²	km² mm²	dm² cm²	hectare (ha), 1 ha = 10^4 m² are (a), 1 a = 10^2 m²	
1–5.1	volume	m³	mm³	dm³ cm³	hectolitre (hl), 1 hl = 10^{-1} m² litre (l), 1 l = 10^{-3} m³ = 1 dm³ centilitre (cl), 1 cl = 10^{-5} m³ millilitre (ml), 1 ml = 10^{-6} m³ = 1 cm³	In 1964 the Conférence Générale des Poids et Mesures adopted the name litre (l) as the synonym for cubic decimetre (dm³) but discouraged the use of the name litre for precision measurements
1–6.1	time	s (second)	ks ms μs ns		day (d), 1 d = 24 h hour (h), 1 h = 60 min minute (min), 1 min = 60 s	Other units such as week, month and year are in common use
1–8.1	angular velocity	rad/s				
1–10.1	velocity	m/s			kilometre per hour (km/h) 1 km/h = $\dfrac{1}{3 \cdot 6}$ m/s *knot (kn) 1 kn = 1 n mile/h = 0·514 444 m/s	
Periodic and related phenomena						
2–3.1	frequency	Hz (hertz)	THz GHz MHz kHz			
2–3.2	rotational frequency	1/s			revolution per minute (rev/min) revolution per second (rev/s)	
Mechanics						
3–1.1	mass	kg (kilogramme)	Mg g mg μg		tonne (t), 1 t = 10^3 kg	The metric carat (1 metric carat = 2×10^{-4} kg) is used for commercial transactions in diamonds, fine pearls and precious stones
3–2.1	density (mass density)	kg/m³	Mg/m³	1 kg/dm³ = 1 g/cm³	1 t/m³ = 1 kg/l = 1 g/ml g/l	For litre (l) see item 1–5.1

Table V List of SI units and a selection of recommended decimal multiples and sub-multiples of the SI units together with other units or other names of units which may be used (from PD 5686) *continued*

An asterisk against a unit means that the unit may be used in the UK but is not yet included in the ISO draft recommendation.

Item No in ISO/R31	Quantity	SI unit	Selection of recommended decimal multiples and sub-multiples of SI unit	Other decimal multiples and sub-multiples of SI unit	Other units or other names of units which may be used	Remarks
3–5.1	momentum	kg m/s				
3–6.1	moment of momentum, angular momentum	kg m²/s				
3–7.1	moment of inertia	kg m²				
3–8.1	force	N (newton)	MN kN mN μN	daN		
3–10.1	moment of force	N m	MN m kN m μN m	daN m		
3–11.1 3–11.2	pressure and stress	N/m²	GN/m² MN/m kN/m² mN/m² μN/m²	daN/mm² N/mm² N/cm²	1 hbar = 10⁷ N/m² 1 bar = 10⁵ N/m² 1 mbar = 10² N/m² 1 μbar = 10⁻¹ N/m²	The hectobar (hbar) is used in certain fields in some countries. The name 'pascal' is given to the newton per square metre in certain countries
3–19.1	viscosity (dynamic)	N s/m²	mN s/m² * μN s/m²		centipoise (cP) 1 cP = 10⁻³ N s/m²	
3–20.1	kinematic viscosity	m²/s	mm²/s		centistokes (cSt) 1 cSt = 10⁻⁶ m²/s	
3–21.1	surface tension	N/m	mN/m			

3–22.1	energy, work	J (joule)	GJ
			
			MJ
			kJ
			
			mJ
					kilowatt hour (kW h)
					1 kW h
					  = 3·6 × 10⁶ J = 3·6 MJ
					
					electronvolt (eV)
					1 eV
					  = (1·602 10 ± 0·000 07)
					    × 10⁻¹⁹ J
						The units W h, kW h, MW h, GW h and TW h are used in the electrical industry.
						The units keV, MeV and GeV are used in accelerator technology

3–23.1	power	W (watt)	GW MW kW mW μW			
	impact strength	J/m²	kJ/m²	daJ/cm² J/cm²		

Heat

Item No in ISO/R31	Quantity	SI unit	Selection of recommended decimal multiples and sub-multiples of SI unit	Other decimal multiples and sub-multiples of SI unit	Other units or other names of units which may be used	Remarks
4–1.1	thermodynamic temperature	K (kelvin)				
4–2.1	Celsius temperature				degree Celsius (°C)	
4–1.1 4–2.1	temperature interval†	K			°C	1°C = 1 K
4–3.1	linear expansion coefficient	1/K			1/°C	

† The abbreviation 'deg' is commonly used to express a temperature interval, but is now regarded as obsolescent by CGPM. However, it may be used in textual matter, eg 'Increase the temperature by 20 degC'.

Table V List of SI units and a selection of recommended decimal multiples and sub-multiples of the SI units together with other units or other names of units which may be used (from PD 5686) *continued*

An asterisk against a unit means that the unit may be used in the UK but is not yet included in the ISO draft recommendation

Item No in ISO/R31	Quantity	SI unit	Selection of recommended decimal multiples and sub-multiples of SI unit	Other decimal multiples and sub-multiple of SI unit	Other units or other names of units which may be used	Remarks
4–4.1	heat, quantity of heat	J	TJ GJ MJ kJ mJ			
4–5.1	heat flow rate	W	kW			
4–6.1	density of heat flow rate	W/m²	MW/m² kW/m²			
4–7.1	thermal conductivity	W/m K	.		W/m °C	
4–8.1	coefficient of heat transfer	W/m² K			W/m² °C	
4–10.1	heat capacity	J/K	kJ/K		kJ/°C J/°C	
4–11.1	specific heat capacity	J/kg K	kJ/kg K		kJ/kg °C J/kg °C	
4–13.1	entropy	J/K	kJ/K			
4–14.1	specific entropy	J/kg K	kJ/kg K			
4–16.1	specific energy	J/kg	MJ/kg kJ/kg			
4–18.1	specific latent heat	J/kg	MJ/kg kJ/kg			

Electricity and magnetism (see notes 1, 2 and 3 below)

Item No in ISO/R31	Quantity	SI unit	Selection of recommended decimal multiples and sub-multiples of SI unit	Other decimal multiples and sub-multiple of SI unit	Other units or other names of units which may be used	Remarks
5–1.1	electric current (intensity of electric current)	A (ampere)	kA mA μA nA pA			
5–2.1	electric charge, quantity of electricity	C (coulomb)	kC μC nC pC			
5–3.1	volume density of charge, charge density	C/m³	MC/m³ kC/m³	C/mm³ C/cm³		
5–4.1	surface density of charge	C/m²	MC/m² kC/m²	C/mm² C/cm²		
5–5.1	electric field strength	V/m	MV/m kV/m mV/m μv/m		V/mm V/cm	

NOTE 1. In electricity and magnetism the SI units assume the rationalized form of the equations between the quantities. See ISO/R 31, Part V.
NOTE 2. THE IEC has not considered the arrangement and the content of the list. In order to give guidance to ISO, IEC/TC 24 provided the list of multiples and sub-multiples used here, but without division into columns.
NOTE 3. This list is a selection only from that given in PD 5686.

Table V List of SI units and a selection of recommended decimal multiples and sub-multiples of the SI units together with other units or other names of units which may be used (from PD 5686) *continued*
An asterisk against a unit means that the unit may be used in the UK but is not yet included in the ISO draft recommendation.

Item No in ISO/R 31	Quantity	SI unit	Selection of recommended decimal multiples and sub-multiples of SI unit	Other decimal multiples and sub-multiples of SI unit	Other units or other names of units which may be used	Remarks
5–6.1	electric potential		MV kV			
5–6.2	potential difference, tension	V (volt)				
5–6.3	electromotive force		mV μV			
5–7.1	displacement	C/m^2	kC/m^2	C/cm^2		
5–41.1	resistance	Ω (ohm)	G Ω M Ω k Ω m Ω μ Ω			
5–42.1	conductance	1/Ω			kS S (siemens) mS μS	1 S = 1/Ω the name 'siemens' and the symbol 'S' are adopted by IEC and ISO, but not so far by CGPM
5–43.1	resistivity	Ω m	G Ω m M Ω m k Ω m m Ω m μ Ω m n Ω m	Ω cm		$\mu\,\Omega$ cm = $10^{-8}\,\Omega$ m $\dfrac{\Omega \text{ mm}^2}{\text{m}} = 10^{-6}\,\Omega$ m = $\mu\,\Omega$ m are also used
5–44.1	conductivity	1/Ω m			MS/m kS/m S/m * μS/m	
5–45.1	reluctance	1/H				
5–46.1	permeance	H				

3 **Design data**

Contents

1 Data contained in this handbook

The metric change will create a state of flux in the technological world of the UK for at least three years and the authors' immediate concern is to put at the architect's disposal a reference book of basic metric design data whose contents will remain, as far as possible, valid and unaffected by forthcoming publication of various metric standards, codes, authoritative information sheets and a welter of manufacturers' literature.

Selected anthropometric and ergonomic data form the core of information contained in this handbook and it is quite unlikely that any minor dimensional variance with other sources will materially affect the validity of the design data. A comparative study of international sources will confirm the existence of marginal differences between the dimensional values given in various publications; slight deviations are of no consequence in the wider context of overall space requirements.

It is not the intention of this handbook to set up or give

rise to any dimensional standards not compatible with prevailing trends and dimensional co-ordination principles, but where in the authors' opinion the situation requires a firm decision on some aspect not covered by precedence this decision has been taken after careful scrutiny of the wider issues involved. In most instances the international field of technical metric information has been consulted and cross-checked to compare the validity of the design data given.

2 Space requirements

In evaluation of space requirements a sensible dimensional approach was aimed at similar to that established by the *Mobleringsplaner* published by Staters Byggeforsknings-institut, Copenhagen; *Woningbouw* published by the Bouwcentrum, Rotterdam and *Voorschriften en wenken voor het ontwepen van woningen 1965* published by the Central Directorate of Housing and Building, The Hague, **3.1**. This approach was also accepted by the Ministry of Housing and Local Government in its Design Bulletin 6 *Space in the home* (metric edition) **3.2** and is generally followed by Danish, Swedish, German and Norwegian architects and designers.

Metric sizes indicated throughout this handbook do not necessarily imply dimensional standards for actual objects or furniture shown, but refer generally to the space required to accommodate them within the overall design.

3 References and sources

MINISTRY OF HOUSING AND LOCAL GOVERNMENT *Design bulletin* 6 Space in the home (metric edition), 1968, HMSO
STATE BUILDING INSTITUTE *Mobleringsplaner* Copenhagen, 19
BUILDING CENTRE *Woningbouw* Rotterdam, 19
CENTRAL DIRECTORATE OF HOUSING AND BUILDING *Voor-schriften en wenken voor het ontwepen van woningen* The Hague, 1965

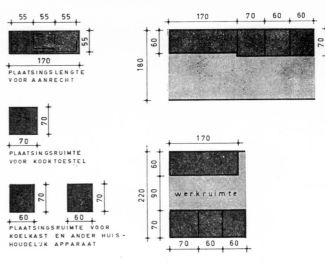

3.1 *Dimensional approach of 'Voorschriften' (cm)*

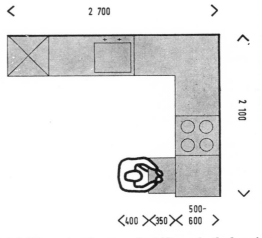

3.2 *Dimensional approach of 'Space in the home' (mm)*

4 Basic materials and products

Contents

1 Introduction

The problems facing the building industry in the change to metric have been outlined in section 1 of this handbook. The most important aspect of the change is the need to produce dimensionally co-ordinated products and to ensure that all manufacturers follow the recommendations contained in the British Standards as they are produced.

1.1 Dimensional co-ordination

The basic standard is BS 4011:1966 *Recommendations for the co-ordination of dimensions in building: basic sizes for building components and assemblies.*

Para 3 of that document states:
'The first selection of basic sizes* for the co-ordinating dimensions of components and assemblies should be, in

* As defined in BS 3536:1963 *Recommendations for a system of tolerances and fits for building*

descending order of preference, as follows (where n is any natural number including unity). The figures are given in decimetres: the SI unit of millimetres has been added in brackets.

FIRST	n × 3 decimetres	(300 millimetres)
SECOND	n × 1 decimetres*	(100 millimetres)
THIRD	n × 0·5 decimetres up to 3 decimetres	
	(50 mm–300 mm)	
FOURTH	n × 0·25 decimetres up to 3 decimetres	
	(25 mm–300 mm)	

The third and fourth preferences should not be used for basic sizes over 3 decimetres unless there are strong economic or functional reasons for doing so. The fourth preference is put forward provisionally. There may be a need for other sizes below 0·5 decimetre, but there is as yet insufficient evidence on which to base a firm recommendation.

Basic sizes for the co-ordinating dimensions of components and assemblies should be chosen from the first selection after consideration of the relevant functional requirements.

Within each category, such as windows, door frames, wall panels, floor slabs, and so on, the preferred sizes should as far as possible be the same in all materials. Account should be taken of the need for different categories of components or assemblies to occupy building spaces of the same size. Where several components are used to build up an assembly, the overall size for the assembly should be a basic size.

The work sizes of components and assemblies should be determined, taking account of space for joint and allowance for tolerances, in accordance with BS 3626:1963.

1.2 Programme for manufacturers

Table 1 on p4 of this handbook shows that the change to metric for products must be done in three stages:

1 Manufacturers provide BSI with information about their present products couched in metric terms by **end of 1968**;
2 BSI produces metric dimensional recommendations or standards for:
(a) products for which dimensional co-ordination is essential by **end of 1971**;
(b) all other products by **end of 1972**;
3 manufacturers must change to full production of their products to metric standards by **end of 1972**.

1.3 Problems of manufacturers

Manufacturers have three main problems:

1 Time allowed for the change, which some claim is too short. A major problem *could* arise if the new British Standards are not produced according to the present programme.
2 Need to invest in new plant, or to change present production lines, without certain knowledge of what the demand for the new metric product is likely to be.
3 Need to produce and stock, for a time, both imperial and metric sizes and to adjust prices so that this situation is economically viable. It is possible that prices may be averaged out so that imperial products subsidise metric products which are bound, initially, to cost more to produce. However by the end of 1971 the balance of advantage should be swinging in favour of the new metric products: rationalised dimensions should, ultimately, lead to reductions in cost.

* This is the 'basic module' in the draft ISO *General regulations governing modular co-ordination*. Use of the basic module is described in *The co-ordination of dimensions for building*. 1965, RIBA

1.4 Revision of product data

The AJ publishes regular news and technical features keeping materials and products information and the publication of new British Standards up to date.

2 Steel

Rationalised metric products will be introduced gradually into the general market and timing will vary according to complexity or cost of production methods. Any product or component formed by means of rolling mills, presses, castings or moulds will most probably be metricised in the last stage of changeover.

2.1 Metric equivalent sizes

Steel joists and beams will fall into this category and the architect will be confronted for some years by the present sizes with metric equivalents shown alongside. Steelwork tables are already available in metric terms, usually included as a separate section. BS 4: *Structural steel sections.* Part I 1962 *Hot-rolled sections* and Part 2 *Hot-rolled hollow sections* also include metric tables. The 6in × 6in × 15·7lb for example becomes 152 mm × 152 mm × 23 kg with 2980 mm² area of section. It must be noted that the tables give the sizes of steel sections in millimetres, but the other properties, like area of section, moment of inertia, radius of gyration, elastic modules and plastic modules are all given in cm², cm⁴, cm, cm³ and cm³ respectively. In trying to operate in millimetres in area dimensions and cubic dimensions it must be remembered that $1 \text{ cm}^2 = 100 \text{ mm}^2$ and $1 \text{ cm}^3 = 1000 \text{ mm}^3$. Thus the sectional area of a beam 6in × 6in × 15·7lb expressed as 29·8 cm² will read in millimetres 2980 mm². The existing steel joists sizes, while not conforming directly to the preferred ranges of dimensional zones (DC series), can be brought into line by forming a column core which keeps station within the allotted dimensional zone and is built up to the required dimension, or forms part of an assembly which is within the dimensionally co-ordinated zone (solid beam, lattice beam, castellated beam and so on). Section 11 of this handbook (p62) shows how this can be done for several building types, with recommended spacing and zone dimensions from BS 4330.

2.2 Future steel beam sizes

The future pattern of rationalised metric sizes of steel beams will most likely follow the international agreement of the International Standards Organisation (ISO). Britain plays a very active role within that body and will most likely be one of the first countries actually producing the internationally agreed metric sizes.

The final formulation of the ISO standards may not take place before 1970 as it is unlikely that the organisation will complete its work earlier. The British Standard will probably accept the ISO standard for its own metric range and therefore the rolling of metric beam and column sizes will be delayed until the 1971 to 1973 period. Some products however, like angles, may be standardised before then.

2.3 Flat steel products

Flat products (mild steel flat plates and sheet material) in metric sizes will not cause the same difficulty in production as beams and columns, and some flat products are already supplied in metric sizes for export. Provision for adjustment to the overall size of flat products in the production process will speed the changeover to metric of those types of products but availability will depend on the demand.

2.4 Steel bars and mesh
Availability and sizes of steel bars and mesh for reinforcement of concrete are dealt with below in para 3.1. See also para 6.2 and Table VI on p172.

2.5 Structural steel calculations
See p171 for design of structural steel.

2.6 Steel sheet and strip
The recommended basic sizes of thickness for sheet and strip (from BS 4391:1969) are given in table I.

Table I Standard thicknesses for steel sheet and strip

Thickness mm		Thickness mm		Thickness mm	
Choice		Choice		Choice	
First (R″10)	Second (R″20)	First (R″10)	Second (R″20)	First (R″10)	Second (R″20)
			0·10	0·10	1·0	1·0
				0·11		1·1
		0·12	0·12		1·2	1·2
			0·14			1·4
		0·16	0·16		1·6	1·6
			0·18			1·8
0·020	0·020	0·20	0·20	2·0	2·0
	0·022		0·22		2·2
0·025	0·025	0·25	0·25	2·5	2·5
	0·028		0·28		2·8
0·030	0·030	0·30	0·30	3·0	3·0
	0·035		0·35		3·5
0·040	0·040	0·40	0·40	4·0	4·0
	0·045		0·45		4·5
0·050	0·050	0·50	0·50	5·0	5·0
	0·055		0·55		5·5
0·060	0·060	0·60	0·60	6·0	6·0
	0·070		0·70		7·0
0·080	0·080	0·80	0·80	8·0	8·0
	0·090		0·90		9·0
0·100	0·100	1·00	1·00	10·0	10·0

NOTE. If additional thicknesses are required, it is recommended that they be selected by interpolation from the second choice (R″20) sizes

3 Concrete
The process of rationalisation in the concrete industry began with the issue of metric sizes for steel bar reinforcement together with a programme of change for the steel reinforcement manufacturers. The change to metric has been used to effect a reduction in the variety of existing sizes by 28 per cent. The following sizes of bars will be available (in mm) 6, 8, 10, 12, 16, 20, 25, 32, 40 and 50 (see table VI p172). The metric units used as far as size (diameter), length, and cross-sectional areas are concerned will be millimetres. The length will be expressed in metres and millimetres to the nearest 5 mm and the standard length will be 12 000 mm. The density of steel will be taken as exactly 0·00 785 kg/mm²/m.

Steel reinforcement manufacturers indicate that the programme will culminate in 1970 in the production solely of metric sized bar reinforcement. The programme is set out below. A similar programme is to be followed for the manufacture of metric sized fabric reinforcement.

3.1 Programme for manufacture of bars and fabrics
After 1st June 1969—metric sizes only will be rolled as standard.

After 1st July 1970—metric sizes only will be supplied as standard.

The programme of change to metric reinforcement indicates that designers will be faced at first with having to use metric reinforcement for schemes designed in imperial values.

This will call for careful procedure in calculations and phasing of jobs, noting particularly that the stocks of old imperial sizes of reinforcement will be diminishing after December, 1969.

The reference coding of metric reinforcing fabric has been conveniently related to the cross sectional areas (per metre width) of main wires (see table XII p175).

3.2 Steel reinforcement metric changes: Contractual implications
Architects are reminded of the Practice Note published in the December 1968 issue of the RIBA Journal advising them on the contractual implications of the programme for manufacture and supply of metric-sized steel reinforcement.

3.3 Reinforced concrete design
A document which gives guidance to the designer of concrete structures is The Technical Report prepared by the Concrete Society and the Institution of Structural Engineers *The detailing of reinforced concrete* (Report of the Joint Committee). The report deals with the problems of detailing reinforced concrete structures in the transitional period and metric data is given alongside the foot/inch values. Notation forms an important part of this document. Care should be taken however in checking with BSI documentation on the correct abbreviations for different types of reinforcement. The BSI preferred abbreviations are as follows:

R = Round mild steel bars in the metric range of areas

Y = High yield bars having a high bond strength in the metric round range of areas

X = A general abbreviation for types not covered by R or Y

The recommended bar spacings should be chosen from the following values in mm:
50, 75, 100, 125, 150, 175, 200, 250, 300. (See table IX p173). The design of reinforced concrete is dealt with at length on p171 to p180.

3.4 Concrete blocks
A number of manufacturers will, during 1969 and onwards, make metric concrete blocks. The first range of sizes is as follows:
200 × 400 × 100 mm
200 × 450 × 100 mm
200 × 500 × 100 mm
200 × 600 × 100 mm
In addition to 100 mm there will be other thicknesses to meet particular performance specifications. Additional overall sizes are also being considered.

3.5 Concrete bricks
The Brick Product section of the British Precast Concrete Federation have recommended to the manufacturers that the present imperial sizes (which include a metrically equivalent format of 225 × 112·5 × 75 mm) be augmented by metrically co-ordinated sizes. The initial suggested range will be as follows:
200 × 100 × 75 mm (basic unit)
200 × 100 × 100 mm*
300 × 100 × 75 mm*
225 × 112·5 × 75 mm†
*anticipated additional sizes
† metric equivalent of present imperial size of brick
The revised version of BS 1180 in metric terms will include a new range of compressive strength values set in SI units.

3.6 Paving flags

Recommended metric sizes for concrete paving flags are:

Type A	600 × 450 mm	⎫
Type B	600 × 600 mm	⎬ tolerances +0 to −4 mm
Type C	600 × 750 mm	⎪
Type D	600 × 900 mm	⎭

All these sizes are to be available in thicknesses of 50 mm and 63 mm. Tolerances ± 3 mm.

3.7 Kerbs

Recommended metric sizes for concrete kerbs are:

Maximum linear dimension: 900 mm

Sections and profiles: As at present contained in BS 340

Radius kerbs: The following radii will be produced:

1·5 metres

3·0 metres

4·5 metres

6·0 metres

7·5 metres

9·0 metres

10·5 metres

12·0 metres

These are the dimensions from the radial point to the face of the kerb, whether for external or internal radii. The industry has been recommended to make the change effective from 1 April 1970, after which date the present imperial sizes would no longer be manufactured.

3.8 Cement

The Cement Makers' Federation announces that the makers will change to the use of metric weights and decimal currency simultaneously on 1 January 1971. From that date all their deliveries will be in metric tonnes in 50 kg bags and invoices will be made out in decimal currency. No deliveries will be made in imperial weights after 31 December 1970 and none will be made in metric weights before 1 January 1971.

4 Timber

There is no opposition to a rapid changeover to metric within the international timber trade and by the time the building industry adopts metric measurement the timber trade will have gone over to metric sizes. A British Standard for metric dimensions for softwood is due to be issued shortly.

Apart from BS Code of Practice 112 *Structural use of timber* there has never been a British Standard for a range of soft-wood dimensions, nor has there been a metric, unified standard in any country in Europe. The existing sizes evolved and proliferated from the commercial considerations of timber sawmills and timber merchants, from transport problems and from the requirements of the building industry.

4.1 New metric timber sizes

The metric change will precipitate adoption of a new rationalised range of softwood dimensions simplifying design, production and stockholding. The sizes proposed are related to, though not identical with, those which have been in common use for many years in imperial measures. The relationship however is close enough in the great majority of uses to permit direct transfer without extensive redesign or increase of cost.

Table II shows the proposed new sizes compared with existing sizes.

Table III shows cross-sectional dimensions and lengths of the new sizes (the full BS should be consulted for tolerances, reductions etc).

Tables IV to VII show these new sizes extracted and listed for various purposes.

4.2 Machined timber

New proposals for metric dimensions of softwood introduce a novel approach to the sizing of machined timber. The nominal and finished dimensions are dealt with on a more rational basis than hitherto, directly helping the manu-facturer. Precision timber is particularly useful for factory assembly in jigs where a uniform dimension is essential, and for structural purposes involving engineering calculations where absence of tolerance and shrinkage allowances leads to economy.

4.3 Softwood flooring

It must be underlined that there is distinction between dimensional metric standards for plain edged timber and that of tongued and grooved (t & g) softwood board and strip flooring. The proposals contained in the revision of BS 1297 *Grading and sizing of softwood flooring* should be noted because of their implications on future practice. The latest revision of the softwood flooring standard excludes from its scope plain edged boards for flooring. In all cases the metric figures quoted apply to finished product after machining. The naming of nominal sizes has been abandoned as having no relevance to the product. The finished trans-verse dimensions are to comply with the following sizes in any combination of thickness and face width:

Finished thicknesses (mm)	16	19	22	29
Finished widths of face	65	90	115	140

(See table V)

Lengths shall be 1·8 m and longer with a minimum average in any delivery of 3·0 m.

Tongue and groove

Details of the tongue and groove are shown in **4.1**.

4.4 Hardwood

The position with regard to the buying and selling of hard-wood in metric sizes is more complicated because the countries supplying hardwood to Britain are not affected by Britain's decision to change to metric.

The Hardwood Metrication Committee has come to the conclusion that the British timber trade should adopt the same range of thicknesses for hardwood as proposed for softwood under the new British standard. But the 16, 22 and 24 mm thicknesses will be omitted from the range. For constructional purposes only, hardwoods will be sold from 125 mm upwards rising by 25 mm increments, with strips or narrows from 50 mm rising in 25 mm steps, and normal random specification from 150 mm rising by 25 mm stages to 300 mm. *Length:* normal length specification from 1·800 m rising by 100 mm stages. *Shorts:* falling from 1·700 m by 100 mm increments. *Width:* 150 mm and up in increments of 10 mm or 25 mm. *Strips and narrows:* 50 mm and up in increments of 10 mm or 25 mm. The unit of sale in the hardwood trade will be the cubic metre (m³).

The critical day for the change to metric in the British timber trade will be 1 April 1970.

See also Table II. Dimensions will normally be expressed in m and mm but the cm may occasionally be used for intermediate measures.

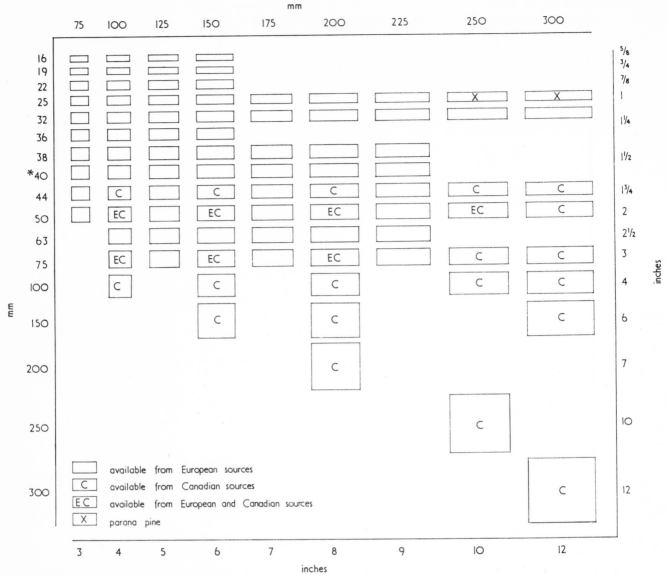

mm

Top axis: 75   100   125   150   175   200   225   250   300

Left axis (mm): 16, 19, 22, 25, 32, 36, 38, *40, 44, 50, 63, 75, 100, 150, 200, 250, 300

Right axis (inches): 5/8, 3/4, 7/8, 1, 1¼, 1½, 1¾, 2, 2½, 3, 4, 6, 7, 10, 12

Bottom axis (inches): 3   4   5   6   7   8   9   10   12

Key:
- available from European sources
- C available from Canadian sources
- EC available from European and Canadian sources
- X parana pine

*not on list of standard sections agreed by importers / exporters conference check on avaiability before specifying

Table II Softwood sawn timber: new metric sizes
Sizes also available in hardwood shown bold

New rationalised metric sizes	Old comparable sizes*
16 mm	⅝in
19 mm	¾in
22 mm	⅞in
25 mm	1in
32 mm	1¼in
38 mm	1½in
44 mm	1¾in
50 mm	2in
63 mm	2½in
75 mm	3in
100 mm†	4in
150 mm	6in
200 mm	8in
250 mm	10in
300 mm	12in

*NOTE. These are not exact translations.
†Hardwood thicknesses above 125 mm (5in) rise in increments of 25 mm.

Table III Cross-sectional dimensions (mm) *above*, lengths (m) below

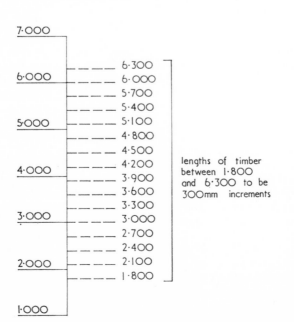

7·000

6·300
6·000   6·000
5·700
5·400
5·000   5·100
4·800
4·500
4·000   4·200
3·900
3·600
3·300
3·000   3·000
2·700
2·400
2·000   2·100
1·800

1·000

lengths of timber between 1·800 and 6·300 to be 300mm increments

Table IV Floor joists, ceiling joists, joists for flat roofs in mm

38 × 75
38 × 100
38 × 150
38 × 175
38 × 200
38 × 225

44 × 75
44 × 100
44 × 150
44 × 175
44 × 200
44 × 225

50 × 75
50 × 100
50 × 125
50 × 150
50 × 175
50 × 200
50 × 300

63 × 150
63 × 175
63 × 200
63 × 225

75 × 200
75 × 300

Table V Floor boards in mm

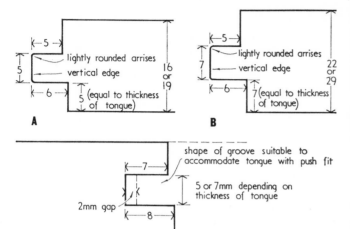

	65	90	115	140
16				
19				
22				
29				

A

k—5—|
5
lightly rounded arrises
vertical edge
16 or 19
k—6—|
5 (equal to thickness of tongue)

B

k—5—|
7
lightly rounded arrises
vertical edge
22 or 29
k—6—|
7 (equal to thickness of tongue)

C

k—7—|
shape of groove suitable to accommodate tongue with push fit
5 or 7mm depending on thickness of tongue
2mm gap
k—8—|

Table VI Timber beams, binders in mm
50 × 75
50 × 100
50 × 125
50 × 150
50 × 175
50 × 200
50 × 300
63 × 150
63 × 175
63 × 200
63 × 225
75 × 175
75 × 200
75 × 300

Common or jack rafters in mm
50 × 75
38 × 100
38 × 125
50 × 75
50 × 100
50 × 125

Table VII purlins in mm
50 × 75
50 × 100
50 × 125
50 × 150
50 × 175
50 × 200
50 × 230
63 × 150
63 × 175
63 × 200
63 × 225
75 × 175
75 × 200
75 × 300

4.1 *Tongued and grooved boards.* **A** *Detail of tongue for 16 or 19 mm boards.* **B** *Detail of tongue for 22 or 29 mm boards.* **C** *Detail of groove for all thicknesses of boards*

5 Bricks

The Brick Development Association announced at the end of 1967 that the advantage of a single standard brick size must not be lost. They therefore proposed metric brick dimensions of 21·5 cm (215 mm) × 10·25 cm (102·5 mm) × 6·5 cm (65 mm) which, with a 1 cm (10 mm) joint produces a working size of 22·5 cm (225 mm) × 11·25 cm (112·5 mm) × 7·5 cm (75 mm). The size is very close to the present dimensions and is achieved by translating the present 9in × 4½in × 3in at the rate of 25 mm = 1in instead of 25·4 mm. Four courses would equal 300 mm in height and four bricks 900 mm in length.

5.1 Modular Society proposals

These proposals have been criticised by many, including the Modular Society which itself proposed four brick sizes:

Decimetres	Millimetres
3 × 1 × 1	(300 × 100 × 100)
2 × 1 × 1	(200 × 100 × 100)
2 × 1 × 0·75	(200 × 100 × 75)
2 × 2 × 1	(200 × 200 × 100)

The society's reasons for proposing these sizes, and its objections to the BDA sizes, are given in a news note 'BDA and MOD SOC: What size the metric brick?' (The Architects' Journal 20.12.67, p1549-1550).
Certain brick manufacturers are able to supply modular bricks to the sizes shown above.

5.2 Latest BDA proposal

BDA have since announced that as from 1 January 1970 the standard brick format becomes 225 × 112·5 × 75 mm. The work size of the standard brick would thus be 215 × 102·5 × 65 mm with a 10 mm morter joint.

This metric format is claimed to provide the following advantages:
1 it maintains the advantages of a single standard size
2 it can be used for brickwork that conforms to the first preference (300 mm) of BS 4011, and will co-ordinate dimensionally with other building components and assemblies also conforming to this first preference
3 it maintains the existing relationship between the length and the width of a brick (ie length = twice the width plus the joint)
4 bricks in the new format can be used to match in with existing brickwork (conversions, extensions, repairs, etc)
5 the new standard bricks will continue to meet the functional requirements of the Building Regulations and Codes of Practice
6 it avoids serious manufacturing complications and consequent increases in cost.

Manufacture of the present imperial standard bricks will cease during the period immediately following 1st January 1970.

5.3 Ultimate brick sizes

Of the future, the BDA reports that 'extensive investigations in many different fields and the co-operation of other sectors of the building industry, including Government departments, will be required before the optimum metric brick size(s) to meet our long-term needs can be identified with certainty. It will be appreciated that the final decisions may well exert considerable influence over the future course of building in this country. They will certainly hold far-reaching implications for the future of the brick industry.'

6 Aluminium

New metric standards relating to plate, sheet, strip, extrusions, tube, wire, forgings and castings (but not finished products) will be issued by the BSI in mid-1969 and materials will then be offered by the Industry to conform with those standards. Explanatory notes relating to the BS General Engineering Specifications concerned will be published by the Aluminium Federation in advance of the issue of the Specifications. It is envisaged that the Industry's change to metric working will be completed by mid-1970.

7 Copper

Thicknesses of copper will be given in millimetres: present dimensions or thicknesses will be rounded off, not merely converted. Metric specifications will be available during 1969 and 1970: the first one probably dealing with light-gauge copper tubes for domestic services, including compression and capillary joints.

The Copper Development Association will amend its literature as metric information becomes available from the BSI or the ISO.

From 1 January 1970 copper products will generally be weighed in kilogrammes, and pricing units will be expressed in £ per 100 kg and not in £sd per long ton or pence per lb as at present. Although decimal currency will not officially come into use for a further year, the new pricing unit will be decimal £ per 100 kilogrammes.

7.1 Copper pipes

Table VIII shows existing pipe sizes compared with ISO recommended sizes. The future edition of BS 864 *Capillary and compression fittings* will be revised in accordance with these recommendations but not all ISO sizes will necessarily be adopted.

Table VIII Comparison of internal pipe diameters (inch sizes) with ISO recommended sizes

Present inch sizes	Proposed ISO sizes (mm)
$\frac{1}{8}$	—
$\frac{3}{16}$	6
$\frac{1}{4}$	8
	8
	10
$\frac{3}{8}$	12
$\frac{1}{2}$	15
	18
$\frac{3}{4}$	22
1	28
$1\frac{1}{4}$	35
$1\frac{1}{2}$	42
2	54

8 Lead

Metric sizes of lead sheet and strip are laid down in the metric version of BS 1178. Thicknesses (below) are similar to existing: BS numbers are the same as present weights expressed in lbs/sq ft.
Standard width will be 2·40 m; length up to 12 m.

BS no	Range of thicknesses			Metric weights			
		Corresponding thickness (in)		Sheets 2·40 m wide and of length			
					3 m	6 m	9 m
	(mm)	Decimal	Fraction	kg/m²	kg to nearest 0·5 kg		
3	1·25	0·049	3/64+	14·18	102·0	204·0	306·5
4	1·80	0·071	5/64—	20·41	147·0	294·0	441·0
5	2·24	0·088	3/32—	25·40	183·0	366·0	548·5
6	2·50	0·098	3/32+	28·36	204·0	408·5	612·5
7	3·15	0·124	⅛—	34·73	250·0	500·0	750·0
8	3·55	0·140	9/64—	40·26	290·0	579·5	869·5

9 Zinc

The thicknesses of zinc sheet and strip for all applications will be expressed in millimetres. The zinc gauge and the standard wire gauge will no longer be used for zinc in the UK. The thicknesses commonly recommended and used in building construction will be expressed as follows. Other thicknesses will continue to be available.

Millimetres	Zinc gauge number	Standard wire gauge
1	16	19
0·8	14	21
0·6	12	23

The length of sheet and strip will be expressed in metres. Whether width will be expressed in millimetres or as decimals of a metre will be determined by the BS.

A completely new and revolutionary continuous casting and rolling mill—the first of its type in Europe, designed to produce zinc sheet in continuous lengths and in any width up to 1000 millimetres—has now started production. This plant will extend the fields of application for rolled zinc, at present restricted to some extent by the existing standard sheet sizes of 8ft × 3ft and 7ft × 3ft.

10 Sheets and boards

Future sizes of rigid flat sheet products used in building are to be contained within the following metric range:
Width: 600, 900, 1200 mm
Length: 1800, 2400, 2700, 3000 mm
Rigid flat sheet products for building include asbestos cement, asbestos insulating board, blockboard, compressed strawboard, expanded polystyrene, glass, gypsum wallboard, hardboard, insulation board, laminated plastics, metal (ferrous and non-ferrous) particle board, plywood and woodwool. The thicknesses are likely to follow the existing pattern of dimensions ie gypsum wallboards and gypsum planks will continue to be made in 9·5 mm (⅜in), 12·7 mm (½in) or 19 mm (¾in). The thickness of compressed straw slabs used for partitioning or wall lining will be n × 25 mm, where n is any natural number, including unity. The Finnish birch blockboard and laminboard will be supplied in:
5-ply: 12, 16, 18, 19, 22, 25 mm
3-ply: 12, 16, 18, 19, 22, 24, 25 mm

10.1 Plasterboard

The major manufacturing source of plasterboard production announced in its metric timetable, that the major change will take place on 1 April 1970. From that date the standard stock sizes in metric will be as follows:
Gypsum wallboard:
Widths: 600, 900, 1200 mm
Lengths: 1800, 2350, 2400, 2700, 3000 mm
Thickness: 9·5 mm and 12·7 mm
Gypsum plank
Widths: 600 mm
Lengths: 2350, 2400, 2700, 3000 mm
Thickness: 19 mm
Industrial Plastic-faced board
From 1 April 1970 these boards will be changed in width from 2ft to 600 mm. The existing imperial sized lengths will continue for the time being and the additional new metric lengths will be 1800, 2400, 2700 and 3000 mm. Thicknesses will remain unchanged.

10.2 Sarking boards

From 1 April 1970 these boards will be changed in width from 2ft to 600 mm. The new metric length of 1800 mm will be manufactured in addition to the existing 6ft length. The thickness will remain unchanged.

10.3 Gypsum lath and baseboards
From 1 April 1970, it is probable that widths will be changed to an absolute metric dimension and a decision on this will be announced no later than September 1969. Other dimensions will be:

Lengths

Lath	New metric 1200 mm
	Imperial 1219 mm (4ft); 1372 mm (4ft 6in)
Baseboard	New metric 800 mm; 1200 mm
	Imperial 813 mm (2ft 8in); 1219 mm (4ft); 1372 mm (4ft 6in)

Thicknesses

| Lath | 9·5 mm ($\frac{3}{8}$in); 12·7 mm ($\frac{1}{2}$in) |
| Baseboard | 9·5 mm ($\frac{3}{8}$in) |

10.4 Dry partitions
From 1 April 1970 cellular plasterboard partitions will only be manufactured in 600, 900 and 1200 mm widths and 1800, 2350, 2400, 2700 and 3000 mm lengths. The present dry partition thicknesses will continue.

No imperial widths of board will be manufactured after 31 March 1971 and the standard lengths in imperial sizes will be progressively withdrawn both before and after this date. All plasterboard and dry partition, whether in metric or imperial dimensions, will be invoiced in square metres from 1 April 1970.

10.5 Fibre building boards
A decision on thicknesses has not yet been taken. Lengths and widths of board will be varied to conform to BS 4011: 1966 required dimensions (see para 1.1 in this section).

Fibre building boards are manufactured by machinery of extremely high capital cost and of fixed dimensions. This implies that sheet sizes can be *reduced* from their present sizes to comply with a metric dimension but, in all probability, can be *increased only slightly*, if at all.

11 Bagged materials
Building and industrial plasters, jointing compounds and gypsum mineral will be packed as follows from 1 February 1971:

Present packing	*From 1971*
1 cwt	50 kg
56lb	25 kg
28lb	12·5 kg
14lb	10 kg
7lb	5 kg
5lb	2·5 kg

All products so changed will be invoiced in metric tonnes or kilogrammes from 1 February 1971.

12 Glass
The British glass industry will complete metrication of flat glass by 1 October 1970. Since 1 January 1969 all flat glass has been made in metric thicknesses. Small changes have been made in some glass thicknesses to bring them into line with internationally agreed standards: all nominal thicknesses will be in exact millimetres. Linear measurement of glass will continue to be in imperial units until October 1970, after which some changes in stock sizes are inevitable n converting imperial sizes to the nearest metric equivalent. All linear dimensions will be expressed in millimetres; areas will be in square metres.

13 Metal windows
The present imperial standard ranges of metal windows will continue to be the main stock ranges until, probably, 1970 but steel and aluminium windows to the new dimensionally co-ordinated metric sizes should be available early in 1970. The basic spaces which these will fit are given below. At the beginning of 1971 the imperial modular ranges may be withdrawn, available thereafter only as 'specials'.

13.1 Basic spaces for steel windows
The Steel Window Association defines the agreed Modular Basic Spaces for windows for:
1 HOUSING (Standard Steel Windows)
2 OTHER BUILDING TYPES (Purpose Made Steel Windows)
The Spaces have been selected from the matrices produced by the BSI Functional Group Panel B/94/4/2 (External Envelope). In turn the matrices were derived from information in BS 4330.

The matrices
1 Housing
The lengths of the basic spaces conform to BS 4011 first preference, in increments of 300 mm. The heights of the basic spaces are derived from the preferred head and cill heights for public sector housing given in DC 7. Spaces larger than those in the basic matrix can be filled by combinations of the unit spaces.

2 Other building types
The lengths of the basic spaces conform to BS 4011 first preference in increments of 300 mm. The heights of the basic spaces are those indicated as first preference in the BSI matrices for basic spaces for health and office buildings, which can be filled by single units. Spaces larger than those in the basic matrix, can be filled by combinations of the single units. Increased flexibility for modular spacings in increments of 100 mm is available by the use of fixed lights and box mullions.

Surrounds
Timber
1 ex-75 × 75 mm timber adding 100 mm to the modular space.
2 ex-50 × 75 mm timber adding 50 mm to the modular space.

Metal
Metal surrounds will also conform to preferred dimensional requirements.

Lintels and cills
These may be contained:
1 within the joint allowance between window and adjacent cladding unit.
2 within the basic space for the adjacent cladding unit. In this case they should penetrate the adjoining space in increments of 25 mm.

Windows and doors: domestic housing types
Outline spaces in table IX will be filled with single or multipane standard window units. Details of types of casement and of work sizes will be published by individual manufacturers. Flexibility of the basic matrix is increased as follows:

Length
1 by including one or more fixed light units 500 mm long in the composite window.

2 by including one or more pressed steel box mullions or partition covers in the assembly. Available in all matrix heights, these add 100 mm to length.

3 by applying a wood surround, adding 100 mm to length and height.

Height

1 by combinations of basic units to fill the intermediate heights

$$eg \frac{500: 500: 500: \quad 500: \quad 500: \quad 200}{500 \quad 700 \quad 900 \quad 1100 \quad 1300 \quad 1500}$$

2 by applying a wood surround, adding 100 mm to height. Thus all modular lengths and heights from 900 mm upwards can be achieved in increments of 100 mm.

Windows and doors: all other building types

Outline spaces in table X are the preferred sizes to be filled with single pane window units. Details of types of casement and of work sizes will be published by individual manufacturers. Flexibility to fill all modular spacings to BS 4011

Table IX Basic spaces for standard steel windows (Housing) in mm

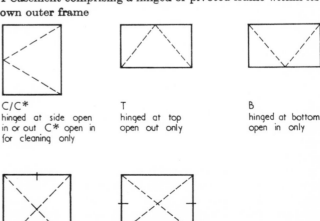

second preference increments of 100 mm is provided as follows:

Length

1 by using, either alone or in combination with the single pane basic units, fixed lights available in increments of 100 mm.

2 by including one or more pressed steel box mullions or partition covers in the assembly. These are available with height increments of 100 mm and in widths which add 100, 200 or 300 mm to the length of assemblies.

Height

1 by combinations of single pane basic units

$$eg \frac{300: \quad 300: \quad 600}{600 \quad 1100 \quad 11\,000}$$

2 by using 100 mm increment fixed lights either alone or in combination with the basic single pane units.

13.2 Basic spaces for aluminium windows

The metric and dimensional co-ordination committee of the Metal Window Federation and technical committee of the Aluminium Window Association have issued sizes of basic spaces and ranges of aluminium windows, illustrated below in table XI. Work is proceeding on joints, tolerances, performance specifications and nomenclature, following which details should be available from individual manufacturers showing sections and work sizes.

Derivation of spaces

The spaces have been derived from the matrices produced by the BSI Functional Group Panel B/94/4/2 (External Envelope). In turn the matrices were derived from the information given in BS 4330.

Those spaces which include an additional • are derived from the recommendations of the Ministry of Housing and Local Government in Design Bulletin 16.

Window types and abbreviations (see table XI)

1 Casement comprising a hinged or pivoted frame within its own outer frame

Table X Basic spaces for purpose made steel windows (other building types) in mm

2 Slider comprising sliding panel(s) within own frame

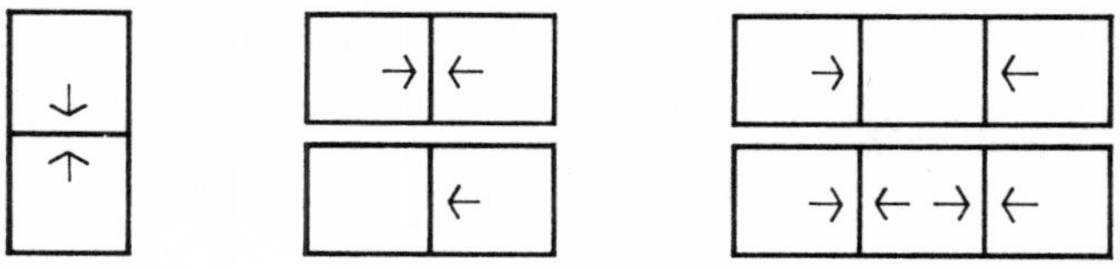

VS
vertical sliding sash
(or double hung)
one pair of vertical
sliding sashes

HS
horizontal sliding sash
two movable - one movable and one fixed
two movable and one fixed or
three movable

3 Doors for occasional use comprising casement and sliders as in sections 1 and 2

D
single leaf
casement

DD
double leaf
casement

SD
single, double or
triple leaf slider

4 Fixed light F, a single frame prepared for glass or panelling. All windows shown in table XI are available as fixed lights.

5 Spaces marked with the above abbreviations can be filled with the appropriate individual windows. All spaces can be filled with one pane fixed lights or with combinations of individual casement and fixed light or individual slider and fixed light. Larger spaces can be filled by combinations of these units.

6 In addition to the sizes shown there is available a range of fixed lights with increments of 100 mm in both height and width together with vertical box mullions (open and closed), 100 mm, 200 mm and 300 mm wide. These will provide a flexibility of 100 mm in both directions.

Table XI Basic spaces for aluminium windows

	600	900	1200	1500	1800	2100	2400
300	T B	T B	T B	T B	T		
500	T B HP	T B HP HS	T B HP HS	T B HP HS	T HS	HS	HS
600	T B HP	T B HP HS	T B HP HS	T B HP HS	T HS	HS	HS
700	T B HP	T B HP HS	T B HP HS	T B HP HS	T HS	HS	HS
900	C T HP VS	C* T VP HP/R VS HS	T HP/R VS HS	T HP/R HS	T HS	HS	HS
1100	C T VP HP VS	C* T VP HP/R VS HS	T VP HP/R VS HS	T HP/R HS	T HS	HS	HS
1300	C VP HP VS	C* VP HP/R VS	VP HP/R VS HS	HP/R VS HS	HS	HS	HS
1500	C VP VS	VP VS	VP VS	VS HS	HS	HS	HS
1800	C VS	VS	VS	VS			
2100	VS	VS D	VS DD	VS DD	VS DD SD	SD	SD

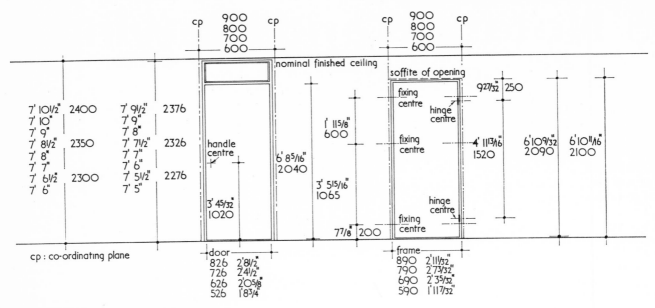

4.2 ISO *standards for doors and door sets*

14 Wood windows and doors

Dimensional requirements of the main components (windows, door sets and so on) should be decided by the BSI by the end of 1969. Production of the new metric components will begin in 1970—availability will depend, to some extent, on the demand forthcoming from architects and builders.

14.1 Windows

British Woodwork Manufacturers' Association proposals for basic spaces for standard wood windows are:
Widths: 600, 900, 1200, 1800, 2400 mm
Heights: 600, 900, 1050, 1200, 1500 mm

14.2 Doors

The British Woodwork Manufacturers' Association has issued proposals **4.2** which have the support of the MOHLG and have been adopted by eleven European countries. They are now in the course of being adopted as an ISO recommendation and will almost certainly be used in the future BS on doors.
These proposals are to fill basic spaces 600, 700, 800 and 900 mm wide × 2100 mm high for door height sets. Where the set is required to ceiling height the same sized units with a separate panel above the door can be used for the heights anticipated in BS 4176:1967. It is anticipated that these are the most likely sizes to be adopted at least for dwellings. For external door sets for all types of buildings the BWMA's proposals are to fill basic spaces 900, 1000, 1500 and 1800 mm wide × 2100, 2400, 2700 and 3000 mm high.

15 Kitchen units

BWMA are currently envisaging widths of 400, 500 and 600 mm for single units and double units 800, 1000, 1200, 1500 and 1800 mm wide. They propose that worktop and sinktop to floor units should be 600 mm deep and wall units 300 mm deep. Normal worktop height will be 850 mm and sinktop height 900 mm.

16 Stairs

The main BWMA dimensional proposals submitted to the BSI are set out below.
They relate particularly to public sector housing where the finished floor to finished floor height is fixed at a mandatory 2600 mm (see fig **26.2** p 139).

Dimension	mm
Total height for 13 risers	2600
Rise per step	200
Going per step	225
Width overall strings	850 or 1750
Basic space: width	900 or 1800
Gap between external faces of strings	100
Clearance 90° to rake	1525
Headroom vertically	2000

17 References and sources

BS 4011:1966 Recommendations for the co-ordination of dimensions in building: basic sizes for building components and assemblies CI/SfB (F4j)
BS 3626:1963 Recommendations for a system of tolerances and fits for building CI/SfB (F4j)
BS 4 Structural steel sections Part 1:1962 Hot-rolled sections and Part 2 Hot-rolled hollow sections CI/SfB Hh2
BS 4330:1968 (Metric units) Recommendations for the co-ordination of dimensions in building: Controlling dimensions CI/SfB (F4j)
BS 1297:1961 Grading and sizing of softwood flooring CI/SfB (43) Hi2
PD 6440 Draft for development: Accuracy in building: part 1 Imperial units, part 2 Metric units [(F6j)]
PD 6444 Recommendations for the co-ordination of dimensions in buildings: basic spaces for structure, external envelope and internal subdivision: part 1 Functional groups 1, 2 and 3 [(F4j)]
PD 6445 Recommendations for the co-ordination of dimensions in building: tolerances and joints, the derivation of building component manufacturing sizes from co-ordinating sizes [(F4j)]
See review of these three PDs in AJ *4.3.70 p 553–556*
BS 449 The use of structural steel in building: part 2 Metric units [(2-) Gh2 (K)]
BS 1178 Milled lead sheet and strip for building purposes [Mh8]
BS 3921 Bricks and blocks of fired brickearth, clay or shale: part 2 Metric units [Fg]
BS 4482 Hard drawn mild steel wire for the reinforcement of concrete, metric units [Jh2]
BS 4483 Steel fabric for the reinforcement of concrete, metric units [Jh 2]

5 Notation and drawing office practice

Contents

1 Notation
1.1 Decimal marker
The decimal marker in this country has been settled definitely as the conventional decimal point—opposite the middle of the figure when printed or handwritten, on the line if typewritten (unless the typewriter contains a special character above the line). When the value to be expressed is less than unity it should always be preceded by zero (0·6 not ·6). Whole numbers may be expressed without a decimal marker.

The appropriate number of decimal places should be chosen depending on the circumstances in which the resulting value is to be used.

Thousand marker
No thousand marker should be used but where legibility needs to be improved a space can be left in large groups of digits at every thousand point. Where only four digits are used a space between the first digit and the others is usually not desirable.

The comma should *never* be used as either decimal marker or thousand marker in this country, although it is in common use as a decimal marker in other metric countries and will be encountered when referring to foreign books and magazines. (See also p37, para 3.1)

1.2 Exclusive use of SI units
The speed with which metric values are assimilated is directly proportional to the means adopted for learning them, and to the method of communication. The more direct the communication, the quicker the response. We in this country are learning a completely new system quite different from the duodecimal system we have grown up with. In countries already using metric the problem is slightly different: they must get used not to a new system but to a revised, SI, system incorporating a number of fairly basic changes. Thus it can be unwise to trust too much to the experience and preferences of these countries whose advice can be tinged with prejudice. This is particularly true in the rivalry of the centimetre and the millimetre as acceptable units of linear measurement.

Centimetres or millimetres
Anyone used to operating the metric system from childhood will experience no difficulty in using a mixture of millimetres, centimetres or metres and will recognize at a glance plan dimensions or sizes of products from a drawing irrespective of whether the dimensions on that drawing are expressed in millimetres, centimetres or metres, or a mixture of all three. It is usual for the dimensions to be indicated by figures alone, without m, cm or mm symbols—and only experience, and the convention associated with the given material, product or scale of drawing tells the reader *how* to interpret these mixed values. Such a situation should not be entertained in this country: the methods of communication must be as direct and unambiguous as possible. In accepting the metre and millimetre exclusively for linear measurement the AJ Metric Handbook conforms to international agreement and to subsequent ratification of that agreement by the BSI. These agreements have been embodied in BSI documents PD 6031 (2nd edition) *The use of the metric system in the construction industry* and BS 1192: 1969 *Recommendations for building drawing practice* metric units.

Need for a standard notation
A standardised approach to notation will make it unnecessary to waste time on checking, even mentally, whether the dimensions are in centimetres or in millimetres, with consequent chances of error due to mistaken identity. The technical and economic success of the metric change depends on the way each professional establishment organises its educational and design practice. Drawings as communication media between the design, production and assembly teams, must be informed, clear and quite unambiguous in their presentation of technical data.

1.3 Specifying imperial and metric sizes
During the changeover period, where necessary also to give equivalent dimensions, it should be done thus:

1 On metric projects, using imperial sized products and components, give metric equivalent sizes, usually to nearest mm, in brackets—eg 4in (102 mm).

2 On imperial projects using metric sized products and components: give imperial equivalent sizes, usually to nearest $\frac{1}{16}$in, in brackets—eg 100 mm ($3\frac{15}{16}$in).

Where both sizes are not required, and where no ambiguity can arise, the first size may be discarded and the second size in brackets retained. Thus the use of brackets indicates that on a metric project an imperial size is required and vice versa. The use of this convention should be noted in the drawing. Examples:

	Sizes of products/components	
Metric projects	*Imperial*	*Metric*
Normally	4in (102 mm)	100 mm
Where no ambiguity can arise	(102 mm)	100 mm
Imperial projects		
Normally	4in	100 mm
		($3\frac{15}{16}$in)
Where no ambiguity can arise	4in	($3\frac{15}{16}$in)

Fig **5.1** shows the difference between imperial and metric modular sizes. From this it will be seen that they are not interchangeable, eg the difference between 3 metres and its modular equivalent of 10ft is nearly 2in or 50 mm.

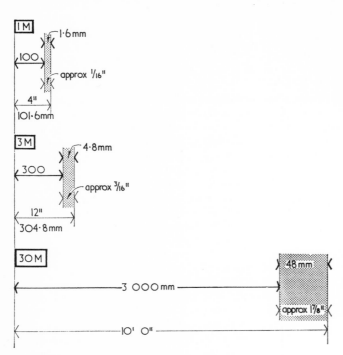

5.1 *Diagram showing the difference between imperial and metric modular sizes (not to scale)*

1.4 Linear measurements on drawings

When both metres and millimetres are used on drawings it will be less confusing if the dimension is always written to three places of decimals, ie 3·450. No unit symbol need be shown unless a lesser number of decimal places is used; ie 3·450 or 3·45 m or, under some circumstances, 3·5 m, are all correct. The first of these (3·450) is to be preferred. 3450 mm can also be used.

Where no ambiguity can arise symbols may be discarded, according to the following three rules:

1 Whole numbers indicate millimetres.

2 Decimalised expressions to three places of decimals indicate metres (and also, by implication, millimetres).

3 All other dimensions must be followed by the unit symbol. Rule 1 could safely be broken in certain circumstances, eg in a large scale detail drawing where all dimensions are in whole number millimetres, except for the occasional one which may be a whole number plus a decimal fraction. There could be no confusion over the unit size in relation to the other dimensions and it would not seem necessary to add mm after the dimension because it was not a whole number.

Table I Examples of the different ways of expressing linear measurements on drawings (from PD 6031 revised)

Millimetres	Metres		Kilometres
3·5 mm			
3			
30			
300	0·300	0·3 m, 0·30 m	
3000	3·000	3 m, 3·0 m, 3·00 m	
3300	3·300	3·3 m, 3·30 m	
3330	3·330	3·3 m, 3·30 m	
3333	3·333		
		3000 m	3 km

2 Paper sizes

The international A-series of paper sizes is to be adopted for all drawings and written material.

2.1 Sizes in the A-series

The basic size of the A range is derived from a rectangle A0 having a surface area of 1 m² on the basis of two rectangles, $x{:}y \doteq 1{:}\sqrt{2}$ and $x \times y = 1$ (ie x = 841 mm; y = 1189 mm). Smaller sizes are obtained by halving the larger dimension and large sizes by doubling the shorter one.

The original basis of the series is shown in **5.2** and its derived sizes in **5.3**.

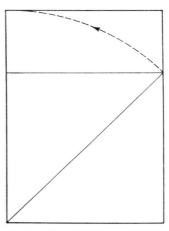

5.2 *Unit square as basis of A series*

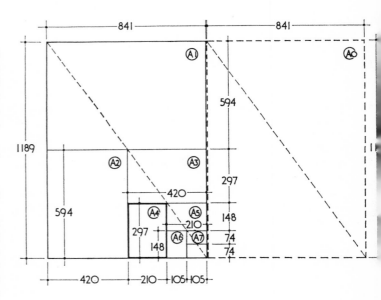

A size	mm
A0	841x1189
A1	594x841
A2	420x594
A3	297x420
A4	210x297
A5	148x210
A6	105x148
A7	74x105
A8	52x74
A9	37x52
A10	26x37

measurements represent trimmed sizes

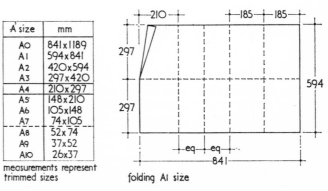

folding A1 size

5.3 *A-series of paper sizes*

2.2 Trimmed sizes

The A formats are trimmed sizes and therefore exact; stubs or tear-off books, index tabs, etc, are always additional to the A dimensions. Printers purchase their paper in A or C sizes allowing for the following tolerances of the trimmed sizes:

for dimensions up to and including 150 mm, $\pm 1\cdot 5$ mm;

for dimensions greater than 150 mm up to and including 600 mm, ± 2 mm;

for dimensions greater than 600 mm, ± 3 mm.

2.3 Drawing boards

As traditional types of working drawing are superseded it is probable that the range of paper sizes will be restricted to A1, A2, A3 and A4. Only rarely will it be necessary to use A0. Thus traditional sizes of drawing boards can accommodate the likely range of paper sizes. See table II.

New drawing boards are available to fit A size paper. There is as yet no agreement on exact sizes but for preliminary planning purposes the sizes given in table III may be used (all of which are available).

Vertical and horizontal filing chests and cabinets will have *internal* dimensions approximately corresponding to the board sizes listed in table III.

Table II Use of A-sized paper on traditional sized drawing boards

Paper size (mm)	Minimum board size (mm)	Comments
A0 1189 × 841	Antiquarian 1372 × 813	Paper needs to be shifted up or down about 50 mm. This paper size probably rarely used
A1 841 × 594	Double elephant 1092× 737	Probably largest size of paper normally required : therefore double elephant board s largest needed. Paper must be shifted if largest paper dimension is to be vertical
A2 594 × 420	Imperial 813 × 584	Or use double elephant board if largest paper dimension is to be vertical
A3 420 × 297	Half imperial 584 × 406	Or use imperial board if largest paper dimension is to be vertical
A4 297 × 210	Half imperial 584 × 406	Half imperial board may be used whether largest paper dimension is horizontal or vertical

Table III New A-sized drawing boards

Paper size (mm)	Board size (mm)
A0 1189 × 841	A0 1270 × 920
A1 841 × 594	A1 920 × 650
A2 594 × 420	A2 650 × 470

3 Envelope sizes

Some years ago the GPO introduced its preferred range of envelope sizes intending a surcharge on any envelope sizes outside that range. The Postmaster General has since announced that because of the response and co-operation of customers it will not be necessary to make a surcharge on envelopes outside the range during 1969 but he reserves the right to make additional charges at some later date if necessary.

Envelopes outside the POP range (Post Office Preferred) may still be used and mail weighing *more* than 4oz (113 grammes) will not be affected by the choice of envelope.

3.1 POP sizes

Envelopes should be at least 90 mm × 140 mm and not longer than 120 mm × 235 mm, oblong, with the longer side at least $1\cdot 414$ × the shorter side, made from paper weighing at least 63 grammes per square metre.

Post cards

Post cards will be treated as letters and conform to the same POP sizes.

Aperture and window envelopes

Aperture envelopes are not popular and irrespective of their size are classed outside the POP range. *Window envelopes* however are allowed provided that they conform to POP specifications.

Unenveloped post

All unenveloped post, except cards, will be classed as outside the POP range.

3.2 Envelopes for unfolded A4 and A5 paper

The RIBA Journal noted in February 1969 that 'while the Preferred range will take A4 and A5 sheets *when folded*, it does not include the C4 and C5 envelopes which must be used for A4 or A5 material sent flat. The penalty for not folding A4 or A5 material to go into 'Preferred' size envelopes is going to be savage. Packets weighting less than 4oz will pay an extra 2d for second class and 4d for first class mail—increases of 50 and 80 per cent respectively. This is the price the building industry in general, and architects in particular, seem likely to have to pay for having introduced standardisation and rationalisation, unless the Postmaster General can be persuaded to have second thoughts. It is not a subsidy that is wanted—only a Preferred system which takes account of the needs of the industry.'

4 Ordnance survey maps

The first Ordnance Survey maps to be based completely on metric measurements are to be published towards the end of 1969. The changeover will be gradual and in the first stage it will be limited to the large-scale OS maps, including the six-inch-to-the-mile series. Metrication of the popular one-inch and smaller scale maps will follow.

4.1 1:1250 and 1:2500 maps

The International Organisation for Standardisation (ISO) recommended the use of 1:1000 and 1:2000 ratios for Block Plans—but because of large costs and labour involved, Great Britain decided not to change immediately to these rational metric scales, but to retain for at least 10 years the existing ratios of 1:1250 and 1:2500.

Method of conversion

In the meantime the new and revised 1:1250 and 1:2500 maps will be converted to metric on the following basis: heights of bench marks will be shown to $0\cdot 01$ m accuracy and spot heights to $0\cdot 1$ m accuracy. On the 1:2500 maps areas of parcels of land will be given to the accuracy of $0\cdot 001$ hectare, and will also be given in acres as at present. The first metric maps at those scales will appear in October

1969 but it will be many years before all 1:1250 and 1:2500 maps (there are some 150 000 of them) are converted to metric form.

Bench marks and levels

The *Recommendations for Building Drawing Practice* contained in BS 1192:1969 (Metric units) state that although Ordnance Survey bench marks are taken in metres to an accuracy of 0·01 m, the levels given in BS 1192:1969 are to 0·001 m. Metric levelling staffs may be read by interpolation to an accuracy of up to 10 mm (ie staff gradations will be in 10 mm increments and accuracy will depend on skill in reading and interpolating values of less than 10 mm. See para 7.4 p31). Temporary bench marks will usually be established for particular projects. Also, some levels will be obtained by direct measurement and three places of decimals (0·001 m) is consistent with the recommended notation for linear dimensions.

4.2 Six-inch and 1:25 000 scales
The six-inch (1:10 560) scale will be replaced by the 1:10 000 with metric contours. The contour interval will be 10 metres in the more mountainous areas and 5 metres in the remainder of the country. The first sheets at the 1:10 000 scale are to be published in December 1969, but it will take a number of years before the country is covered with a homogeneous series at this scale.

Publication of the 1:25 000 map with metric contours will follow the 1:10 000 maps. The contour intervals will be consistent over the whole of the 1:25 000 sheets.

5 Bench mark lists
The heights of bench marks will be given in both metres and feet in Bench Mark Lists obtainable from Ordnance Survey Headquarters, Romsey Road, Maybush, Southampton SO9 4DH.

6 Levels
Levels record the distance of a position above or below a defined datum.

6.1 Datum
As a minus sign is easily misread, a suitable fixed point should be taken as TBM (temporary bench mark) such that all other levels are positive. Thus datum should be clearly indicated or described on the drawings and all levels and vertical dimensions should be expressed in metres to three places of decimals above this datum. Particularly on large jobs, it is usually necessary to relate the job datum to the Ordnance Survey datum. It is important to state clearly whether the Newlyn or Liverpool Ordnance Datum is being used.

6.2 Levels on plan
It is important to differentiate on site layout drawings between existing levels and intended levels, thus:

Existing level: × 58·210

Intended level: × 60·255

The exact position to which the level applies should be indicated by ' × '.
Finished floor levels should be indicated by the letters FFL followed by the figures of the level, thus:

FFL   12·335

6.3 Levels on section and elevation
The same method should be used as for levels on plan except that the level should be projected beyond the drawing with an arrowhead indicating the appropriate line, as in **5.4**.

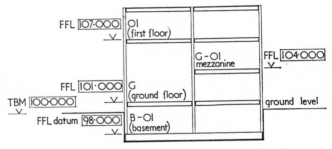

5.4 *Method of indicating levels on sections and elevations*

7 Instruments for metric measurement
A new British Standard BS 4484 has now been published on *Measuring instruments for constructional works. Metric graduation and figuring of instruments for linear measurement.* The following notes are based on that standard.

7.1 Folding rules and rods, laths, and pocket tape rules
Lengths of instruments will be as follows:
(a) Folding rules: 1 m
(b) Laths: 1 m, 1·5 m or 2 m
(c) Folding and multi-purpose rods: 2 m
(d) Pocket tape rules: 1 m, 2 m, 3 m, or 5 m
The form of graduation is shown in **5.5**. The instruments are graduated in millimetres along one edge with 5 m and 10 m graduation marks. Along the other edge the millimetre graduations are omitted.

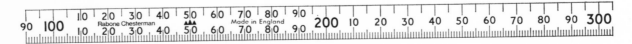

5.5 *Graduation markings for folding rules and rods, laths and pocket tape rules*

5.6 *Graduation markings for steel tapes*

7.2 Steel and synthetic tapes

Lengths will be: 10 m, 20 m or 30 m long. Etched steel bands are available in 30 m and 50 m lengths.

Tapes will be graduated at intervals of 100 mm, 10 mm (with the 50 mm centre graduation mark 'arrowed') and 5 mm. The first and last metre of the tape will be further subdivided into minor graduation marks at 1 mm intervals. See **5.6**. Note that synthetic material tapes however will *not* be subdivided into millimetres over the first and last metre.

7.3 Chains

Studded steel band chains will be in lengths of 20 metres and will be divided by brass studs at every 200 mm position and figured at every five metres. The first and last metre will be further divided into 10 mm intervals by smaller brass studs with a small washer or other identification at half metre intervals. The markings will appear on both sides of the band.

Land chains will also be in lengths of 20 metres, made up of links, which from centre to centre of each middle connecting link, will measure 200 mm. Tally markers will be attached to the middle connecting ring at every whole-metre position. Red markers will be used for 5 m positions, with raised numerals; yellow markers of a different shape and with no markings will be used for the rest, **5.7**.

7.4 Levelling staffs

Lengths will be 3 m, 4 m or 5 m long with a reading face not less than 38 mm wide.

Graduation marks will be 10 mm deep and spaced at 10 mm intervals. At every 100 mm the graduation marks will be offset to the left and right of centre **5.8**. The outside edges of the lower three graduation marks will be joined together to form an 'E' shape. Different colours will be used to distinguish graduation marks in alternate metres.

Staffs will be figured at every 100 mm intervals with metre numbers (small numerals) followed by the decimal point and first decimal part of the metre (large numerals).

7.5 Ranging rods

Lengths will be 2 m, 2·5 m or 3 m painted in either 200 mm or 500 mm bands alternating red and white. A rod of 2 m length painted in 200 mm bands is shown in **5.9**.

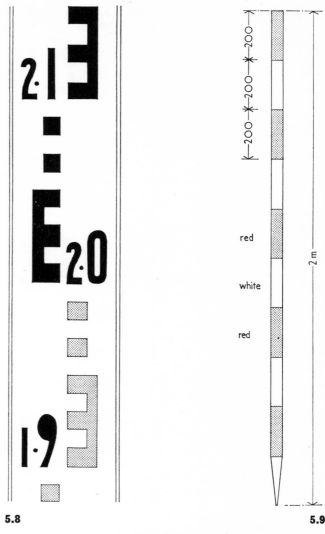

5.8

5.9

5.8 *Levelling staff marked in 10 mm increments*

5.9 *Metric ranging rod*

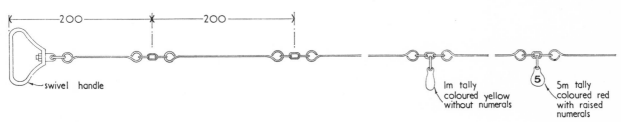

5.7 *Land chain markings*

8 Metric scales

The metric changeover will introduce a range of new scales with ratios similar but not quite the same as foot/inch units. The internationally agreed and recommended range of scales for use with the metric system is given in table IV. The simplicity of the metric system is reflected in the ease with which metric ratios can be operated **5.10**. The decimal subdivision of the scale 1:1 and 1:100 allows for almost all drawings to be done with one metric scale. However to acquire metric knowledge quickly, it is suggested that the recommended ratios of the metric scales should be used.

A comparison of metric and imperial scales giis ven in table V.

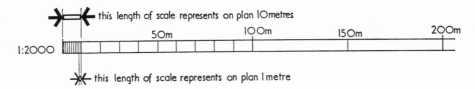

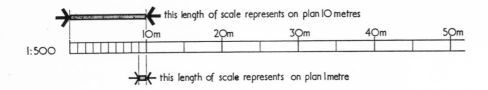

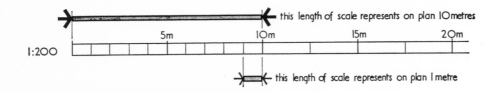

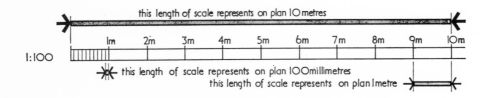

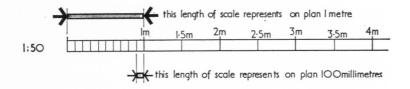

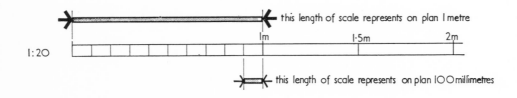

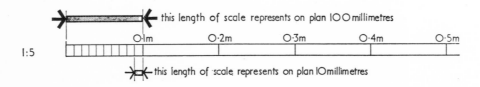

5.10 *Representation of metric lengths (to scale). This drawing may be used to check the correct interpretation of a scale.*

Table IV Preferred scales commonly used for different types of drawing

Type of drawing	Scales for use with		Notes
	Metric system	**Foot-inch system**	
DESIGN STAGE Sketch Drawings			Scales will vary but it is recommended that preference be given to those used in production stage
PRODUCTION STAGE **Location drawings** Block plan	1:2500 (0·4 mm to 1 m) 1:1250 (0·8 mm to 1 m)	1:2500 (1in to 208·33ft) 1:1250 (1in to 104·17ft)	Discussions with Ordnance Survey indicate that due to the large costs and labour involved these scales will probably have to be retained for the next 10 years
Site plan	1:500 (2 mm to 1 m) 1:200 (5 mm to 1 m)	1:500 (1in to 41·6ft) 1:192 ($\frac{1}{16}$in to 1ft)	
General location	1:200 (5 mm to 1 m) 1:100 (10 mm to 1 m) 1:50 (20 mm to 1 m)	1:192 ($\frac{1}{16}$in to 1ft) 1:96 ($\frac{1}{8}$in to 1ft) 1:48 ($\frac{1}{4}$in to 1ft)	
Component drawings Ranges	1:100 (10 mm to 1 m) 1:50 (20 mm to 1 m) 1:20 (50 mm to 1 m)	1:96 ($\frac{1}{8}$in to 1ft) 1:48 ($\frac{1}{4}$in to 1ft) 1:24 ($\frac{1}{2}$in to 1ft)	
Details	1:10 (100 mm to 1 m) 1:5 (200 mm to 1 m) 1:1 (full size)	1:12 (1in to 1ft) 1:4 (3in to 1ft) 1:1 (full size)	
Assembly	1:20 (50 mm to 1 m) 1:10 (100 mm to 1 m) 1:5 (200 mm to 1 m)	1:24 ($\frac{1}{2}$in to 1ft) 1:12 (1in to 1ft) 1:4 (3in to 1ft)	

Table V Comparison of scales

Scales for use with metric system	Scales for use with foot/inch system	
1:1 000 000	1:1 000 000	$\frac{1}{16}$in to 1 mile approx
1:500 000	1:625 000	1/10in to 1 mile approx
1:200 000	1:250 000	$\frac{1}{4}$in to 1 mile approx
1:100 000	1:126 720	$\frac{1}{2}$in to 1 mile
1:50 000	1:63 360	1in to 1 mile
1:20 000	1:25 000	2$\frac{1}{2}$in to 1 mile approx
1:10 000	1:10 560	6in to 1 mile
1:5000		
1:2000	1:2500	1in to 208·33ft
1:1000	1:1250	1in to 104·17ft
1:500	1:500	1in to 41·6ft
	1:384	$\frac{1}{32}$in to 1ft
1:200	1:192	$\frac{1}{16}$in to 1ft
1:100	1:96	$\frac{1}{8}$in to 1ft
1:50	1:48	$\frac{1}{4}$in to 1ft
1:20	1:24	$\frac{1}{2}$in to 1ft
1:10	1:12	1in to 1ft
1:5	1:4	3in to 1ft

8.1 Types of drawings

Types of drawings done to the most suitable scales are shown in **5.11** to **5.17**. Note that in **5.15**, **5.16** and **5.17** alternative dimensional units are shown for comparison. The method of expressing dimensions as shown in the shaded drawings is *not* recommended.

Scales on drawings

The scale should be stated on every drawing; drawings which are intended to be read by the non-specialist (eg sketch drawings), or which are to be microfilmed or published, should in addition have a drawn scale. Where two or more scales are used on the same sheet these should be particularly clearly indicated.

8.2 RIBA Metric scales

A number of metric scales are now on the market. Details of the scale being sold by the RIBA are given in table VI.

Table VI RIBA metric scale

Description		Metric ratios									Sizes
		Full size (1:1)	**1:5**	**1:10**	**1:20**	**1:50**	**1:100**	**1:200**	**1:1250**	**1:2500**	
RIBA Metric Scale For use in fully metric work Uses metric ratios and gives dimensions in metric	Edge 1	m*		m			m				
	Edge 2				m			m			
	Edge 3		m			m					300 mm 150 mm
	Edge 4								m	m	(Also in plastic)

m Gives dimensions in metric. * Full size (1:1) is read off the 1:100 edge, which is calibrated millimetre dividings.

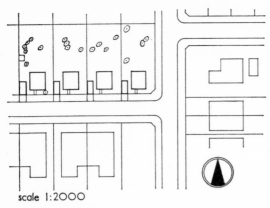

scale 1:2000

5.11 *Layout plan (note that the scale may be 1:2500 for some years even though not strictly an accepted scale. See para* **4.1** *p29)*

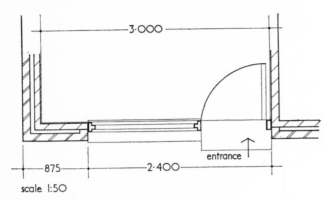

scale 1:50

5.14 *Location drawing*

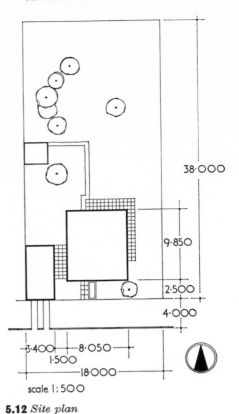

scale 1:500

5.12 *Site plan*

Note: These illustrations are drawn to the scales indicated

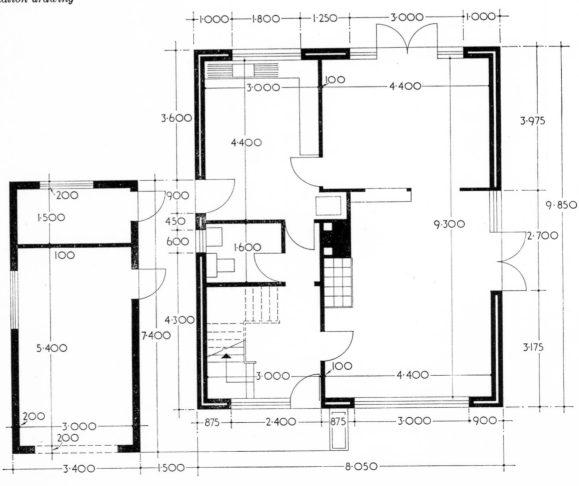

scale 1:100

5.13 *Location drawing and sketch plan*

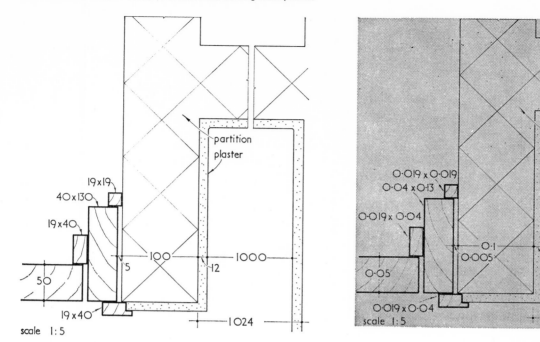

5.15 *Assembly detail drawing*

Note: These illustrations are drawn to the scales indicated. For the larger scale drawings (eg those illustrated in **5.15** to **5.17**) dimensions should be shown in millimetres and not in metres (as in the shaded drawings). There can be no confusion about unit size as the dimensions, in this context, could not be other than millimetres.

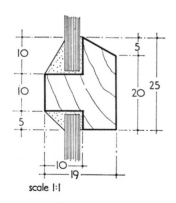

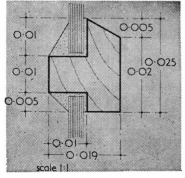

5.16 *Full size detail*

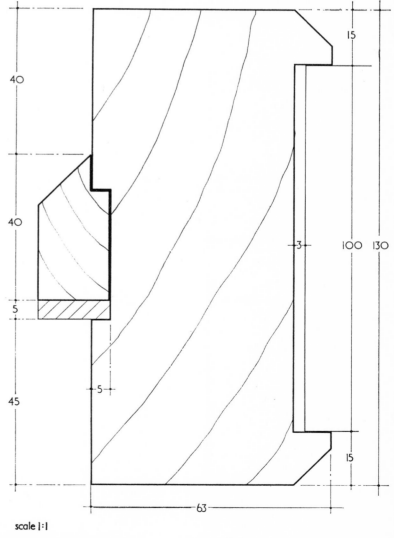

5.17 *Full size detail*

9 Guidance on metric style

The use of symbols and notation is explained for the benefit of secretarial staff and others concerned with written and typed documents and letters. This page could be photocopied and distributed throughout the office.

9.1 Symbols

1 The main symbols should be used as shown on the summary sheet, table VII. The same symbol ie m mm kg should be used for singular and plural values (1 kg, 10 kg) and no full stops or other punctuation marks should be used after the symbol unless it occurs at the end of a sentence. The sloping line as a separator between numerator and denominator should be used as it is used in the imperial system, ie 3 kg/m³ or 3 kg/cu m (three kilogrammes per cubic metre).

2 The unit should be written in full if there is any doubt about the use of a symbol. For example, the recognised unit symbol l for the unit litre can be confused with the number 1 and it will often be less confusing to write litre in full. Also during the changeover period, the unit symbol t for tonne should be avoided in situations where it may be confused with the imperial ton, and the unit tonne should be written in full.

3 When symbols are raised to various powers, it is *only* the symbol which is involved and not the number attached to it. Thus 3 m³ equals 3 (m)³ and *not* 3 m × 3 m × 3 m (ie the answer is 3 cubic metres and not 27 cubic metres).

4 During the changeover period, difficulty may be experienced with some office machinery and computers in current use when reproducing the squaring and cubing indices m² or mm², and m³ or mm³. In such cases the abbreviations 'sq' and 'cu' may be used temporarily (sq m and cu m) but the use of these should be limited to the UK as these terms may not be understood internationally.

5 Hyphenating a unit is wrong (milli-metres) and hyphenating a unit because it will overrun a margin should be avoided (eg millimetres, not milli—next line—metres).

6 A single space should separate figures from symbols: 10 m not 10m.

9.2 Notation

1 Care should be taken when expressing sizes of units to keep descriptions consistent within themselves and not mix different units, eg 1500 mm × 600 mm × 25 mm thick and *not* 1·5 m × 600 mm × 25 mm thick. Occasionally, as with timber sizes, it may be necessary to mix the units, eg 100 mm × 75 mm × 10 m long, but in any event, the latter description is preferred to: 0·100 × 0·075 × 10 m. An exception to this rule is when certain mathematical formulae must contain similar units for calculation purposes in which case these figures may be written as 0·1 × 0·075 × 10 (if all in metres) or 100 × 75 × 10 000 (if all in millimetres).

2 It is important to distinguish clearly between the (**metric**) tonne and the imperial ton. The tonne is equivalent to 2204·6lb whilst the ton is equal to 2240lb—a difference of 1·6 per cent.

3 The interval of temperature should be referred to as degree Celsius (°C) and not as centigrade. The word centigrade is used by the continental metric countries as a measure of plane angle and equals 1/10 000th part of a right angle.

Examples

Correct use	Incorrect use
33 m	33m 33m.
10·100 m	10 m 100 mm *
33 mm	33mm 33m.m.
50·750 kg	50 kg 750 g

* NOTE. Some metric values are expressed differently in certain metric countries. The value 10·100 m for example could mean ten thousand one hundred metres and *not* ten metres one hundred millimetres, as in UK.

10 References and sources

PD 6031:1968 *The use of the metric system in the construction industry* (2nd edition)

BS 1192:1969 *Recommendations for building drawing practice (metric units)*

BS 4484:1969 *Measuring instruments for constructional works. Part 1 Metric graduation and figuring of instruments for linear measurement*

Table VII Summary of symbols and notation

Quantity	Description	Correct unit symbol	Acceptable alternatives	Incorrect use	Notes
Numerical values		0·1 0·01 0·001		.1 ˙1 ·1 ·01 ·001	When the value is *less* than unity, the decimal point should be *preceded* by zero
Length	metre millimetre	m mm		m. M meter m.m. mm. MM M.M.	
Area	square metre	m²	sq m	m.sq sm sq.m. sq m.	
Volume	cubic metre cubic millimetre	m³ mm³	cu m cu mm	cu.m. m.cu. cu.mm. mm.cub. mm.cu.	
	litre (liquid volume)	l		l. lit.	Preferably write *litre* in full to avoid 'l' being taken for figure 'one'
Mass (weight)	tonne	t		ton	Preferably write *tonne* in full to avoid being mistaken for imperial ton
	kilogramme	kg		Kg kG kg. kilogram	
	gramme	g		g. G.	
Force	newton	N		N. n	Note that when used in written text, the unit of newton is spelled out in full and begins with a lower case letter 'n'. When used as unit symbol, in calculations or in a formula it is then expressed as capital letter 'N'

6 Decimal currency

Contents

1 Introduction
The metrication of English money was first envisaged in 1849 by the introduction of the florin, valued at one-tenth of a pound sterling. This action was a point in an evolution which began about 800 AD with a coinage based upon the *denarius* (d), of which 240 weighed one pound or *libra* (£).

1.1 Decimal Currency Act
The Decimal Currency Act 1967 has implemented a decision that British currency shall revert to a pound-penny system but have £1 divided into 100 parts. The Act comes into force on 15 February 1971 following a timed programme beginning in 1968 and ending some eighteen months after changeover D-day. The building industry is to go metric from 1 January 1970. It is therefore essential that drawings, bills of quantities and contracts that will be effective then should be prepared in metric sizes and decimal money.

2 New decimal coinage
Details of the new coins are given in table I. The illustrations are actual size.

3 Decimal currency notation
During the changeover period there will be a time when dual currency notation will be in use. This will be expressed as £sd and equivalents in decimal £p. The 'p' represents the new penny while the £ sign will remain unchanged. When writing or printing £ no full stop should be used unless a sentence ends with £. There are 100 new pennies to the £1, therefore sums containing pounds and pence must always be expressed to two places of decimals to avoid mistakes, eg

£1·01   £29·07   £105·32   £20 370·01

Where no new pence are involved amounts of money may be printed or written:
£1, £50, or alternatively **£1·00** and **£50·00**

3.1 Decimal points and the use of commas
When typewriters lacking a decimal point are used a *full-stop* in the baseline may be substituted. *Commas* should not be used for separating groups of digits; a space should be left as when expressing linear dimensions.*

* The Decimal Currency Board recommends the use of a comma as a thousand marker but for all normal documentation in the construction industry, including Bills of Quantities and other documents and letters where money is expressed, a space should be left between groups of digits. It may occasionally be desirable to include a comma for handwritten cheques to minimise forgery.

Table I Value and appearance of new decimal coins

Size and appearance	Name	Decimal value	£sd value
	New halfpenny	½p	1·2d
	New penny	1p	2·4d
	Two new pence	2p	4·8d
	Five new pence	5p	1/-
	Ten new pence	10p	2/-
	Fifty new pence	50p	10/- replacing 10/- notes

Bank notes to the values of £10, £5 and £1 remain as now

3.2 Sums less than one pound

Amounts in pence only may be printed or written in either of two ways as commonsense or convenience dictates:

1p, 10p, 40p   or   **£0·01, £0·10, £0·40**

The two symbols £ and p should never be used at the same time for writing a sum of money; either the one or the other must serve. An initial *zero* should always be included when the £ symbol is employed for sums amounting to less than one pound. The new halfpenny is always to be printed or written as a vulgar fraction:

½p, 3½p, 20½p   or   **£0·00½, £0·03½, £0·20½**

3.3 Cheques

New halfpennies written on cheques will be ignored by banks and should not, therefore, be included in total accounts or invoices. Cheques should be made out with *hyphens* replacing decimal points. The British Bankers Association recommend this and also that the following conventions should be adopted:

Handwritten cheques

figures	words
£29-00	Twenty-nine pounds
£29-26	Twenty-nine pounds 26
£29-08	Twenty-nine pounds 08
£0-26	Twenty-six pence

Typed or printed cheques
£29.08 Twenty-nine pounds 08 (*typed*)
£29·08 Twenty-nine pounds 08 (*printed*)

Guineas

The use of guineas is to be discontinued and only equivalent amounts should be charged or paid in future transactions if they are based upon £1·05.

4 Changeover programme

The decimal currency programme is given in table II. Old £sd currency and new £p currency will both be legal tender for not more than eighteen months after 15 February 1971.

5 Checklist for management

Anticipating administrative action in offices will be necessary well before 15 February 1971 if difficulties as regards clients, contractors and office routines are to be avoided. Final settlements, fee accounts, current bills of quantities, specifications, schedules of rates for both incompleted and new contracts, certification, salaries and taxes, will all be affected. The following checklist for necessary action is derived from suggestions made by the Decimal Currency Board.

5.1 Organisation

Appoint someone to assume chief executive responsibility for programming changes and maintaining liaison in:
Costs and budgetting
Client accounting
Staff training
Machines and stationery
RIBA, RICS and other professional bodies

Table II Programme of the change to decimal currency

Year	Date or period	Action
1967	14 July	Decimal Currency Act passed
1968	Spring	5p and 10p coins go into circulation
1969	August	½d demonetised *
	October	50p coin replaces 10/- note
1970	1 January	Half-crown (2/6) demonetised
1971	11-14 February	Clearing banks closed for normal public business
	15 February	**Changeover D-day** ½p, 1p and 2p coins go into circulation
		From this day banks will work wholly in the new currency. Accounts must be kept in decimal and all cheques and other bank documents have to be written decimally.
		Special arrangements will be made to handle £sd cheques not already cleared.
		Income tax and social security benefits and payments decimalised.
		Postage rates, postal orders and GIRO decimalised.
	15 February to (approx) August	1d, 3d and 6d coins demonetised: £sd coins will have to remain in use for some purposes until completion of decimal conversion of accounting equipment and coin-operated machines etc.

* Demonetisation: the act of depriving a coin of its status as legal tender. Authorised by Royal Proclamation.

5.2 Accounting

Establish effects of decimalisation upon:
Salaries and wages
Inland revenue
Social security payments etc
Customs and Excise
Banking arrangements and practice
Post Office stamps, orders, GIRO etc
Petty cash account
Management costing
Job certification
Tax concessions

5.3 Business machines

Make arrangements for conversion or replacements as soon as possible for:
Accounting machines
Adders and calculators
Postal franking
Cheque printing
Cost computing
Typewriting

5.4 Printed materials

Check and ensure availability of:
Stationery with cash rulings
Ready-reckoners
Price lists and catalogues

5.5 Administration

Survey effects of decimalisation on administration as regards:

Updating and keeping records

Legislative requirements

Staff training

Professional relationships with quantity surveyors, specialists, suppliers and contractors

Ministry pro formas and returns

The Decimal Currency Board gives information and advice on changeover problems.

6 Conversion £sd to £p

In the new decimal currency system £1 = 100p whereas in the £sd system £1 = 240d. It follows that only those £sd amounts which end in sixpence or multiples of sixpence will convert precisely into £p. Rounding off to the nearest decimal value will be necessary and so two standard tables have been drawn up to ensure that debtor and creditor gains and losses balance out, while inconsistencies are avoided. Table III is intended mainly for shopping but may have other applications, eg catalogue pricing.

Table III New halfpenny conversions for cash transactions

£sd	£p	£sd	£p
1d	½p	7d	3p
2d 3d	1p	8d	3½p
4d	1½p	9d 10d	4p
5d	2p	11d	4½p
6d equals	2½p	12d (1/–) equals	5p

Table IV Whole new penny conversions for banking and accounting purposes

£sd	£p	£sd	£p
1d	0	1/2 1/3	6p
2d 3d	1p	1/4 1/5 1/6	7p
4d 5d	2p		
6d 7d 8d	3p	1/7 1/8	8p
9d 10d	4p	1/9 1/10	9p
11d 1/– 1/1	5p	1/11 2/–	10p

6.1 Conversion factors and tables

Conversions of money references in contracts and agreements, or with reference to wage rates, schedule rates, low value unit prices and so on should not have tables III and IV applied to them. In such cases conversion should be made arithmetically on the basis that one £sd penny (1d) = 5/12 new penny; ie by applying a conversion factor = 0·4167. The Decimal Currency Board proposes to issue conversion tables of the ready-reckoner type but has not officially approved any tables so far published. Table XIX on p204 may, however, be used with reasonable confidence.

7 Cost analysis and estimates

The analysis of building costs and the preparation of rough estimates by price per square foot or cubic foot must be replaced by costs per square or cubic metre. A £4000 house with a floor area of 1000 sq ft (93 m²) will cost £43/m². The same house with a cube (for estimation) of 13 000 cu ft (368 m³) would therefore cost about £1·09/m³. Metric values in £p/m² lie between 10 and 11 times (actually 10·764) the equivalent £ per sq ft values. Similarly, costs reckoned in £/m³ are between 30 to 40 times (actually 35·315) the £ per cubic foot values. These rough figures should only be used to provide a quick check against miscalculation. Typical £p/m² cost analyses have been published in the Architect's Journal since 10 November 1960. See also tables XX and XXI on p204 and 205.

8 Specification and bills of quantities

The fifth edition of the RICS Standard Method of measurement is in metric terms. It will influence specification writing and require the use of conversion tables when comparisons are made with priced bills of quantities in imperial terms. Tables III and IV (this page) and XIX to XXI p204 to 205 will be relevant for this but table XXII p205 may also be needed.

8.1 Conversion from old bills of quantities

Caution should be exercised when converting prices from old bills of quantities to metric because the new standard method is not a direct translation of the old one into metric. Figures have been rounded off. Eg 'Voids not exceeding 4 sq ft' (0·372 m²) now becomes 'Voids not exceeding 0·50 m²' (5½ sq ft approx). Such alterations are not always directly translatable because contractors will now have to apply differing rates in pricing. An analysis of the clauses affected is given in notes appended to the fifth edition.

9 References and sources

DECIMAL CURRENCY BOARD *Decimal currency, three years to go* HMSO 1968 1s CI/sfB (Y)

DECIMAL CURRENCY BOARD *Expression of amounts in printing, writing and speech.* HMSO 1968 1s CI/sfB(Y)

DECIMAL CURRENCY BOARD *Britain's new coins* HMSO 1968 1s CI/sfB (Y)

DECIMAL CURRENCY BOARD *Points for businessmen* HMSO 1969 1s CI/sfB (Y)

DECIMAL CURRENCY BOARD *Cash transactions during the changeover* HMSO 1969 1s CI/sfB (Y)

DECIMAL CURRENCY BOARD *Facts and forecasts. Second edition* 1969 HMSO 1s. CI/sfB (Y)

ROYAL INSTITUTE OF CHARTERED SURVEYORS *Standard method of measurement of building works* London 19 (5th edition metric) The Institution CI/sfB (A4s)

7 Public sector and other official bodies

Contents

1 Introduction

This section summarises the work being done by the chief ministries and some other official bodies concerned with the construction industry.

The main interest is centred on dimensional co-ordination and dimensionally co-ordinated components. All the ministries have collaborated (as described in para 2 below) in producing, so far, seven main publications* on this theme:

DC 4 and DC 5 deal with the main controlling dimensions;
DC 6 explains how these dimensions are to be applied;
DC 7 deals with intermediate controlling dimensions;
DC 8 and DC 10 deal with spaces allocated for selected components and assemblies;
DC 9 gives advice on performance specification writing for building components.

This dimensional framework has been incorporated in British Standard 4330:1968 Recommendations for the co-ordination of dimensions in building: Controlling dimensions.

2 Ministry of Public Building and Works

To achieve greater standardisation between departments an interdepartmental Building Liaison Committee was set up and MOPBW was made responsible for co-ordinating development work in individual central government departments. An interdepartmental working party prepared recommendations for co-ordinated dimensions for schools, offices, hospitals, and housing which resulted in the original publications DC 1, DC 2 and DC 3. These were later put into metric terms and expanded, and now constitute documents DC 4, DC 5, DC 6, DC 7, DC 8, DC 9 and DC 10 (see para 1 above).

This work was done by a reconstituted working party known as the Interdepartmental Subcommittee for Component

Co-ordination working under the Building Liaison Committee. The subcommittee is now reinforced by a full-time Component Co-ordination Group whose job it is to carry out research and studies involved in co-ordinating the recommendations from individual departments.

Departmental bulletins are now being prepared to show in more detail how the recommendations in the DC documents can be used for specific building types. The programme of work of the subcommittee is:

1 preparation of revisions, in metric terms, of the original DC documents (now complete);
2 identification of six functional groups to assist the work of BSI in producing new metric standards for individual components in the correct order of priority (now complete)— see table II p5,6.
3 standardisation of other performance requirements after agreement on dimensional co-ordination of components has been reached;
4 study of the constructional and jointing problems involved in assembling individual components into a wide variety of building situations.

MOPBW is also to act as a clearing house for all government departmental building work information to be given to BSI and industry generally.

2.1 Publications

Apart from the DC documents mentioned above, two bulletins have been published so far in an MOPBW series: Going metric in the Construction industry. The first deals with 'why and when', and the second with dimensional co-ordination. A metric bibliography is also published, with supplements to keep it up to date.

Other bulletins deal with education and training, a building case study and application of dimensional co-ordination. See bibliography p189.

2.2 Metrication officer

The supervision of the implementation of the changeover to the metric system within the Ministry is the responsibility of the Metrication officer L. J. F. STONE. His office also issues metric instructions and metric memorandums which are not generally available but which are held by all MOPBW libraries.

2.3 Demonstration building

A metric controlling dimensions demonstration building has been erected by MOPBW at the Building Research Station to give a better understanding of the dimensional relationships between components, as well as providing a full-scale model as guidance for designers in the problems they have to face in producing co-ordinated metric designs. The demonstration consists of a series of building frames representing the five main types of construction used in Ministry buildings: load bearing brickwork, in situ concrete frame, Public Building frame, composite construction, and Nenk system.

---

*For full titles see section 31, Selective bibliography p185.

3 Ministry of Housing and Local Government

As in the case of other ministries, MOHLG's aim is to establish a dimensional framework within which it will be possible to make maximum use of dimensionally co-ordinated standard components and so reduce building costs and provide better value for money.

The department's policy is outlined in MOHLG Circular 1/68 (HMSO). The Research and Development Group has published Design Bulletin 16 *Co-ordination of components in housing 'The metric dimensional framework'* This incorporates the aspects of BS 4330 *Controlling dimensions* considered relevant to Public Sector Housing, together with examples of application. The bulletin replaces Design Bulletin 8. Recent work has been directed towards establishing the dimensional and functional performance of components. The dimensional information being the department's contribution to the BSI Functional Group panels. Work is now proceeding on the functional aspects, together with a working party consisting of representatives of Local Authorities and Consortia. Test pilot schemes are being carried out to examine practical implications.

Metric versions are also now available of the following Design Bulletins:

DB 1 Some aspects of designing for old people
DB 2 Group flatlets for old people
DB 3 Space in the Home

Bulletins are at present under preparation covering user requirements for bathrooms and kitchens.

3.1 Timing the change to metric

Authorities will not be expected to submit metric plans at layout stage before **1 January 1969** or at tender stage before **1 January 1970,** unless there is some special justification. All schemes submitted after **1 January 1972** *must* be in metric dimensions.

From this date, the single floor-to-floor height for housing in the public sector is fixed at a mandatory height of 2600 mm (Circular 31/67). When schemes are designed in metric, the associated site development works should also be in metric.

3.2 The transitional period

For the benefit of contractors, manufacturers and local authorities, departments of regional offices of the MOHLG, in collaboration with the National Building Agency, will establish a clearing house of information on numbers and types of dwellings to be designed in metric and imperial dimensions, types of construction, tender dates and starting dates.

3.3 Standard metric components

All authorities have been asked to make the fullest use of the opportunity provided by metrication for rationalising design, production and site operations by using standard housing components.

The departments will be associated with the production of metric standards for all components: manufacturers can then design components to these standards advised by the departments and the NBA.

Once a British Standard has been promulgated, components esigned in accordance with the standard should be used. In principle for all schemes submitted at tender stage on or after **1 January 1972,** authorities will be required to specify metric British Standards for all components for which such standards exist. Metric schemes submitted before this date should similarly specify metric standards, where possible. Until metric components conforming to a new metric British Standard are available, authorities should continue to specify the existing components which, by the end of 1968, will be described also in metric terms. Authorities have been urged not to develop their own local standards for metric components. Any metric scheme submitted at layout stage will be expected to conform to the recommendations in Design Bulletin 16, and approval may be withheld if it departs unreasonably from those recommendations.

Metric schemes will be subject to the minimum standards and cost limits set out in Circular 36/37 (in Wales, 28/67). These standards are set out in the revisions to appendixes I and II of that circular, which will apply to metric schemes, and are included in section 26 of this handbook*.

3.4 Metrication of Building Regulations

This work is being done in three stages:
Stage I Exact metric equivalents:
(a) by regulation;
(b) by dimensions
Already published.

Stage 2. Proposed rationalised metric dimensions for comment. *Publication September* **1969.**†
Stage 3 The final and complete revision in metric will be published *towards the end of the metric changeover period.*

3.5 Metric Liaison officer

J. A. FOWLER

4 Department of Health and Social Security

This department also stresses the importance of dimensional co-ordination and of achieving standard ranges of co-ordinated components.

Considerable work has already been done by the department, in conjunction with Interboard Study Groups, in developing feet/inch co-ordinated components‡. This work is now being revised in metric terms.

Design notes published so far deal in metric with basic anthropometric data, user activities and room data. Work is continuing on establishing a dimensional framework for health buildings; on techniques for incorporating within that framework, co-ordinated components; on providing component performance specifications; on identifying and selecting basic dimensional and other properties of the ranges of components to be used.

Hospital Design note 5 *Co-ordination of Components for health buildings* will be published Autumn 1969. This gives advice on principles of co-ordination; design techniques and the co-ordinated components to be adopted for health buildings. Until new metric standards are developed for ranges of components, boards have been asked to specify those contained in the Design Note. A metric guidance handbook for architects for limited distribution to regional architects will be published late 1969.

4.1 Metric programme for hospital and health buildings

All projects starting on the ground before **1 January 1970** will be in feet/inches. All projects starting on the ground after **1 January 1973** will be in metric terms. Certain

---

*Full details are contained in a joint circular issued by the Ministry of Housing and Local Government and the Welsh Office on 1 January 1968: Circular 1/68 *Metrication of housebuilding* HMSO 1s 6d

†Also a circular is issued for guidance to local authorities. Circular 48/69 (MOHLG) or 43/69 (Welsh office)

‡These have been summarised in a compendium. Vol I covers the years 1965-1967; Vol II covers 1967-1970. See also Hospital Design Note 3 *The* IHB *programme; components and assemblies.*

projects starting on the ground between **1 January 1970** and **1 January 1973** will be carried out in metric terms.

4.2 Metrication officer
D. PILKINGTON (For complete department)
J. CALDERHEAD (Architects' Branch)

5 Department of Education and Science

The most important work being undertaken by DES at present is the development of a common pool of dimensionally related components: components capable of being used without modification in more than one educational system. For future development projects DES will be using systems already evolved by the educational building consortia and will, therefore, follow whatever arrangements the various consortia are making for the changeover to metric.

Building Bulletin 42 *Co-ordination of components for educational buildings* is one of the most important contributions in the field of rationalisation of building components to come from the public sector. This contains, as an appendix, metric analogues for the dimensions employed in Building Bulletin 24 *Controlling dimensions for educational building*. All further Building Bulletins not in draft form will employ metric measurements. Metric equivalents for most basic reference documents have now been published including Administrative memorandum 14/68 which gives metric analogues of the *Standards for school premises regulations* (Statutory Instrument 890:1959). See also bibliography p 189.

Projects which are expected to begin construction on or after 1 April 1972 must be designed in metric measurements.

5.1 Metrication officer
S. M. SMITH

6 Greater London Council

Supplements to the *London Building Acts* and the current *London Building (Constructional) By-Laws* have been published setting out the metric equivalents to all measurements and values included in those Acts and By-Laws. Documents dealing with means of escape in case of fire have also been issued.

Similar supplements will follow in connection with other regulations, codes and documents published by the council in connection with building.

6.1 Metrication officer
W. F. A. BEESTON metrication officer for the Department of Architecture and civic design.

7 Scottish Development Department

Metric equivalents of all dimensions referred to in the Building Standards (Scotland) Regulations 1963-67 have been published, and work is proceeding on a final and complete revision of the Regulations.

Advice on internal design standards for dwellings (in metric) is available in the new Scottish Housing Handbook Bulletin 1, which replaces the imperial standards set out in the Scottish Housing Handbook part 3 (1956 edition).

7.1 Metrication officer
A. M. GRAHAM

8 Scottish Education Department

In due course, the School Premises (General Requirements and Standards) Regulations 1967, will be amended to take account of the changeover to the metric system. Meanwhile, it will be necessary to apply metric equivalents in order to determine whether projects expected to start after **1 January 1970** and designed in metric terms comply with them or not. The School Building Code will also require amendment. The opportunity may also be taken to make such alterations in standards as may be required. Metric equivalents of the dimensions in the current Regulations and Code are given in Appendix II to Circular 701 (VII). The Further Education Building Code already includes metric equivalents.

8.1 Standards for Social Work Projects
There are, at present, no statutory standards equivalent to the School Premises Regulations for buildings in the social work field, for example, children's homes, remand homes, old people's homes, etc. Future guidance material on space standards in this field will be in metric terms or indicate metric equivalents.

8.2 Metrication officer
W. A. P. WEATHERSTONE

9 National Building Agency

The NBA are now providing an advisory service to designers, contractors and manufacturers to answer queries about the application of design disciplines and sizing and availability of components.

In addition the NBA is monitoring a random sample of local authority housing schemes and contracts, suppliers and sub-contractors associated with these schemes. The information obtained will be collated and summarised periodically.

9.1 Appraisal of Metric Housing Systems
Those system sponsors who wish to do so may now obtain an appraisal certificate for low rise housing systems in metric. Where system sponsors already hold an appraisal certificate, the metric version of the system will be included in the same certificate. A summary of the metric dimensional disciplines will be included within the certificate.

9.2 Training
Following the thirty-four half-day seminars which the NBA have given for local authorities in England and Wales, a series of six one-day seminars have been held for architects in contractors' offices and consultant architects. Two seminars were held in London and four in provincial cities. Lectures cover—programme for change to metric; SI units; dimensional co-ordination in housing; and availability of metric materials and components.

9.3 Publications
Metric Housing—What it Means. A booklet summarising one of the talks given in a series of seminars for local authority housing designers. The booklet deals with, the programme, when to change, how to prepare for metric change and SI units.

The Standard Method of Measurement in Metric—A comparison with Imperial. Leaflet produced as an aid to Quantity Surveyors. It shows, by comparison, the changes that have been made in converting the Standard Method of Measurement to metric.

Wall Charts for Designers. A wall chart for housing designers showing the co-ordinated dimensions for housing, together with useful metric tables and conversions.

Metric component file. Contains data sheets on dimensionally co-ordinated metric products and components. Costs £6 a year for 6 issues.

10 Building Research Station

BRS is directly involved with the BSI in preparing and implementing the metric programme, especially in the provision of specialist advice and information. The station is also strongly represented on the BSI Functional Group Panels and other BSI committees. BRS publications now contain only metric SI units.

11 British Standards Institution

The British Standards Institution has been entrusted with the overall planning of the metric change in British industry. In the construction field, the coming change has provided the opportunity of obtaining dimensional co-ordination over a wide range of components and assemblies, and the revision of standards relating to building products is proceeding along these lines.

There is a complete programme for metric change in the industry as a whole already published by BSI (PD 6030). A later programme (PD 6249) deals with metric change for actual components. This is related directly to BSI work on revision of building standards as divided among "functional group panels" and shows when metric standards can be expected. The groups are: structure; external envelope; internal envelope; services/drainage; and fixtures. The latest stage is the completion of detailed lists of components and assemblies in the first four of these groups (PD 6432) which enables everyone concerned to see (a) the relative importance of any one item for the purpose of co-ordination (b) the dimensions which are to be co-ordinated and (c) a general grouping of the construction materials to cover all components.

Meanwhile architects and designers are able to prepare drawings on the basis of published controlling dimensions for various types of building issued as recommendations by BSI in 1968. This document (BS 4330) is the basis of co-ordination work within the functional groups mentioned. A published list of metric standards issued or in progress is published periodically.

12 Royal Institute of British Architects

The RIBA is concerned that the profession should participate to the full in the change to metric and the concurrent emphasis on dimensional co-ordination. It has supported the BSI programme for the change and considers that it is now for members to make themselves aware of the implications so that they may take their proper place in implementing the change.

The RIBA is represented on all the BSI committees preparing the revised British Standards and other documents concerned with the change and has a member on each of the functional groups set up by BSI to apply in detail the BSI recommendations on metric dimensional co-ordination. At the institute, staff members, assistant secretary Kenneth Claxton and Mrs Cassandra Bhatia, both of whom are architects, are responsible for ensuring that there is liaison between those members representing the RIBA on BSI, and that BSI drafts are appraised by the Metric Change Liaison members in the Allied Societies. Recently the reorganizations of the BSI representatives into RIBA liaison committees has been conducted to ensure that an appropriate and consistent RIBA comment can be made throughout the respective BSI committees.

The RIBA are also conducting some six seminars on the change to metric which are designed to provide feedback based upon experience derived from actually having built in metric. The results of these seminars are being reported in the newly introduced Technical Papers produced by the RIBA. In addition to this the RIBA are producing two Metric Teaching Texts in this same series, and have also specially selected packs of Metric information for ready use by architects.

In addition the RIBA staff give advice on essential reading and sources of speakers for local bodies of architects.

Further courses on metric drawing practice are also being considered. Information on the availability of metric products has become a major concern of the RIBA who are collaborating with the Building Centre to encourage manufacturers on the production of the new co-ordinated products required. In this context the RIBA is also collaborating with the NCC on the provision of statistics related to projects being conducted in metric.

13 Construction Industry Metric Change Liaison Group (CIMCLG)

Several bodies from within the construction industry have formed a Metric Change Liaison Group. They are:

The Association of Industrialised Building Component Manufacturers Ltd.	(AIBCM)
The Modular Society	
The National Federation of Builders and Plumbers Merchants	(NFBPM)
The National Federation of Building Trades Employers	(NFBTE)
The National Council of Building Material Producers	(NCBMP)
The Royal Institute of British Architects	(RIBA)
The Royal Institution of Chartered Surveyors	(RICS)
The Institute of Structural Engineers	(ISE)

The formation of the group will enable the member bodies to discuss jointly the problems associated with change to metric, with a view to ensuring a collective understanding and to initiate joint action on agreed issues.

The current chairman is Peter Cocke ARIBA, and further information is available from Kenneth Claxton, assistant secretary, Professional Services at the RIBA, 66 Portland Place, London W1 (telephone 580 5533 ext 223).

14 Metrication Board

The job of the Board is to guide, stimulate and co-ordinate the planning for the change to metric for the various sectors of the economy. It is an *advisory* board with members drawn from industry, the distributive trades, education, the general public and consumers.

Its membership is as follows:

Chairman: Lord Ritchie-Calder

Deputy Chairman: Lord Bessborough

Director: Gordon Bowen

Secretary: F. Howard Whitaker

Members: Ailsa Stanley, Herbert J. Cruickshank, Alan G. Dawtry, Edward F. Knight, Professor M. L. McGlashan, Sir Thomas Padmore, Dr F. Lincoln Ralphs and Mark Abrams,

Further appointments are to be made.

Headquarters: 22 Kingsway, London WC2

8 **Anthropometric data**

Contents

1 Principles of anthropometrics

The body and reach characteristics of people have a direct influence on design. Average (mean) dimensions of some of the more important of these characteristics for men and women are shown in **8.1** and **8.2**.

Although in certain situations it is appropriate to use the average as a criterion it must be emphasised that averages should be treated with caution. Where account is taken of average dimensions only, the likelihood is that in any specific circumstance only about half of the population under consideration will be satisfied. To ensure that the broad range of the population is accommodated account must be taken of people whose dimensional characteristics deviate from the mean. In tables I and II and in **8.4** and **8.5** values for each characteristic are given for the 5th percentile, i.e. the position at or below which measures for 5 per cent of the total population are found, and for the 95th percentile, the position at or below which measures for 95 per cent are found. It is not always economic or practicable to cover 100 per cent of the population by

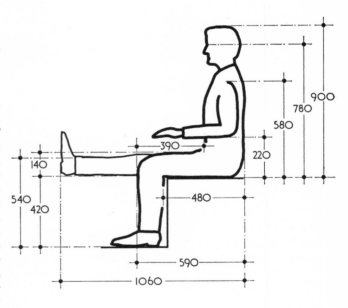

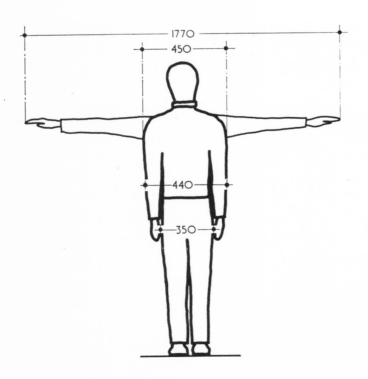

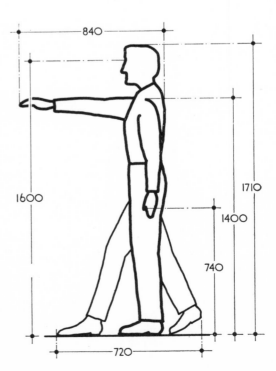

8.1 *Mean average (50th percentile) dimensions of adult British males*

catering for people at the extremes, and attempts to do so can compromise the convenience of solutions for the broad range of normal people. It may not for example be possible to obtain a solution to a specific design problem which is equally efficient for a typical ambulant person and a person in a wheelchair.

2 Application of anthropometric data

As an example, it can be observed from the data in table II that 95 per cent of men are taller than 1628mm. When data are applied to design problems it is usually found that there is a limiting factor in one direction only, eg if the problem relates to obstructions at head height the measures of short people are not significant. In applying data from table II the architect should inquire in which direction the dimension is critical. It is not the case that whenever the value for the 95th percentile is observed 95 per cent of the population will be accommodated; if the critical dimension is in the opposite direction only 5 per cent will be accommodated, and the correct course is to apply the 5th percentile instead.

3 Body clearances and maintenance access

Typical allowances for body clearances are given in **8.6 to 8.11**. Space requirements for maintenance are given in **8.12 to 8.19**. Minimum entries for one man are as follows:

330mm to 450mm: difficult

450mm to 610mm: fair

610mm to 920mm: good

(See also section **9**: Internal circulation.)

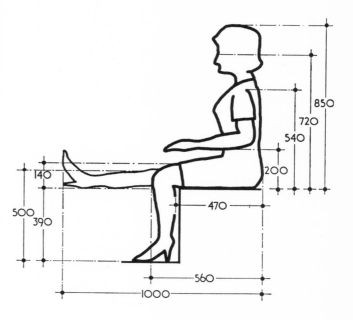

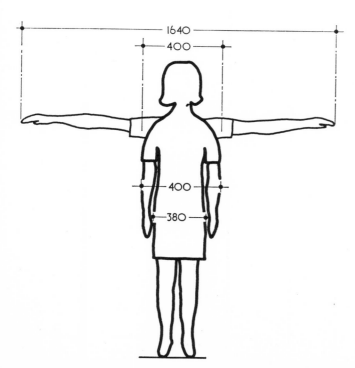

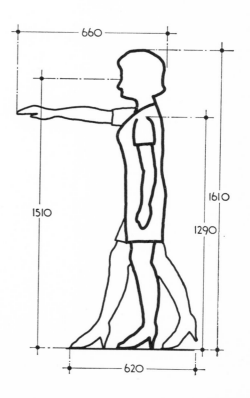

8.2 *Mean average (50th percentile) dimensions of adult British females*

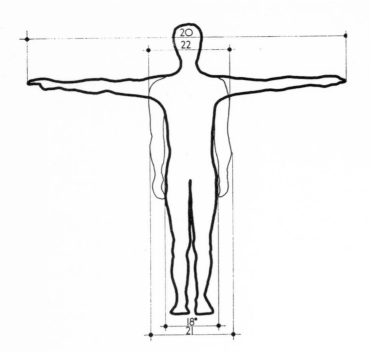

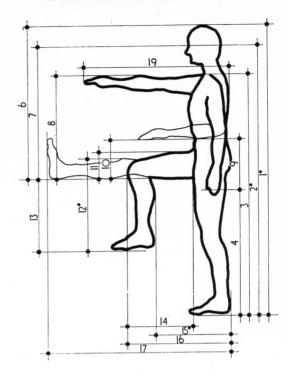

8.3 *Key dimensions listed in table II. To assist interpretation these figures are shown unclothed, though allowances have been made in table II for the wearing of clothes and shoes. Dimensions marked ● are most commonly used.*

4 References and sources

Sources of anthropometric data reproduced here are listed in the references below. For original sources of data used in table II reference should be made to AJ information sheet 1185. Any apparent inconsistencies between dimensions in table II, such as distances between eyes and top of head for standing and seated people, are because data were drawn from several studies.

AJ information sheet 1185 Anthropometric data and their application, AJ, 1963, February 13 CI/SfB (E2d)

AJ information sheet 1194 Internal circulation, AJ, 1963, March 20 CI/SfB 91 (E2p)

GOLDSMITH, S. Designing for the disabled, London, 1967, Royal Institute of British Architects, 2nd edition £3 10s CI/SfB (E3p)

FLOYD, W. F. and others. A study of the space requirements of wheelchair users, Paraplegia journal, May 1966 p24 CI/SfB (E3p)

BRITISH STANDARDS INSTITUTION BS 4467: 1969. Anthropometric and ergonomic recommendations for dimensions in designing for the elderly, London, 1969, The Institution CI/SfB (E2d)

Table I   Estimated heights, in millimetres, of children at various ages

Age	Boys			Girls		
	Percentiles			Percentiles		
	5th	50th	95th	5th	50th	95th
3	879	942	1005	876	930	984
6	1068	1143	1218	1059	1138	1217
9	1215	1311	1407	1204	1300	1396
12	1345	1458	1571	1355	1468	1581
15	1504	1633	1762	1507	1603	1699
18	1651	1755	1859	1534	1626	1718

Table II Estimated dimensions, in millimetres, of body and reach characteristics of the British population

Because in nearly all situations to which the architect applies anthropometric data users will be clothed, the data in this table includes allowances for clothing and shoes. The allowances for footwear are 28mm for men, 40mm for women and 31mm for elderly women. The allowances for clothing, affecting most of the dimensions from item 6 on, range according to circumstance from 3mm to 20mm. In situations where clothes are not worn, eg bathrooms and shower rooms, appropriate deductions should be made.

Key dimension	MEN aged 18 to 40			WOMEN aged 18 to 40			ELDERLY women aged 60 to 90			Examples of applications to design problems
	Percentiles 5th	50th	95th	Percentiles 5th	50th	95th	Percentiles 5th	50th	95th	
STANDING										
1 **Stature**	**1628**	**1737**	**1846**	**1538**	**1647**	**1756**	**1454**	**1558**	**1662**	95th: Minimum floor to roof clearance; allow for headgear, say 100mm, in appropriate situations.
2 **Eye height**	**1524**	**1633**	**1742**	**1437**	**1546**	**1655**	**1338**	**1451**	**1564**	50th: Height of visual devices, transoms, notices, etc.
3 Shoulder height	1328	1428	1428	1237	1333	1429	1201	1288	1375	5th: Height for maximum forward reach
4 Hand (knuckle) height	703	770	837	—	—	—	653	732	811	95th: Maximum height of grasp points for lifting.
5 **Reach upwards**	**1972**	**2108**	**2264**	—	—	—	**1710**	**1852**	**1994**	5th: Maximum height of controls; subtract 40mm to allow for full grasp.
SITTING										
6 Height above seat level	841	900	959	790	849	908	739	798	857	95th: Minimum seat to roof clearance; allow for headgear (men 75mm, women 100mm) in appropriate situations.
7 Eye height above seat level	726	785	844	676	735	794	621	684	747	50th: Height of visual devices above seat level.
8 Shoulder height above seat level	537	587	637	494	544	594	479	529	579	5th: Height above seat level for maximum forward reach.
9 Lumbar height	—	254	—	—	—	—	—	—	—	50th: Height of table above seat.
10 Elbow above seat level	178	224	270	157	203	249	143	193	243	50th: Height above seat of armrests or desk tops.
11 Thigh clearance	124	149	174	121	146	171	93	131	169	95th: Space under tables.
12 **Top of knees, height above floor**	**506**	**552**	**598**	**473**	**519**	**565**	**460**	**498**	**536**	95th: Clearance under tables above floor or footrest
13 Underside thigh, height above floor	402	435	468	385	418	551	366	404	442	50th: Height of seat above floor or footrest
14 Front of abdomen to front of knees, distance	336	386	436	—	—	—	—	—	—	95th: Minimum forward clearance at thigh level from front of body or from obstruction, eg desk top.
15 **Rear of buttocks to back of calf, distance**	**436**	**478**	**520**	**423**	**465**	**507**	**424**	**470**	**516**	5th: Length of seat surface from backrest to front edge.
16 Rear of buttocks to front of knees, distance	568	614	660	542	584	626	520	579	638	95th: Minimum forward clearance from seat back at height for highest seating posture.
17 Extended leg length	998	1090	1182	—	—	—	892	967	1042	5th (less than): Maximum distance of foot controls, footrest etc from seat back.
18 **Seat width**	**328**	**366**	**404**	**353**	**391**	**429**	**321**	**388**	**455**	95th: Width of seats, minimum distance between armrests.
SITTING AND STANDING										
19 Forward reach	773	848	923	600	675	750	665	736	807	5th: Maximum comfortable forward reach at shoulder level.
20 Sideways reach	1634	1768	1902	1509	1643	1777	—	—	—	5th: Limits of lateral finger tip reach; subtract 130mm to allow for full grasp.
21 Width over elbows	389	456	523	351	418	485	—	—	—	95th: Lateral clearance in work space.
22 Shoulder width	420	462	504	376	418	460	381	431	481	95th: Minimum lateral clearance in work space above waist.

NOTE: Dashes indicate that no data were given in the sources used, or that data were unreliable.
Figures printed in bold type are the most commonly used dimensions (see also **8.3**)

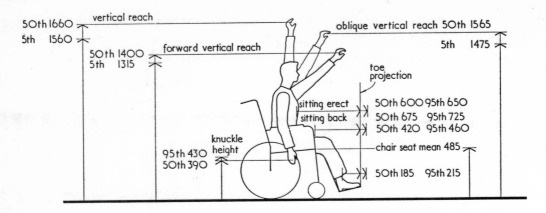

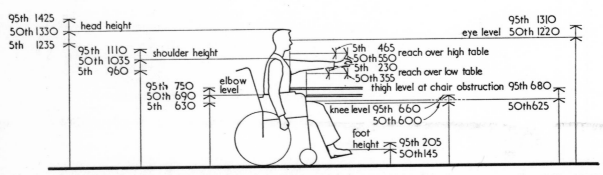

8.4 *Dimensions of adult male wheelchair users. These dimensions and those in* **8.5** *relate to people who use standard wheelchairs and who have no major impairment of upper limbs.*

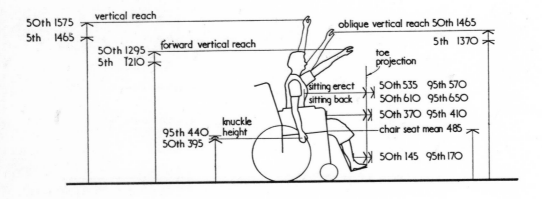

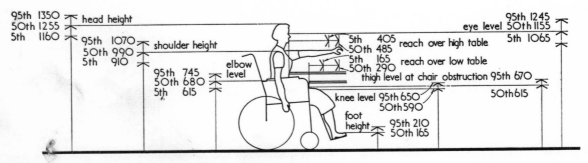

8.5 *Dimensions of adult female wheelchair users*

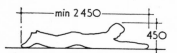

8.6 *Body clearance: prone*

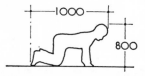

8.7 *Body clearance: crawl*

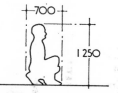

8.8 *Body clearance: squat*

8.9 *Body clearance: stoop*

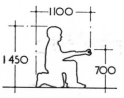

8.10 *Body clearance: kneel*

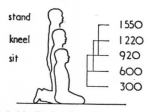

stand	1550
kneel	1220
sit	920
	600
	300

8.11 *Body clearance: maintenance reach levels*

650 diam
(800 square preferable)

8.12 *Services access: crawlway*

min 450 diam or square

8.13 *Service access: hatch*

min 600 high x 400mm wide

8.14 *Service access: access panel*

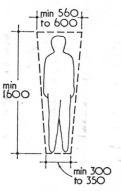

min 560 to 600

min 1600

min 300 to 350

8.15 *Service access: catwalk*

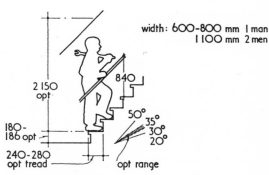

width: 600–800 mm 1 man
1100 mm 2 men

2150 opt

840

50° 35° 30° 20°

180–186 opt

240–280 opt tread

opt range

8.16 *Service access: stairs*

min entries for one man (mm):
330–450 difficult
450–610 fair
610–920 good

width: min 800 mm
opt 1100 mm

2150 opt

850 (910 at 0°)

max diam 40 mm

20° 10° 0°

opt range

8.17 *Service access: ramps*

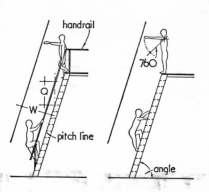

recommended for angles 50° to 75°
handrails are required on both sides if risers are not
left open or if there are no side walls
widths: 500mm to 600mm with handrails
600mm min between side walls

angle	W(mm)	Q(mm)
50°–55°	1620–1570	880
57°–60°	1500–1450	900
63°–66°	1370–1320	910
69°–72°	1270–1200	920
74°–77°	1150–1050	950

recommended riser 180mm to 250mm
tread 75 mm to 150 mm
45mm diam max for handrail

8.18 *Service access: step ladders*

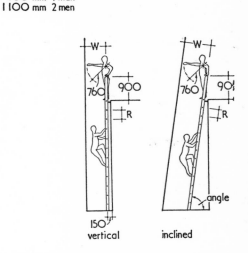

vertical   inclined

generally suitable for vertical movements from 75° to 90°
ladder frame should extend 900mm above platform
widths: 380mm min, 450mm desirable
600mm min between side walls
150mm toe space

angle	R(mm)	W(mm)
75·0°	330	1150
78·0°	335	1050
80·5°	340	1000
83·0°	350	950
85·0°	360	900
87·5°	370	850
90·0°	380max	800
	300min	

provide back guard over 6 000 mm high

8.19 *Service access: rung ladders*

9 Internal circulation

Contents

1 Fire regulations

Relevant fire regulations are important in planning any internal circulation.

Tables I and II summarise requirements of the GLC as set out in the LCC Code of Practice *Means of escape in case of fire*. These figures provide a useful guide to capacities of passages and stairs of various widths whether intended for escape purposes or not. Where a passage or staircase is intended as a means of escape, it is advisable to have discussions with the local fire authority at an early stage. The code divides buildings into nine classes, and while exit widths will be the same for some classes, the maximum travel distances (ie the distance to be traversed in order to reach a place of safety such as the open air or a protected staircase) will differ. The nine classes are set out in table III. Note that in tables I, II and III exact metric equivalents are given as required by the code, with suggested possible rounded dimensions in brackets.

2 Space allowances in buildings

As a guide to assessing circulation, the areas listed in table IV may be used. These areas are overall space requirements for both the activity and its associated circulation.

Table I Minimum widths of escape passages and staircases in one-staircase buildings, classes I to IV, VII and VIII (GLC)

Number of persons	Persons evenly distributed on all floors (mm)	Persons mainly concentrated on one floor (mm)
Up to 50	762 (760)	762 (760)
75	762 (760)	914 (920)
100	914 (920)	1066 (1070)
150	1066 (1070)	1219 (1220)
200	1219 (1220)	1371 (1370)
250	1371 (1370)	1524 (1520)
300	1524 (1520)	

Table II Minimum widths of escape passages and staircases in two-staircase buildings, classes V and IX (GLC)

Number of persons	Persons evenly distributed on all floors (mm)	Persons mainly concentrated on one floor (mm)
Up to 200	914 (920)	1066 (1070)
300	1066 (1070)	1219 (1220)
400	1219 (1220)	1371 (1370)
500	1371 (1370)	1524 (1520)
650	1524 (1520)	2134 (2130)

Table III Classes of building according to LCC Code of Practice *Means of escape in case of fire*

Building class	Construction	Maximum travel distances (ie distance to be traversed to reach place of safety) in metres
I Up to 92·9m² (93m²) at first floor level	Wood floor with plaster or other suitably protected ceiling	18·3 (18·0)
II Up to 185·8m² (186m²) at first floor level	Incombustible floors	30·5 (30·0) of which 18·3 (18·0) must be within a protected corridor
III Up to 325·2m² (325m²) at first floor level	Incombustible floors	As for class II
IV Small buildings in single occupation with no floor more than 12·8m above ground level		18·3 (18·0) on each floor
V Any building not falling into classes I to IV (because of larger area or construction of floors) must have at least two protected or external staircases	(a) Incombustible floors (b) Combustible floors	(a) 30·5 (30·0) (b) 18·3 (18·0)
VI Buildings without limitation of floor area and not more than one storey in height		18·3 (18·0)
New residential buildings VIII Dwellings in single occupation VIII Blocks of flats and maisonettes		30·5 (30·0)
IX Blocks of flats or maisonettes with two or more staircases		27·4 (27·0)

Table IV Minimum areas per person in various types of buildings

Occupancy	Area per person (m²)
Assembly halls (closely seated)	0·46m² (based on movable seats, usually armless, 450mm centre to centre; with fixed seating at 500mm centre to centre will increase to about 0·6m²)
Dance halls	0·55m² to 0·9m²
Restaurants (dining-area)	0·9m² to 1·1m²
Retail shops and showrooms	4·6m² to 7·0m² (including upper floors of department stores except special sales areas)
Department stores, bazaars or bargain sales areas	0·9m² (including counters, etc) 0·46m² (gangway areas only)
Offices	9·3m² (excluding stairs and lavatories)
Factories	7m²

Table V Area per person to be allowed in various circulation areas

Occupancy	Area per person (m²)
Overall allowance for public areas in public-handling buildings	2·3 to 2·8
Waiting areas, allowing 50 per cent seating, 50 per cent standing without baggage, allowing cross-flows (eg airport lounge)	1·1 to 1·4
Waiting areas, 25 per cent seating, 75 per cent standing, without serious cross-flows (eg waiting rooms, single access)	0·65 to 0·9
Waiting areas, 100 per cent standing, no cross-flows (eg lift lobby)	0·5 to 0·65
Circulating people in corridors, reduced to halt by obstruction	0·2
Standing people under very crowded conditions—acceptable temporary densities	Lift car capacities: 0·2m² (four-person car) ; 0·3m² (thirty-three-person car)

Table VI Capacities in persons per hour and per minute, for various widths of corridors and stairs, based on optimum concentrations

Width in mm	Corridor		Stairs	
	Persons per hour	Persons per minute	Persons per hour	Persons per minute
1200	6480	108	4500	90
1800	9780	163	6900	115
2450	12900	215	9240	156
3050	16200	270	11220	187

3 Space required for waiting areas

Table v gives areas to be allowed for specific circulation areas.

4 Flow capacities of corridors and staircases

Average space allowances per person in circulation areas may be based on the following:

General design purposes $\qquad$ 0·8 m²

People moving at over 1·3 m/s (good walking pace) $\qquad$ 3·7 m²

People moving at 0·4 to 0·9 m/s (shuffle) 0·27 to 0·37 m²

People at standstill caused by obstruction $\qquad$ 0·3 m²

Dimensions of stairs for ambulant disabled and elderly people are shown in **9.1**. A formula for suitable relationship of riser to going for most normal staircases is $2R + G = 600$ to 630mm where R = riser; G = going.

R should be max 190mm

G should be min 250mm

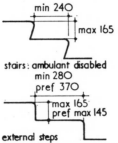

9.1 *Stairs for ambulant disabled and elderly people*

5 Lifts

Table VII Approximate dimensions of lift cars and wells for preliminary planning purposes

Type	Application	No of persons	Load (kg)	Speed (m/s)	A × B (mm)	C × D (mm)	E (mm)	Car size (m²)	Area per person mm²
	Small offices and flats	4	280	0·5	1120 × 940	1730 × 1170	685	0·8	200
		6	410	0·5	1270 × 1120	1900 × 1350	760	1·1	185
		8•	550	0·5 and 0·75	1420 × 1250	2060 × 1480	835	1·5	180
	Perambulators	8•	550	0·5 : 0·75 : 1·0 : 1·5	1140 × 1580	2060 × 1830	835	1·5	180
	General purpose	8•	550	0·75 : 1·0 : 1·5	1520 × 1220	1930 × 1780	910	1·4	175
		10	680	0·75 : 1·0 : 1·5	1830 × 1220	2240 × 1730	910	1·8	180
		13	910	0·75 : 1·0 : 1·5	1930 × 1420	2340 × 1930	910	2·0	150
		16	1140	0·75 : 1·0 : 1·5	2130 × 1520	2540 × 2030	1060	2·8	175
		20	1370	0·75 : 1·0 : 1·5	2130 × 1670	2540 × 2180	1060	3·0	150
	Intensive traffic: offices, hotels, etc	16	1140	1·5 : 2·5 : 3·5 : 4·0 : 5·0	2130 × 1520	2540 × 2080	1060	2·4	150
		20	1370	1·5 : 2·5 : 3·5 : 4·0 : 5·0	2130 × 1680	2540 × 2240	1060	3·0	150
		23	1590	1·5 : 2·5 : 3·5 : 4·0 : 5·0	2130 × 1880	2540 × 2440	1060	3·3	140
		26	1820	1·5 : 2·5 : 3·5 : 4·0 : 5·0	2440 × 1830	2870 × 2440	1520	3·7	140
	Intensive traffic: departmental stores, concert halls, etc	23	1590	1·5 : 2·5	2440 × 1680	2850 × 2290	1520	3·3	140
		26	1800	1·5 : 2·5	2440 × 1830	2870 × 2440	1520	3·7	140
		33	2250	1·5 : 2·5	2740 × 1830	3180 × 2440	1670	4·3	130

Lifts to accommodate wheelchairs are listed on p95 of *Designing for the disabled*

• Suitable for use as firemen's lift (GLC) subject to speed being sufficient to reach topmost storey of the building in one minute

Calculation of the sizes of lift lobbies

The most critical space is at the principal entrance or exit level, and calculations must be based on the peak use period which will usually be that time when the occupants of the building arrive. To find the space required by people waiting in main entrance lift lobby:

1 Calculate lift waiting interval (WI) in minutes

$$WI = \frac{\text{Round trip time}}{\text{No of lifts}}$$

2 Assess filling period (FP) in minutes.
This is the time taken to move the total population of the building; thirty minutes is a usual allowance.
3 The number of people in the lobby will be:

$$\frac{\text{Total users} \times WI}{FP}$$

4 Let X represent the allowance of m²/person in lobby.
5 Allow factor (say 2) for uneven arrivals.
A higher factor must be used if very uneven flow is expected eg because of proximity of stations, bus stops etc. (The selection of this factor will also be influenced by policies specific to the particular design such as type of building and space available for circulation.)

6 Size of lobby will then be: $\dfrac{\text{Total users} \times WI \times X \times 2}{FP}$

Worked example

Assume 800 people on floors served by lifts with FP of 30 minutes, WI of ½-minute and a space allowance of 0·65m² per person.

$$\text{Lobby area} = \frac{800 \times 0\cdot5 \times 0\cdot65 \times 2}{30} = 17\cdot3m^2$$

6 Stairs

Table VIII shows critical dimensions for stairs as required by the LCC Code of Practice *Means of escape in case of fire.* This code is still in force in Inner London Boroughs.

7 Circulation

Some common sizes affecting circulation are shown in **9.2** and certain corridor widths in **9.3**. German examples are shown in **9.4** for comparison.
Common obstructions which may affect circulation flow are shown in **9.5** (See also sections 12 to 26 of this handbook for circulation allowances in particular building types). Circulation areas for disabled people in wheelchairs and on crutches are given in **9.6** to **9.15** (from *Designing for the disabled*).

8 References and sources

AJ information sheet 1194 Internal circulation. AJ, 1963, March 20 CI/SfB 91
AJ information sheet 1195 Sizes of lifts and lift lobbies. AJ, 1963, March 20 CI/SfB (66)
AJ information sheet 1053 Staircases: required sizes 1. AJ, 1962, January 10 CI/SfB (24) (F)
GOLDSMITH, S. Designing for the disabled. London, 1967, Royal Institute of British Architects, 2nd edition £3 10s CI/SfB (E3p)
NEUFERT, E. Bauentwurfslehre. Berlin, 1967, Ullstein CI/SfB (E1)
Manuale dell' architetto. Rome, 1962, Consiglio Nazionale delle Ricerche (3rd edition) CI/SfB (E1)
LONDON COUNTY COUNCIL Document 3868 Code of Practice Means of escape in case of fire. London, 1956, The Council (now Greater London Council) CI/SfB (Ajn)

Table VIII Critical dimensions for stairs (GLC)

Class of building (see also table III)	Number of persons (calculated from floor area per person as laid down in table IV of the code) (a) evenly distributed on all floors (b) mainly concentrated on one floor	Clear width of flight (mm)	Number of risers		Max height of riser (mm)	Min depth of tread (mm)	Depth of landing from first nosing	Max gradient of ramps (where constituting a means of escape)	Remarks
			Min	Max					
Halls and places of assembly	Up to 200 300 500 750 1000	1066 (1070) 1219 (1220) 1524 (1520) 1524 (1520) 1524 (1520)	3	16	152 (150)	279 (280)	Not less than width of flight	1 in 10	Straight flights only permitted

Classes I–IV and VII–VIII (buildings requiring only one staircase)	Up to 50 a
	          b
	       75 a
	          b
	      100 a
	          b
	      150 a
	          b
	Up to 200 a
	          b
	      250 a
	          b
	      300 a
	          b
		762 (760)
		762 (760)
		762 (760)
		914 (910)
		914 (910)
		1066 (1070)
		1066 (1070)
		1219 (1220)
		1219 (1220)
		1372 (1370)
		1371 (1370)
		1524 (1520)
		1524 (1520)
		1524 (1520)
			—	16	190·5 (190)	254 (250)	Not less than width of flight	1 in 10	Straight flights only permitted

Classes V and IX (buildings requiring two or more staircases)	Up to 200 a
	          b
	      300 a
	          b
	      400 a
	          b
	      500 a
	          b
	      600 a
	          b
		914 (910)
		1066 (1070)
		1066 (1070)
		1219 (1220)
		1219 (1220)
		1371 (1370)
		1371 (1370)
		1524 (1520)
		1524 (1520)
		2286 (2290)
			—	16	190·5 (190)	254 (250)	Not less than width of flight	1 in 10	Straight flights only permitted

Figures in brackets are suggested possible metric roundings. All other dimensions are exact metric equivalents to the nearest mm

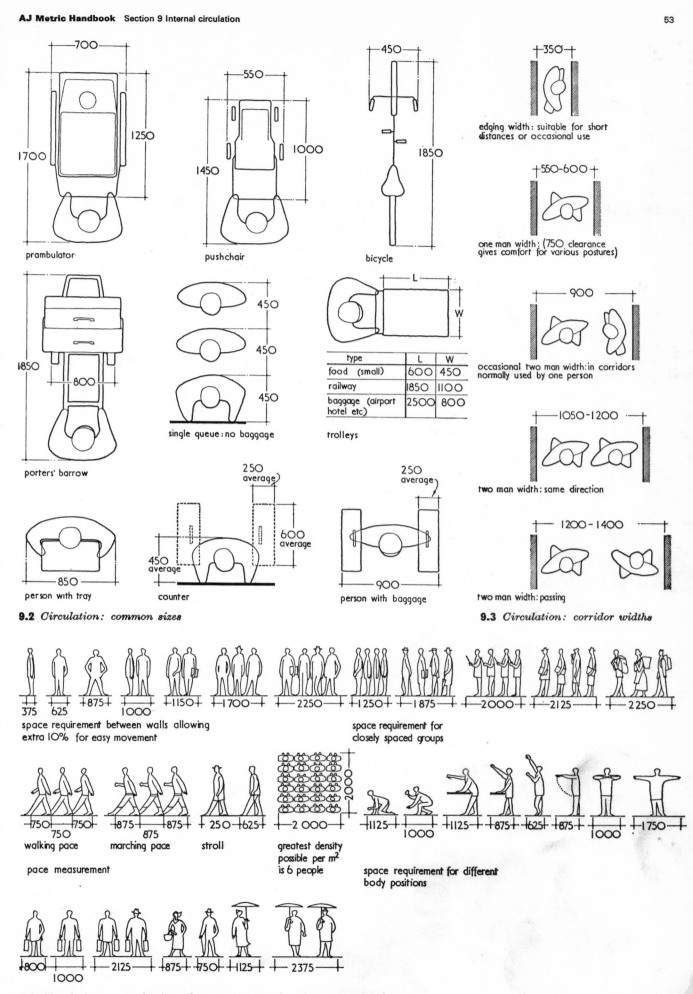

700
1700
1250
prambulator

550
1450
1000
pushchair

450
1850
bicycle

350
edging width: suitable for short distances or occasional use

550-600
one man width: (750 clearance gives comfort for various postures)

1850
800
porters' barrow

450
450
450
single queue: no baggage

L
W
trolleys

type	L	W
food (small)	600	450
railway	1850	1100
baggage (airport hotel etc)	2500	800

900
occasional two man width: in corridors normally used by one person

850
person with tray

250 average)
450 average
600 average
counter

250 average)
900
person with baggage

1050-1200
two man width: same direction

1200-1400
two man width: passing

9.2 *Circulation: common sizes*

9.3 *Circulation: corridor widths*

375 625 +875+ 1000 +1150+ 1700 2250 1250+ 1875
space requirement between walls allowing extra 10% for easy movement

1250+ 1875 2000 2125 2250
space requirement for closely spaced groups

+750+ +750+
750
walking pace

+875+ +875+
875
marching pace

+250+625+
stroll

pace measurement

2000
2000
greatest density possible per m² is 6 people

1125+
1000
1125+ 875+ 625+ 875+
1000 1750
space requirement for different body positions

+800+
1000
2125
+875+ 750+ 1125+
2375

9.4 *Circulation: examples from German sources of various space requirements*

accepted fire hand
appliances:
45·720 reel (150 ft) :P= 320mm
30·500 reel (100 ft) :P = 260mm
22·860 reel (75 ft) :P =240mm

350

fire bucket

250

2 gallon (9·1 litres)
extinguisher

170

4" (1016mm)id. c.i.pipe

100

remote control gear
(wheel type)

60

surface switch

P

door on retainer:
P= door thickness+80mm
(note furniture on door
=further protrusion)

P

radiators on wall brackets

type	P(mm)
2 column	130
3 column	160
4 column	230
5 column	290
7 column	350
3½" hospital (90)	130
5" hospital (165)	170
7" hospital (180)	230

9.5 *Circulation: common obstructions*

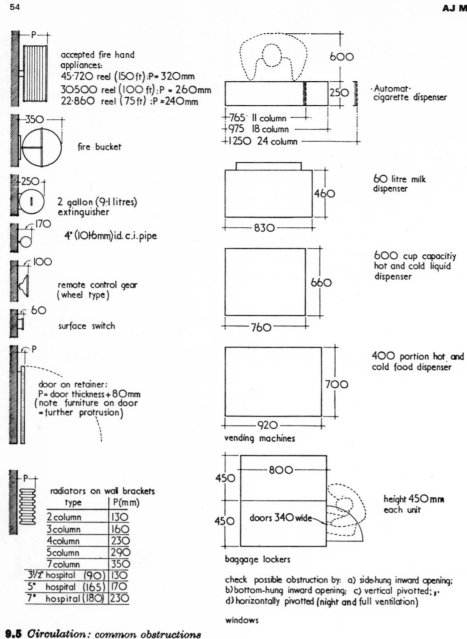

600

250

·Automat·
cigarette dispenser

765 11 column
975 18 column
1250 24 column

60 litre milk
dispenser

460

830

600 cup capacitiy
hot and cold liquid
dispenser

660

760

400 portion hot and
cold food dispenser

700

920

vending machines

800

450

450

doors 340 wide

height 450 mm
each unit

baggage lockers

check possible obstruction by: a) side-hung inward opening;
b) bottom-hung inward opening; c) vertical pivotted; ,
d) horizontally pivotted (night and full ventilation)

windows

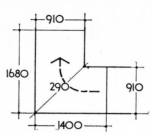

910

1680

290

910

1400

9.10 *Small wheelchairs:*
forward turn through 90deg
Preferred minimum space

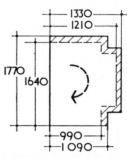

1330
1210

1770

1640

990
1090

9.11 *Small wheelchairs: turn*
through 180deg

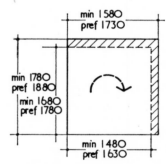

min 1580
pref 1730

min 1780
pref 1880

min 1680
pref 1780

min 1480
pref 1630

9.12 *Small wheelchairs:*
turn through 180deg
for hemiplegics and so on

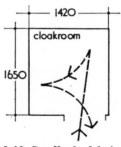

1420

cloakroom

1650

9.13 *Small wheelchairs:*
three-point turn in cloakroom

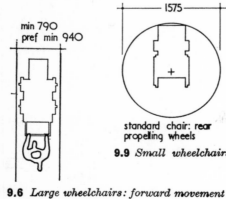

min 790
pref min 940

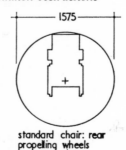

1575

standard chair: rear
propelling wheels

1500

standard chair: front
propelling wheels

1245

indoor chair: model 1
front propelling wheels

9.9 *Small wheelchairs: comparative turning space requirements*

9.6 *Large wheelchairs: forward movement*

790

1520

9.7 *Large wheelchairs: forward turn through 90deg*

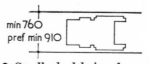

min 760
pref min 910

9.8 *Small wheelchairs: forward movement*

660

9.14 *Stick user*

840

9.15 *Crutch user*

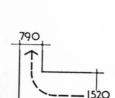
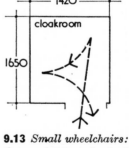

10 External circulation

Contents

1 Car sizes and manoeuvrability

A representative selection of car widths and lengths is shown in **10.1**. Over 90 per cent of cars are under 4570mm long and over 50 per cent are under 4270mm. It is unlikely that cars over 4900mm will ever be more than a minority in this country. Minimum turning circles between kerbs and walls are given in **10.2** and greatest height, width, length and wheelbase in **10.3**. The maximum length of a motor cycle is about 2340mm (motor scooter 1830mm). Maximum width is 710mm (1600mm with side car). Average height is 790mm, excluding windscreen (motor cycle combination 1270mm).

2 Parking arrangements

Critical dimensions of parking stalls are given in **10.4** and **10.5**. A stall size of 4600mm × 2200mm to 2300mm is shown in **10.4** (minimum allowance). A stall size of 5500mm × 2400mm is shown in **10.5** (average allowance for outdoor parking).

Parking spaces for disabled people are shown in **10.7**.

3 Garages and garage courtyards

Typical layouts for garage courts are shown in **10.6**.

4 Private garages

Critical dimensions for private garages are given in **10.9** and for disabled people in **10.8**.

5 Widths of roads and footways

Tables I to V give widths of roads, footways and cycle tracks for urban and rural roads. Tables VI and VII give details of pedestrian bridges and subways.

note: wall-to-wall figures not available for Rolls Royce, Bentley, Jaguar, estimation, add 900mm to kerb-to-kerb dimensions

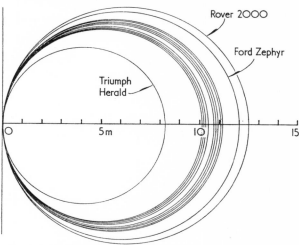

wall-to-wall dimensions

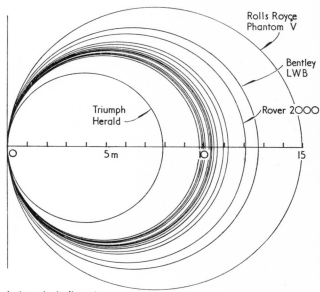

kerb-to-kerb dimensions

10.2 *Minimum turning circles between walls (above) and kerbs*

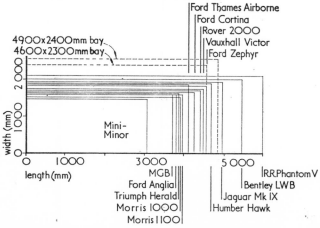

10.1 *Representative selection of car widths and lengths*

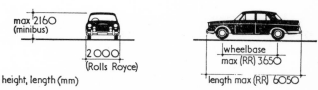

10.3 *Greatest height, width, length and wheelbase*

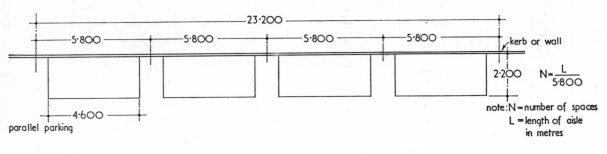

parallel parking

$$N = \frac{L}{5 \cdot 800}$$

note: N = number of spaces
L = length of aisle
in metres

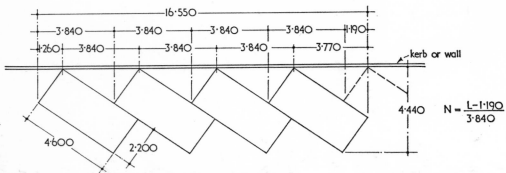

35° parking

$$N = \frac{L - 1 \cdot 190}{3 \cdot 840}$$

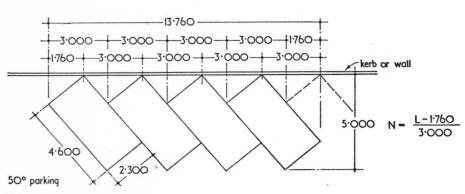

50° parking

$$N = \frac{L - 1 \cdot 760}{3 \cdot 000}$$

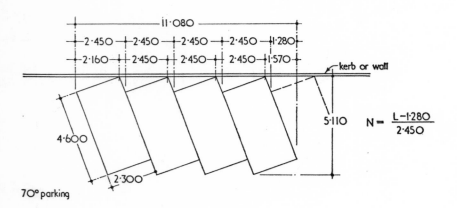

70° parking

$$N = \frac{L - 1 \cdot 280}{2 \cdot 450}$$

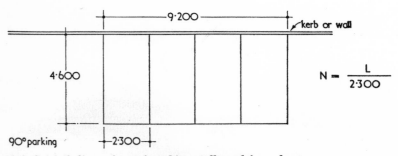

90° parking

$$N = \frac{L}{2 \cdot 300}$$

10.4 *Critical dimensions of parking stalls and formulae to determine number of cars that can be parked in a given aisle length. Stall size 4·600 × 2·200 (to 2·300). These dimensions are suitable for use in car parking buildings.*

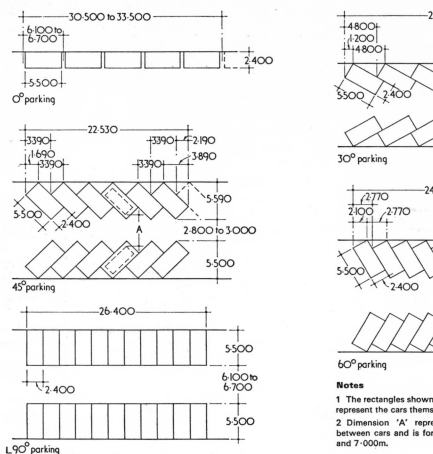

0° parking

45° parking

L90° parking

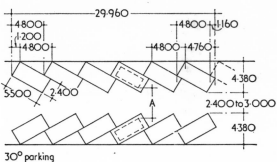

30° parking

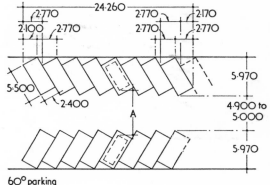

60° parking

Notes

1 The rectangles shown are in all cases the space allotted to the car and do not represent the cars themselves.

2 Dimension 'A' represents the assumed clear distance across the aisle between cars and is for 30°, 45° and 60° parking, respectively 4·570, 4·870 and 7·000m.

10.5 *Critical dimensions of parking stalls with stall size of 5·5m × 2·4m. For use in outdoor parking*

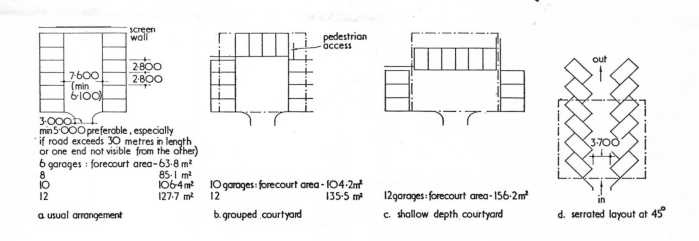

screen wall

7·600 (min 6·100)

2·800
2·800

3·000
min 5·000 preferable , especially if road exceeds 30 metres in length or one end not visible from the other)

| 6 garages : forecourt area–63·8 m² |
| 8   85·1 m² |
| 10   106·4 m² |
| 12   127·7 m² |

a. usual arrangement

pedestrian access

10 garages: forecourt area–104·2 m²
12   135·5 m²

b. grouped courtyard

12 garages: forecourt area–156·2 m²

c. shallow depth courtyard

out

3·700

in

d. serrated layout at 45°

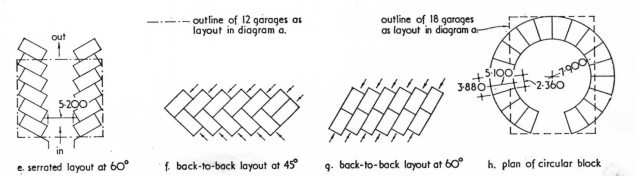

out

5·200

in

e. serrated layout at 60°

—·—·— outline of 12 garages as layout in diagram a.

f. back-to-back layout at 45°

outline of 18 garages as layout in diagram a.

g. back-to-back layout at 60°

5·100   7·900
3·880   2·360

h. plan of circular block

10.6 *Garages and garage courtyards: Space requirements and layout*

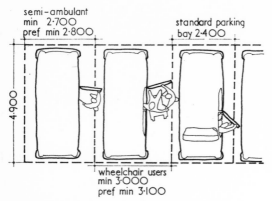

10.7 *Parking spaces for disabled people*

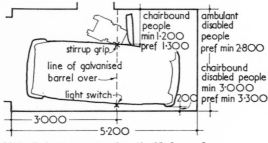

10.8 *Private garages for disabled people*

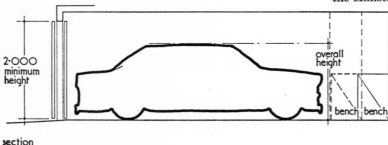

section

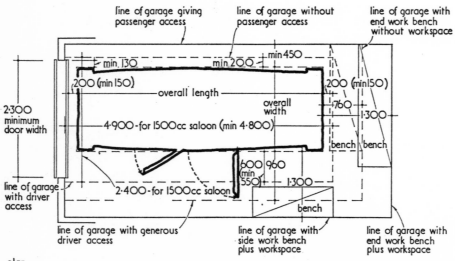

plan

10.9 *Individual garage dimensions*

6 Pedestrians and disabled people

Where bollards are used to separate people from cars, the minimum dimensions shown in **10.11** should be used. Pavement widths for disabled people are given in **10.12** and **10.13**; external ramps (also internal ramps) in **10.14**

7 Cycle stands and stores

Various arrangements are shown in **10.15** and **10.16**. All stands are proprietary products and dimensions given here are approximate, but enough for preliminary design purposes.

8 References and sources

AJ information sheet 1397 Car parking buildings: Basic data. AJ, 1966, June 29 CI/SfB 223

AJ information sheet 1398 Car parking buildings: Ramp systems. AJ, 1966, June 29 CI/SfB 223 (24)

AJ information sheet 1010 Parking arrangements: Cars and cycles. AJ, 1961, October 18 CI/SfB 122

AJ information sheets 1222 and 1225 Garages and garage courtyards: Space requirements and layouts. AJ, 1963, November 6 CI/SfB 97

AJ information sheet 1228 External fixtures: Bollards. AJ, 1963, December 25 CI/SfB (90·7)

GOLDSMITH, S. Designing for the disabled. London, 1967, Royal Institute of British Architects, 2nd edition £3 10s CI/SfB (E3p)

NEUFERT, E. Bauentwurfslehre. Berlin, 1967, Ullstein CI/SfB (E1)

BEAZLEY, E. Design and detail of the space between buildings. London, 1962, Architectural Press, 2nd impression £2 2s CI/SfB 08:90

MINISTRY OF TRANSPORT. Technical memorandum No T8/68: Metrication-Highway Design. Available free from the Ministry CI/SfB 12 (F7)

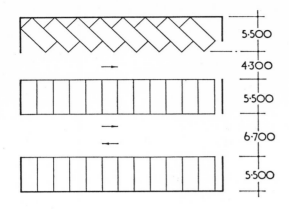

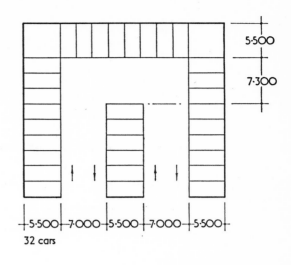

32 (number of cars which can be accommodated)

32 cars

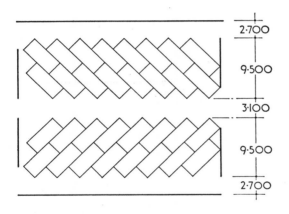

32 cars

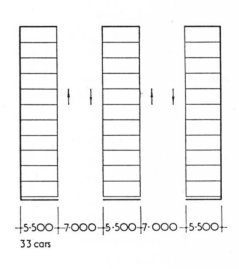

33 cars

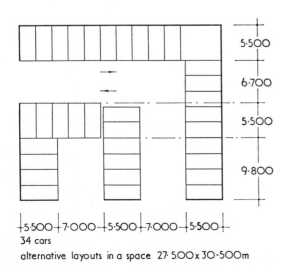

34 cars

alternative layouts in a space 27.500 x 30.500m

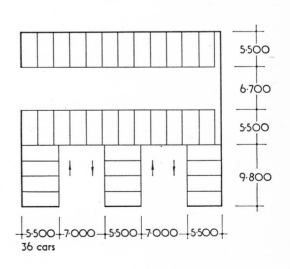

36 cars

10.10 *Various parking arrangements for cars occupying a space of 27.500 × 30.500m translated from* Space between buildings

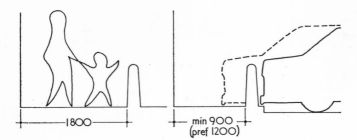

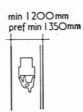

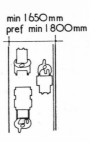

10.11 *Use of bollards to separate vehicles from pedestrians*

10.12 *Pavement widths for disabled people (single)*

10.13 *Pavement widths to allow passing wheelchairs*

Table I Recommended carriageway widths for urban roads*

Road type	Recommended carriageway widths m
Primary distributor	
Dual 4 lane carriageway	14·60
Overall width for 4 lane divided carriageway with central refuges	14·60
Single 4 lane carriageway with no refuges	13·50
Dual 3 lane carriageway	11·00
Single 3 lane carriageway (recommended only for tidal flows)	9·00
Dual 2 lane carriageway	7·30
District distributor	
Single 2 lane carriageway	7·30
Dual 2 lane carriageway	7·30
Dual 2 lane carriageway if the proportion of heavy commercial traffic is fairly low	6·75
Local distributor	
Single 2 lane carriageway in industrial districts	7·30
Single 2 lane carriageway in principal business districts	6·75
Minimum single 2 lane carriageway in residential districts used by heavy vehicles	6·00
Access roads	
In industrial and principal business districts use the dimensions stated above for local distributors	
Minimum width for single 2 lane carriageways in residential districts	5·50
2 lane width for back or service roads used occasionally for heavy vehicles	5·00
Minimum 2 lane width for back roads in residential districts if use is limited to cars	4·00

Table II Recommended carriageway widths for rural roads

Road type	Recommended carriageway widths m
Single lane carriageway used principally in Scotland and Wales	3·5
Minimum width of carriageway in rural junctions	4·5
Minimum width for single 2 lane carriageway	5·5
Motorway slip road width	6·0
Single or dual 2 lane carriageway	7·3
Single 3 lane carriageway	10·0
Dual 3 lane carriageway	11·0
Dual 4 lane carriageway	14·6

Table III Urban and rural roads: lay-bys, bus bays and passing bays

Type of lay-by	Width m
Minimum standard lay-by	2·5
General standard lay-by	3·0
Bus bay standard	3·25
Maximum standard for single carriageway lay-by	3·5
Standard passing bay	2·25

* Tables I to VII are taken from Ministry of Transport Technical Memorandum T8/68.

Table IV Recommended footway widths for urban and rural roads

Type of road	Recommended minimum footway widths m
Primary distributor:	
Urban motorway	No footways
All purpose road	3·00*
District distributor	3·00 in principal business and industrial districts*
	2·50 in residential districts*
Local distributor	3·00 in principal business and industrial districts*
	2·00 in residential districts*
Access roads	Principal means of access:
	3·00 in principal business districts*
	2·00 in industrial districts*
	2·00 normally in residential districts*
	3·50–4·50 adjoining shopping frontages
	Secondary means of access:
	1·00 verge instead of footway on roads in principal business and industrial districts
	0·60 verge instead of footway on roads in residential districts

* If no footway is required provide verge at least 1m wide. Where slabs are used widths are net paved areas (excluding kerbs).

Table V Cycle tracks and cycle ways for urban and rural roads

Type of traffic	Standard width m	Minimum width m
One-way traffic	2·75	1·80
Two way-traffic	3·60	—

Table VI Pedestrian bridges and subways for urban and rural roads

Bridges, subways	Width m	Height m
Pedestrian bridges	1·80 (min for permanent structures)	see table VII below
	1·50 (min for temporary structures)	
Subways	2·30	2·25

Table VII Vertical clearances for urban and rural roads

Obstruction	Minimum vertical clearance including allowance for resurfacing m
Overbridges	5·1
Sign gantries	5·5
Minimum maintained clearance	5·0

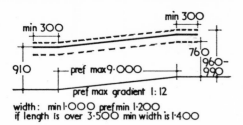

width: min 1·000 pref min 1·200
if length is over 3·500 min width is 1·400

10.14 *General purpose ramps for disabled people*

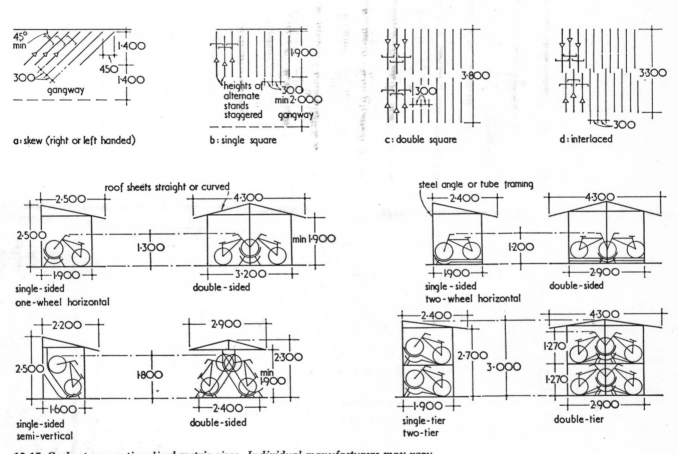

a: skew (right or left handed)   b: single square   c: double square   d: interlaced

single-sided
one-wheel horizontal

double-sided

single-sided
two-wheel horizontal

double-sided

single-sided
semi-vertical

double-sided

single-tier
two-tier

double-tier

10.15 *Cycle stores: rationalised* **metric** *sizes. Individual manufacturers* **may** *vary*

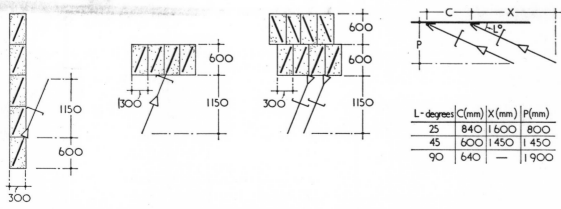

L – degrees	C (mm)	X (mm)	P (mm)
25	840	1600	800
45	600	1450	1450
90	640	—	1900

10.16 *Cycle stands*

11 Dimensional co-ordination

Contents

1 Dimensional grids

Dimensional co-ordination relies on establishment of rectangular and three-dimensional grids of basic modules into which components can be introduced in an interrelated pattern of sizes **11.1**. The modular grid network delineates the space *into* which the component fits. This is the most important facet of dimensional co-ordination—the component must always be *undersized* in relation to the space grid.

In the engineering world the piston and cylinder principle establishes the size relationship between dimensional space grid and component **11.2**. The size of the cylinder must allow for the right degree of accuracy and tolerance to enable the piston to move up and down.

The degree of accuracy to be allowed for in the building process is related to the economics of jointing and adequate space must be allowed for size of component plus joint. Transgressing the rules of locating components within the allotted space contained by grid lines will cause considerable difficulty in site assembly.

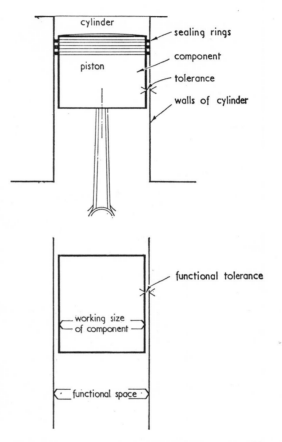

11.2 *Analogy between piston/cylinder relationship and working size/functional space relationship of building component*

2 Arrangement of components in a grid
The basic arrangement of components within the grid layout shows them fitting into the spaces allocated to them: dimensionally they are co-ordinated, thus allowing the designer maximum use of standard components **11.3**, **11.4**.

3 Modular theory and dimensional co-ordination
A dual nomenclature exists in the building industry when referring to modular co-ordination. It may not be clear why it is sometimes referred to as modular co-ordination while in other cases it is called dimensional co-ordination. The difference between the basic modular theory and the modified version of dimensional co-ordination of preferred sized components lies in the fact that the pure *modular theory* uses the module (100 mm) as the *only* common dimensional denominator, with no preferential treatment for any particular number of increments. This theory gives the designer unrestricted flexibility within the pattern of

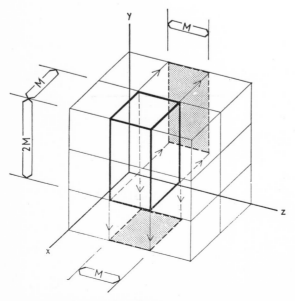

11.1 *Three-dimensional grid of basic modules*

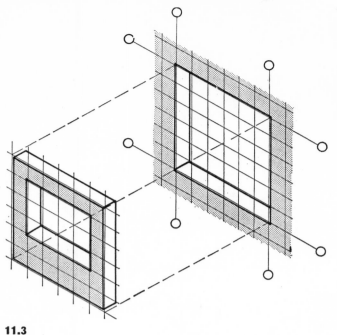

11.3

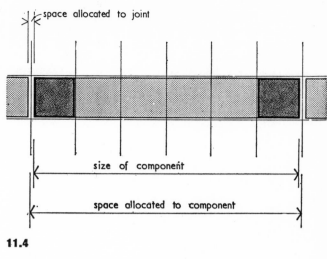

11.4

11.3, 11.4 *Fitting a component into a dimensionally co-ordinated grid*

100 mm multiples **11.5**. It is claimed that the same three-dimensional relationship of the 100 mm grid helps the manufacturer as well as the designer to produce dimensionally co-ordinated components without putting too many restrictions on the aesthetics of the design.

The *dimensional co-ordination theory* though using the basic unit of 100 mm (called here 'increment') establishes an hierarchy of preferred units. The first preference is 300 mm; second preference is 100 mm; third preference is 50 mm up to a maximum of 300 mm; and the fourth preference is 25 mm up to a maximum of 300 mm. These preferences for sizing were embodied in BS 4011:1966 *Basic sizes for building components and assemblies* on which the later BS 4330 *Controlling dimensions (metric units)* was based **11.6, 11.7**.

4 Variety reduction

The reasoning behind the concept of preferred sized components based on 300 mm multiples lies in what is called the next stage of modular co-ordination: *variety reduction*. It is maintained by the exponents of this theory that in order to help the manufacturing and assembling branches of the building industry, sizes of components have to be rationalised still further. The full modular principle—though a vast step from the previous array of sizes—did not go far enough in reducing the variety. Thus with the official backing of the public sector, dimensionally co-ordinated components based on the ranges of *preferred* metric sizes will dictate the future of the metric changeover in the building industry.

5 Controlling lines and dimensions

Introduction of dimensional preferences for sizing of components formed the basis for the network of larger grids related to the functional needs of the building. The 300 mm basic sizes selected for components established the pattern of grids which co-ordinated the structure and infilling components **11.8**. The functional hierarchy of fits established the need for delineating the datum lines in respect to load-bearing walls and columns on the horizontal plane, and floor and roof zones in the vertical plane. These datum lines became known as *controlling lines* while the dimensions between them became the *controlling dimensions*.

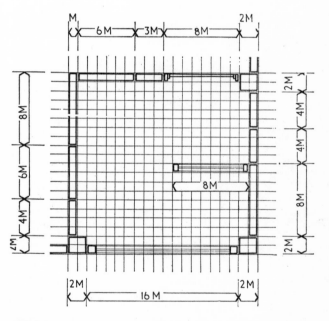

11.5 *Relating modular planning principle to sizing of components*

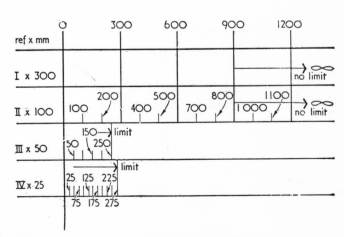

11.6 *Diagram representing preferences for sizing building components and assemblies according to* BS *4011:1966*

5.1 Axial lines and boundaries of zones

The most significant point is the creation of two ways of placing components in relation to structure **11.9**. In one case the structure is located by axial controlling lines, in the second case by boundaries of zones. Both systems impose differing uses of building components. It is sometimes believed that the axial controlling lines imply use of a framed structure with a predominance of structure over other components. In fact this is not the case and framed structures or loadbearing walls can be used by either of the types.

Further complications which must be taken into account when designing within the grid controlling lines involve junctions of components.

6 Building Costs

Location of components within zones will also affect overall cost of building. Sizes of components will be influenced by functional considerations of the job and they will be selected or designed to suit the plan. The designer of future metric components must be aware of the adverse ratio of size and cost in relation to dimensional flexibility. A large number of small units offers a better chance of filling the allocated space without adjustment of their size —allowing for the collective effect of joints **11.10**. In other words the components can be moved about. At the same time the number and cost of joints make small units an expensive item in the elemental cost breakdown of components. On the other hand, large units, with fewer joints, offer better value per square metre—but need different handling because of the smaller allowances for joints.

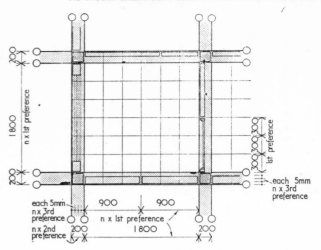

11.8 *Example of dimensional co-ordination of components using preferences for sizing from* BS *4011*

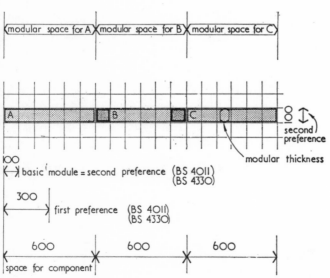

11.7 *Comparison between modular spaces and dimensionally co-ordinated spaces according to* BS *4011 and* BS *4330*

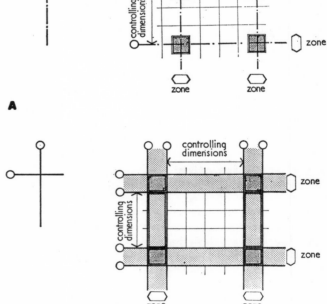

11.9 *Controlling lines: method* A *controlling lines on axial lines of loadbearing walls and columns (axial grid); method* B *controlling lines on boundaries of zones (face grid)*

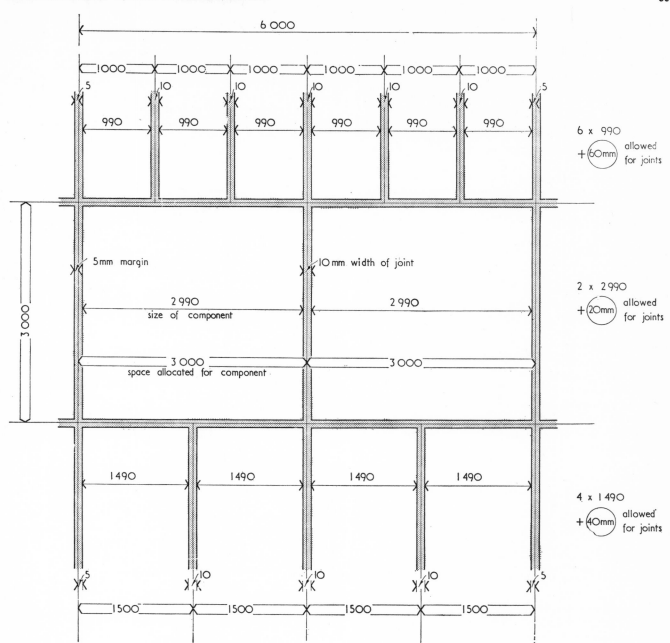

11.10 *Diagrammatic representation of widths of wall cladding units showing effect of sizes of components on standardisation of joint widths*

Table I Floor to ceiling heights

Heights in multiples of 300 mm	100 mm
mm	mm
1500*	
1800*	
2100†	
	2300
2400	
	2500
	2600
2700	
	2800
	2900
3000	

Greater heights in multiples of 300 from 3000 to 6600 and thereafter in multiples of 600

NOTE. In addition to the values in the table, 2350 mm may be used, for housing only, in conjunction with a floor to floor height of 2600 mm.
* Applies only to farm buildings
† Applies only to domestic and lock-up garages, multi-storey car parks, and farm buildings

Table II Heights of zones for floors and roofs

Heights in multiples of 300 mm	100 mm	50 mm
mm	mm	mm
	100	
	200	
		250*
300		
	400	
	500	
600		
900		
1200		
1500		
1800		
2100		

Greater heights in multiples of 300 from 2100

* Applies only to housing for use in conjunction with the floor to floor height of 2600 mm and floor to ceiling height of 2350 mm

Table III Floor to floor and floor to roof heights

Heights in multiples of 300 mm	100 mm
mm	mm
	2600*
2700	

Greater heights in multiples of 300 from 2700 to 8400 and thereafter in multiples of 600

* Applies only to housing to accord with Circular No 31/67 of the Ministry of Housing and Local Government

Table IV Changes in level

	Range	Heights to be in multiples of
mm		mm
From 300 to 2400		300
Above 2400 for the smaller dimension and for the larger when numerically possible		600

Changes in level of 1300, 1400, 1700, 2000 and 2300 may be used for housing in conjunction with a floor to floor height of 2600

7 Recommendations

The practical side of dimensional co-ordination must now be related to the recommendations contained in BS 4330:1968 *Recommendations for the co-ordination of dimensions in building: Controlling dimensions (metric units)*

7.1 Vertical controlling dimensions

Recommendations from BS 4330 are contained in tables I to IV.
See also sections 12, 13, 15, 16, 17, 23, 25 and 26 of this handbook for vertical controlling dimensions applied to particular building types.

7.2 Horizontal controlling dimensions

Recommendations from BS 4330 are contained in tables V and VI. The sources, by building types, from which these recommendations were derived are shown in table VII, which should be read in conjunction with **11.11**.

See also sections 12, 13, 15, 16, 17, 23, 25 and 26 of this handbook for horizontal controlling dimensions applied to particular building types.

Table V Widths of zones for columns and loadbearing walls

Widths in multiples of 300 mm	100 mm
mm	mm
	100
	200
300	
	400
	500
600	

If greater widths are required they should be in multiples of 300 as first preference, or of 100 as second preference, in accordance with BS 4011

Table VI Spacing of zones for columns and loadbearing walls

Range	Sizes of spacings to be in multiples of
mm	mm
From 900	300

NOTE. 800 mm may be used for housing only

8 References and sources

BS 4330:1968 *Recommendations for the co-ordination of dimensions in building: Controlling dimensions* (Metric units) (15s)
BS 4011:1966 *Basic sizes for building components and assemblies* (4s)
DC 4, 'Recommended vertical dimensions for educational, health, housing, office and single-storey general purpose industrial buildings'.
DC 5, 'Recommended horizontal dimensions for educational, health, housing, office and single-storey general purpose industrial buildings'.
DC 6, 'Guidance on the application of recommended vertical and horizontal dimensions for educational, health, housing, office and single-storey general purpose industrial buildings'.
DC 7, 'Recommended intermediate vertical controlling dimensions for educational, health, housing and office buildings, and guidance on their application'.
All published by HMSO.

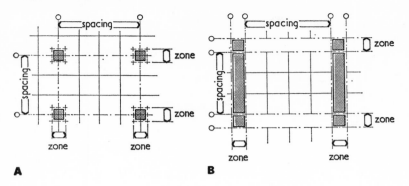

11.11 *Dimensional co-ordination: diagram to show definition of spacing and zone (to be read in conjunction with table VII)* **A** *Axial lines,* **B** *Boundaries*

Table VII Spacing of zones for columns and loadbearing walls located by: A Axial lines; B Boundaries of zones (to be read in conjunction with 11.11)

Education A Spacing	Education A Zone	Health B Spacing	Health B Zone	Housing A & B Spacing	Housing A & B Zone	Offices A & B Spacing	Offices A & B Zone	Industrial A & B Spacing	Industrial A & B Zone	Hotels A & B Spacing	Hotels A & B Zone	Shops A & B Spacing	Shops A & B Zone	Farm buildings A & B Spacing	Farm buildings A & B Zone
	100		100		100		100		100		100		100		
	150														
	200		200		200		200		200		200		200		200
	250														
	300		300		300		300		300		300		300		300
							400		400		400		400		400
							500		500		500		500		
	600								600		600		600		600
				800											
				900		900									
				1 200		1 200									
				1 500		1 500									
1 800				1 800		1 800				1 800					
				2 100		2 100				2 100					
2 400				2 400		2 400				2 400				2 400	
2 700		2 700		2 700		2 700				2 700		2 700			
3 000		3 000		3 000		3 000		3 000		3 000		3 000		3 000	
				3 300		3 300				3 300		3 300			
3 600				3 600		3 600				3 600		3 600		3 600	
				3 900		3 900				3 900		3 900			
4 200		4 200		4 200		4 200				4 200		4 200			
4 500		4 500		4 500		4 500		4 500		4 500		4 500		4 500	
4 800				4 800		4 800				4 800		4 800			
				5 100		5 100				5 100		5 100			
5 400		5 400		5 400		5 400				5 400		5 400			
		5 700		5 700		5 700				5 700		5 700			
6 000		6 000		6 000		6 000		6 000		6 000		6 000		6 000	
6 300		6 300		6 300		6 300				6 300		6 300			
6 600				6 600		6 600				6 600		6 600			
				6 900		6 900				6 900		6 900			
7 200				7 200		7 200				7 200		7 200			
				7 500		7 500		7 500		7 500		7 500		7 500	
				7 800		7 800									
8 100				8 100		8 100									
8 400				8 400		8 400									
		8 700		8 700		8 700									
9 000		9 000		9 000		9 000		9 000		9 000		9 000		9 000	
		9 300		9 300		9 300									
9 600		9 600		9 600		9 600									
9 900				9 900		9 900									
				10 200		10 200									
				10 500		10 500		10 500		10 500		10 500			
10 800				10 800		10 800									
				11 100											
				11 400											
11 700		11 700													
12 000		12 000				12 000		12 000		12 000		12 000			
		12 300													
12 600		12 600													
		12 900													
		13 200		13 200											
						13 500		13 500							
14 400				14 400											
						15 000		15 000							
16 200															
						16 500		16 500							
		17 400													
18 000						18 000		18 000				18 000			
		18 600													
		19 800													

* Sizes of spacings for housing greater than 7200 mm apply to pitched roof construction

12 Offices

Contents

1 Controlling dimensions

Recommendations for horizontal and vertical controlling dimensions for office buildings are contained in *Dimensional co-ordination for building* documents DC 4, DC 5, DC 6 and DC 7 and in BS 4330:1968 (metric units). These recommendations are illustrated below. Horizontal controlling dimensions are shown in **12.1** and table I. Vertical controlling dimensions are shown in **12.2**.

Table I Horizontal controlling dimensions (to be read in conjunction with 12.1). Dimensions are between boundaries of centre lines of loadbearing walls and columns.

Spacing (mm)	Zone (mm)
900	100
1 200	200
1 500	300
1 800	400
2 100	500
2 400	
2 700	
3 000	
3 300	
3 600	
3 900	
4 200	
4 500	
4 800	
5 100	
5 400	
5 700	
6 000	
6 300	
6 600	
6 900	
7 200	
7 500	
7 800	
8 100	
8 400	
8 700	
9 000	
9 300	
9 600	
9 900	
10 200	
10 500	
10 800	
12 000	
13 500	
15 000	
16 500	
18 000	

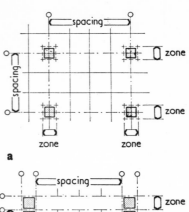

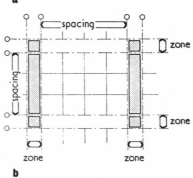

12.1 *Horizontal controlling dimensions. Diagram to define meaning of spacing and zone (to be read in conjunction with table I)* **a** *centre lines* **b** *boundaries of loadbearing walls and columns*

Preferences in millimetres showing increments for first, second and third preferred dimensions (BS 4011)

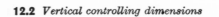

12.2 *Vertical controlling dimensions*

2 Desks, tables and chairs

Swedish Standards

Swedish recommendations for office flat top desks, based on A4 paper, are shown in **12.3** to **12.9**.

Suitable typing desks are shown in **12.10**.

Other possible desk sizes based on the Swedish standards are listed in table II.

Table II Possible desk sizes based on 12.3 to 12.10

Executive/manager	1500 × 750mm plus extension 750 × 500mm
Clerk	1300 × 700mm or 1000 × 700mm plus extension 1000 × 500mm
Secretary/typist	1300 × 700mm plus extension 1000 × 500mm
Typist	1200 × 700mm

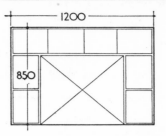

12.8 *Generous amount of space for paper*

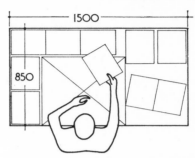

12.9 *Space for papers plus area for references*

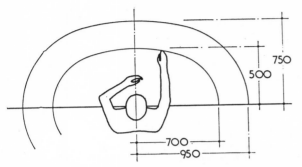

12.3 *Average reach of person sitting at desk. To reach outer area, the user will have to bend but not stand up*

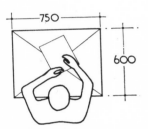

12.4 *Basic space for writing and typing*

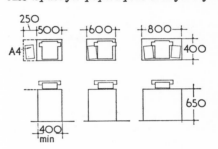

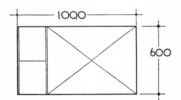

12.5 *With space for paper on one side*

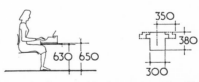

12.10 *Desks suitable for typing (Swedish)*

British Standards

Specifications for office desks, tables and seating given in BS 3893:1965 are summarised in tables III to V below with the recommended range given in exact metric equivalents. Suggested rationalised metric dimensions, not contained in the BS, are shown in brackets.

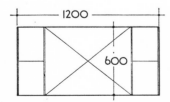

12.6 *With paper on both sides*

Table III Office desk sizes (translated from BS 3893)

Work surface			Use
Minimum depth (mm)	Minimum width (mm)	Height (mm)	
686 (700 or 750)	1067 (1050 or 1100) 1143 (1150 or 1200) 1372 (1400) 1524 (1500 or 1550)	711 (710)	General purpose and executive
610 (600 or 650)	1067 (1050 or 1100) 1372 (1400)	648-711 (650-710)	Typists: single double pedestal
762 (750 or 800)	1219 (1200 or 1250)	711-762 (710-760)	Machine operator

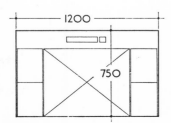

12.7 *Paper plus space for pens and telephones*

Table IV Office table sizes (translated from BS 3893)

Minimum depth (mm)	Minimum width (mm)	Height (mm)
533 (550)	914 (900 or 950)	
610 (600)	1067 (1050 or 1100)	
610 (600)	1143 (1150)	
686 (700 or 750)	1219 (1200 or 1250)	711 to 762
686 (700 or 750)	1372 (1400)	(710 to 760)
762 (750 or 800)	1524 (1500 or 1550)	710 recommended
762 (750 or 800)	1676 (1700)	
838 (800 or 850)	1829 (1800)	

Table V Office chairs (translated from BS 3893) Sizes in mm

(a) Executive and clerical
Heights: 432 desirable for fixed chairs (430)
432–508 for adjustable chairs (430 to 510)
Widths: min 406 (400)
Depths: 356–470; recommended 381 (380)

(b) Typists and machine operators
Adjustable through 102 (100) in one of the following ranges:
406 to 508 (400 to 500)
483 to 584 (480 to 580)
559 to 660 (560 to 660)
635 to 737 (640 to 740)

3 Filing cabinets

Space requirements for drawer cabinets and lateral filing are shown in **12.11** and **12.12**, with circulation requirements in **12.13**. See also minimum space requirements listed in table VIII.

4 Space requirements

Space required per employee will depend on type of work; use of equipment or machinery; degree of privacy; and storage needs. As a general guide see tables VI and VII.

Table VI Space standards

Absolute minimum area per person	3·7m² to 4·2m² (including aisles, filing cabinets and desk space). But this is too low for individual offices
Minimum area for reasonable conditions	4·2m² to 6·0m² (14m² to 17m²). But allow more for individual offices with single occupation. (See table VII)
Requirements of Offices, Shops and Railway Premises Act 1963 (see para 5)	3·7m² minimum floor area per person 11·3m² minimum room capacity per person
Additional space for visitors	Allow minimum 1·8m² extra for visitors
Completely integrated office with all activities under one roof	Minimum average of 9·3m² to 11·6m² per person excluding circulation, wcs and so on
Completely integrated office plus eating and lounge facilities	Minimum 14·0m² per person
Proportion of total floor area to be aimed at for office working space	70 per cent of gross internal floor area

Table VII Appropriate space allowances

	Area per head Including normal furniture (m²)	Type of office
Director	37 to 42	Individual
Assistant director	23 to 33	Individual
Department manager	18 to 23	Individual
Assistant department manager	14 to 18	Individual
Senior executive officer Senior professional	14 to 18	Shared
Senior manager Section supervisor	8 to 11	Shared
Secretary/shorthand typist (includes space for confidential files)	9	Shared
Junior professional/technician	9	Shared
Senior clerk	7	Open
Junior clerk	5·5	Open
Copy typist	4·5 to 5·5	Open
Draughtsman	7	Open

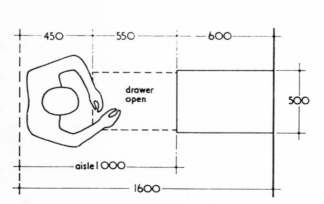

12.11 *Space requirements of drawer filing cabinet*

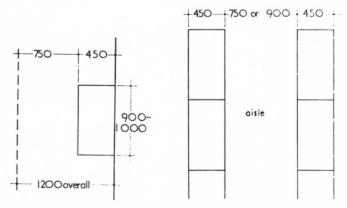

12.12 *Space requirements of lateral filing units*

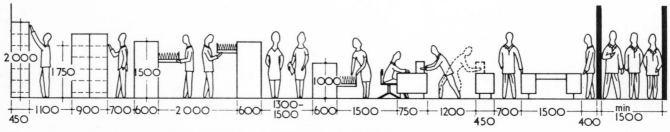

12.13 *Typical space and circulation requirements of filing and other office equipment*

5 Work spaces for individuals

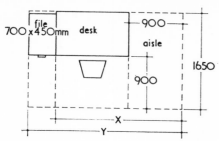

12.14 *Desk and file*

If desk   = 1500 × 750mm
    x = 2400mm   and area = 3·96m²
with file Y = 2850mm   and area = 4·71m²
If desk   = 1200 × 750mm
    x = 2100mm   and area = 3·47m²
with file Y = 2550mm   and area = 4·20m²

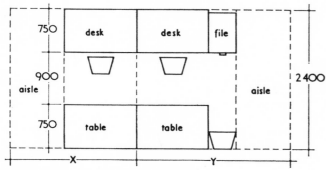

12.15 *Desks with tables, file and chair*

If desk and table = 1500 × 750mm
    x = 2400mm   and area = 5·76m²
with file Y = 2850mm   and area = 6·84m²
If desk and table are 1200 × 750mm
    x = 2100mm   and area = 5·04m²
with file Y = 2550mm   and area = 6·12m²

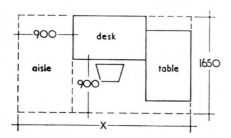

12.16 *Desk with adjacent table*

If desk   = 1500 × 750mm
    x = 3150mm   and area = 5·20m²
If desk   = 1200 × 750mm
    x = 2850mm   and area = 4·70m²

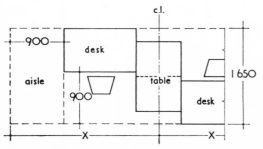

12.17 *Desk with shared table*

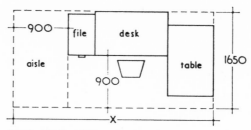

12.18 *Desk, table and file*

If desk   = 1500 × 750mm
    x = 3600mm   and area = 5·94m²
If desk   = 1200 × 750mm
    x = 3300mm   and area = 5·45m²

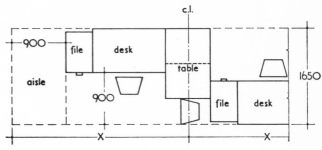

12.19 *Desk, shared table and file*

If desk = 1500 × 750mm
    x = 3225mm   and area = 5·33m²
If desk = 1200 × 750mm
    x = 2925mm   and area = 4·83m²

6 Desk spacings and layouts

Table VIII gives recommendations for spacing office furniture and **12.20** to **12.27** illustrate some examples of typical office layouts. (All desks shown are 1200 × 750mm.)
The arrangement with a screened desk **12.28** gives a working area of 3·6m² of which 90 per cent can be reached from normal sitting position.

Table VIII Recommendations for minimum spacing of office furniture (mm)

(a) Distance from back to front of desks in a row (chair space)	
When each desk is on an aisle	900
When each desk is not on an aisle	900 to 1370
(b) Aisle widths	
Major aisle (large general office)	1500
Normal general office aisle	900
Minor general office aisle	750
(c) Distance required in front of filing cabinets	
Single row	900
Two rows facing	1220
When filing cabinets face an aisle the width of the cabinets when open (1370mm) should be added to the normal aisle width	
(d) Distance required in front of shelving	
Two rows facing	750 to 900
When shelving faces an aisle the width of the shelving should be added to the normal aisle width	

Note Desks should not be placed tightly against and facing a solid wall or opaque glass partition

If desk = 1500 × 750mm (**12.17**)
    x = 2775mm   and area = 4·58m²
If desk = 1200 × 750mm
    x = 2475mm   and area = 4·10m²

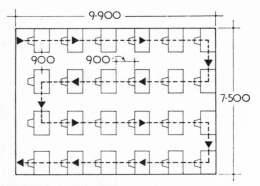

12.20 *Area per person 3·1m²*

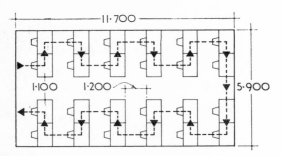

12.21 *Area per person 2·9m²*

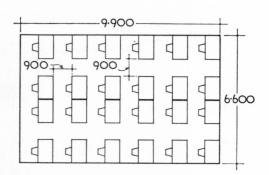

12.22 *Area per person 2·7m²*

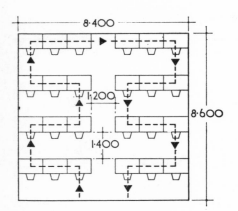

12.23 *Area per person 3·0m²*

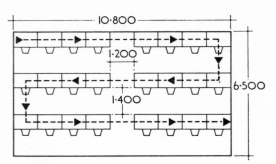

12.24 *Area per person 2·9m²*

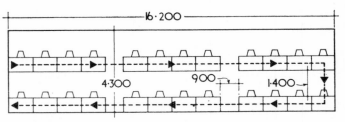

12.25 *Area per person 2·9m²*

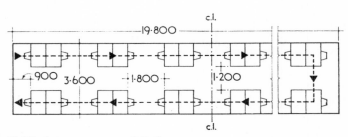

12.26 *Area per person 3·0m²*

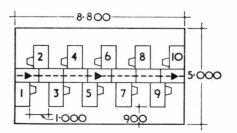

12.27 *Area per person 4·4m²*

Typical office layouts

Illustrated on this page are some typical office layouts
using desk size of 1200 × 750mm (1500 × 750mm in the
case of **12.28**).

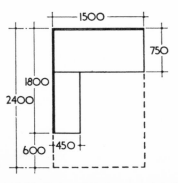

12.28 *Screened desk: area 3·6m²*

7 Drawing offices

BRS Digest 20 recommends a space allowance of about 7·0m² per draughtsman assuming standard equipment and some storage at work position:

Using 1500 × 900mm benches and tables with a 1800mm aisle the arrangements shown in **12.29** to **12.31** are possible. A more economical solution is shown in **12.32**. Layouts of complete drawing offices are shown in **12.33**.

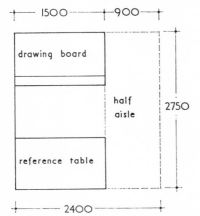

12.29 *Drawing board with front reference: area 7·0m²*

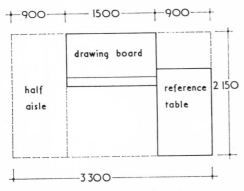

12.30 *Drawing board with back reference: area 6·6m²*

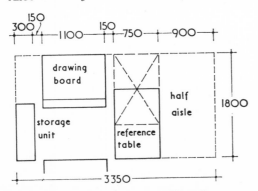

12.31 *Drawing board with side reference: area 7·1m²*

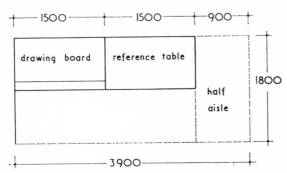

12.32 *Drawing board with mobile reference: area 6·0m²*
(*Building Design Partnership design*)

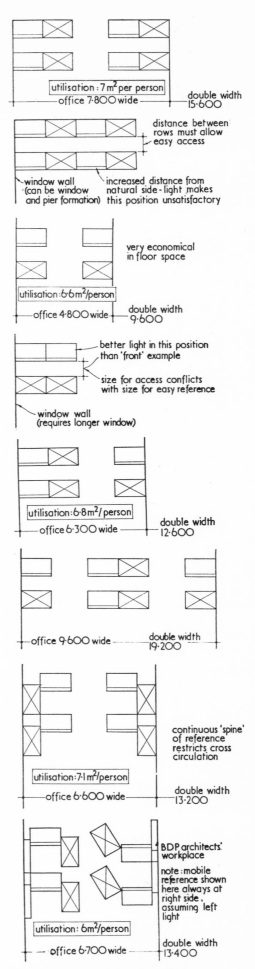

12.33 *Layout of drawing offices*

8 Conference rooms

Typical conference room layouts are shown in **12.34**

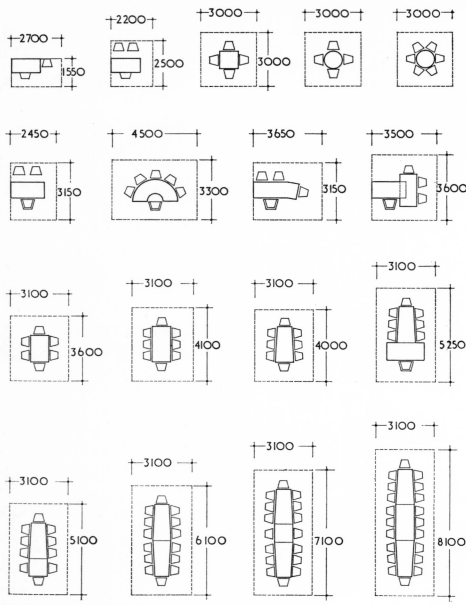

12.34 *Conference room layouts*

9 References and sources

AJ information sheet 1230 Office equipment. AJ, 1964, February 5 CI/SfB 32 (7–)

AJ information sheet 1231 Office storage and filing. AJ, 1964, February 5 CI/SfB 32 (76)

AJ information sheet 1233 Space and planning in offices, AJ, 1964, February 5 CI/SfB 32

AJ information sheet 1239 Offices: Conference rooms. AJ, 1964, February 12 CI/SfB 32

AJ information sheet 1242 Drawing offices. AJ, 1964, February 19 CI/SfB 32

BERGLUND, E. Bord (tables) för måltider och arbete i hemmet. Stockholm, 1957, Svenska Slöjdföreningen (Swedish Society of Industrial Design) CI/SfB (72·2)

BRITISH STANDARDS INSTITUTION BS 3893:1965 Office desks, tables and seating. London 1965, The Institution CI/SfB 32 (72)

MINISTRY OF PUBLIC BUILDING AND WORKS DC4 Dimensional co-ordination for building: Recommended vertical dimensions for educational, health, housing, office and single-storey general purpose industrial buildings. 1967, HMSO (F4j)

MINISTRY OF PUBLIC BUILDING AND WORKS DC5 Dimensional co-ordination for building: Recommended horizontal dimensions for educational, health, housing, office and single-storey general purpose industrial buildings. 1967, HMSO (F4j)

MINISTRY OF PUBLIC BUILDING AND WORKS DC6 Dimensional co-ordination for building: Guidance on the application of recommended vertical and horizontal dimensions for educational, health, housing, office and single-storey general purpose industrial buildings. 1967, HMSO (F4j)

MINISTRY OF PUBLIC BUILDING AND WORKS DC7 Dimensional co-ordination for building: Recommended vertical controlling dimensions for educational, health, housing and office buildings, and guidance on their application. 1967 HMSO (F4j)

BUILDING RESEARCH STATION Digest 20 (second series) Layout and equipment of drawing offices. 1962, HMSO

BRITISH STANDARDS INSTITUTION BS 4330:1968 (metric units) Recommendations for the co-ordination of dimensions in building: Controlling dimensions (F4j)

13 **Shops**

Contents

1 Controlling dimensions

Recommendations for horizontal and vertical controlling dimensions for shop buildings are contained in *Dimensional co-ordination for building* documents DC 4, DC 5, DC 6 and DC 7 and in BS 4330:1968 (metric units). These recommendations are illustrated below. Horizontal controlling dimensions are shown in **13.1** and table I. Vertical controlling dimensions are shown in **13.2**.

The planning dimensions given in this section are those required by the activity taking place within the spaces or rooms. They are not necessarily related to any recommended controlling dimensions as these will depend to some extent on overall planning. Adjustments may therefore need to be made in the dimensions of individual spaces to conform to a dimensional grid.

Table I Horizontal controlling dimensions (to be read in conjunction with 13.1). Dimensions are between boundaries or centre lines of loadbearing walls and columns.

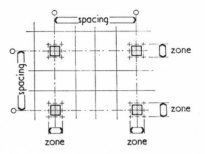

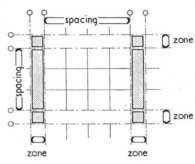

13.1 *Horizontal controlling dimensions. Diagram to define meaning of spacing and zone (to be read in conjunction with table I)*

Spacing (mm)	Zone (mm)
2 700	100
3 000	200
3 300	300
3 600	400
3 900	500
4 200	600
4 500	
4 800	
5 100	
5 400	
5 700	
6 000	
6 300	
6 600	
6 900	
7 200	
7 500	
9 000	
10 500	
12 000	
18 000	

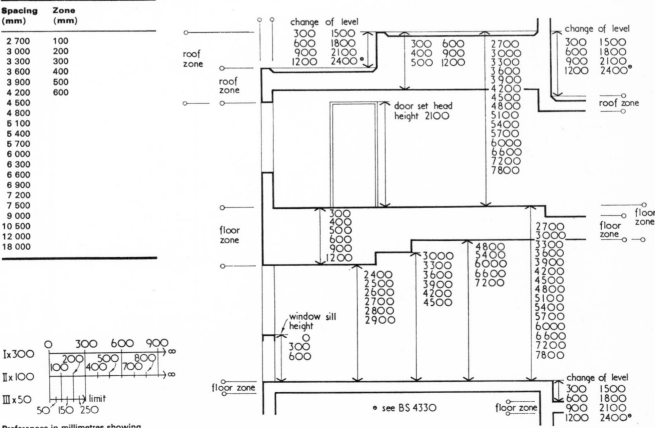

13.2 *Vertical controlling dimensions*

Preferences in millimetres showing increments for first, second and third preferred dimensions (BS 4011)

Foodshops

Traditional layouts for personal service foodshops are shown in **13.3** to **13.7**. Minimum recommended total width of shops is 5·4 metres.

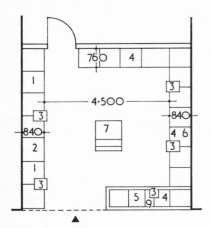

13.3 *Greengrocers shop*
Overall width: nearest controlling dimension (BS 4330) *is 6m or 6·3m*

Key to 13.3

1 Open bins for root vegetables; 2 Slab or mesh shelves for greens (cabbage, etc) over vegetable bins, about 910mm above floor level; 3 Scales; 4 Open bins, racks or display equipment for fruit and flowers; 5 Suspended shelf for choice fruits. A high level rail for bananas is often provided here or over 4; 6 Shelf for flowers; 7 pay desk, sometimes replaced by a short counter, off centre, to facilitate wrapping and storage of wrapping material; 8 The entrance may be closed by a roller shutter and left open by day or by a conventional door which can be fixed open in good weather; 9 The shop window is often arranged so 4 and 5 make a display visible both inside and outside the shop.

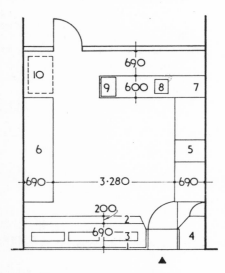

13.4 *Bakers or pastry shop*
Overall width: nearest controlling dimensions (BS 4330) *is 4·5m or 4·8m*

Key to 13.4

The traditional layout shown here is still widely used although frequently combined with a tea-room or restaurant in the back quarters or at first floor. Refrigerated display cases are commonly used instead of the older kind of showcases for cakes and confectionery. 1 Shop window with marble or cooled display bed about 760mm above floor level; 2180mm wide shelf set 150mm above display bed and used for boxing cakes, etc, selected from window display; 3 Pierced or slotted standards supporting adjustable brackets for forward-tilted glass shelves for cakes and 'fancies'; 4 Enclosed window for special display (wedding cakes, etc); 5 serving counter with bread bins or trays below; 6 service counter with biscuit tins below and display cases over; 7 Counter with paper and cardboard box store below; 8 Scales; 9 Cash register; 10 Tubular rack for stock trays from bakehouse or van delivery.

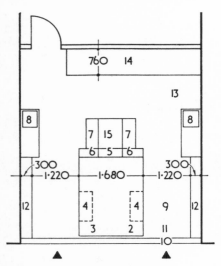

13.5 *Fishmongers shop*
Overall width: nearest controlling dimension (BS 4330) *is 4·5m or 4·8m*

Key to 13.5

The traditional form shown here closely resembles modern specialist shops. An almost invariable addition now is a frozen food refrigerated display cabinet which might replace one of the sinks shown 8. Cold storage and uncrating space is not shown but would have to be planned round a yard at the back. 1 Marble finished concrete slab or refrigerated display bed. Slabs are arranged to fall 150mm from back to front. Front height 610mm above floor level; 2 Gutter with central trapped gulley; 3 Marble curb to traditional slab; 4 Shelves under for wrapping paper, etc; 5 Live fish tank, if included; 6 Scales; 7 Wrapping shelves; 8 Sink, draining and cutting board; 9 Floor laid to fall to street; 10 Gutter cover; 11 Hanging rail for dried or smoked fish or poultry; 12 Display shelves for sundries; 13 Dumping space for crates or customer space for access to frozen food display; 14 Packing bench; 15 Pay desk or cash register.

Key to 13.6

If to be self-service perimeter shelves 6 are usually replaced by dry goods display fittings; counters removed and a central gondola installed. A frozen food display cabinet is essential and may replace one window display 2. 1 Double entrance door; 2 Show window with display bed 760mm above pavement and half height screen at the back; 3 Bacon window with marble display bed, one step shelf and half height screen at back; 4 Hanging rail for bacon, etc. above; 5 Display shelving; 6 Marble shelves at 450mm centres; 7 Counter with shelf below; 8 Marble counter with shelf below; 9 Scales; 10 Cash register; 11 Semi-recess to receive biscuit tins (240mm high × 220mm × 230mm); 12 Cheese cutter; 13 Bacon slicer (requires 920 × 760mm); 14 Marble butter slab and screen; 15 Coffee grinder; 16 Egg display; 17 Bacon or ham hanging rail.

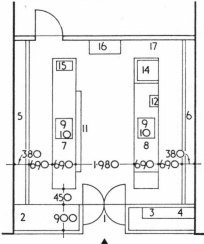

13.6 *Grocery or provisions shop*
Overall width: nearest controlling dimension (BS 4330) *is 5·4m or 5·7m*

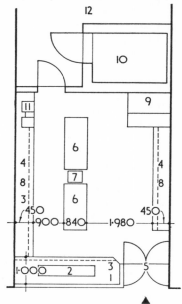

13.7 *Butchers shop*
Overall width: nearest controlling dimension (BS 4330) *is 4·5m or 4·8m*

Key to 13.7

1 Marble finished concrete slab 920mm above floor level laid to fall to a small curb at the window end; 2 12·7mm plate glass shelf suspended at eye level or, alternatively, hanging rails to display joints; 3 Hanging rails; 4 Extra hanging rails, although these are not always included; 5 Entrance doors large enough to admit carcass meat if deliveries are made to the front of the shop; 6 Chopping tables or butchers blocks; 7 Scales; 8 Fixed shelf at counter height; 9 Pay desk; 10 Cold store; 11 Sausage machine; 12 General storage, etc.

3 Supermarkets

Planning

Fittings should be arranged lengthwise to form 'corridors' or gangways in runs of between 3·7m and 7·3m long and widths of at least 1·4m. Where shopping trolleys are used or where an assistant is in attendance minimum width becomes 2m.

Equipment

Collection baskets are normally about 430 × 280 × 200mm and some 30 baskets per 46·5m² of selling floor area are required.

Collection trolleys or carts are about 890mm high × 400mm wide with baskets 610mm long × 305mm deep.

Dry goods display

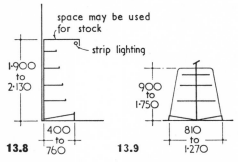

13.8 *Section through a multi-tier display unit with typical dimensions*

13.9 *Section through a gondola display fitting with typical dimensions*

Refrigerated display

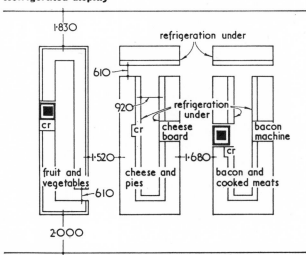

13.10 *Plan showing typical layout for refrigerated display cases where counter service is given*

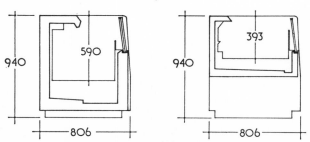

13.11 *Hussmann refrigerated display units for dairy products, perishable provisions, delicatessen, fresh meat. Cabinets may be up to double the width and height shown here. Lengths vary and may be up to 4 metres per unit. Consult manufacturers*

Checkouts

Allow one checkout per 47m² to 60m² of sales floor area, but some supermarket operators may require one checkout fewer than is calculated by these figures for economic reasons.

Checkout counters are usually between 300 and 460mm wide and from 1000mm to 2400mm long. Anything over 1800mm long calls for some kind of conveyor or sloping chute. Height to working top is about 800mm.

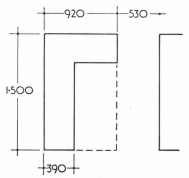

13.12 *Typical plan dimensions for a checkout counter unit. The broken line indicates a screen to protect the cashier*

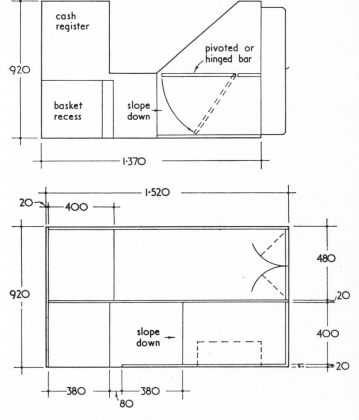

13.13 *Typical checkout units*

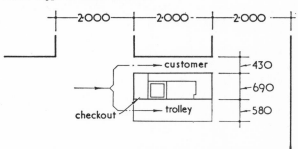

13.14 *Checkout system employed by the French firm of Tirlet*

4 Women's wear

Living models

Dressing rooms should be on a minimum basis of 3·7m².
A small stage may be 6m × 3m deep × 800 mm high.
Sometimes an apron stage about 1·2m wide and projecting
6m to 12m and between 200mm and 800mm high is used.

Display models

Models are about 1630mm high placed on platforms about
230mm high. Allow space of 990mm × 710mm each.

Individual stands

These are usually between 1500mm and 1850mm high and
require about 530mm × 530mm floor area each.

Racks

Table II shows the amount of space required for garment
display. Basic dimensions and types of display unit are
shown in **13.15**, **13.16** and **13.18**.

Table II The use of space for garment display

	Number of customers able to inspect 100 garments at one time	Relative area needed for display and inspection of 100 garments (m²)	Area per customer per ten garments including circulation (m²)	Area per customer per ten garments excluding circulation (m²)
1220mm diam circular rail *	22 (11)	22·85 (11·43)	0·1	0·05
1830mm diam circular rail *	19 (17)	18·12 (15·98)	0·09	0·04
Single rail, one tier	19	13·47	0·07	0·04
Double rail, one tier	10	8·13	0·08	0·05
Wall ftting, one tier	10	8·23	0·08	0·05
Wall fittiing, two tiers	5	4·11	0·08	0·05

* Figures in brackets represent the number of customers able to inspect garments
on one circular display rail (col 1) and the area required by a single display
(col 2). The other figures have been adjusted proportionally to make them
comparable with those for straight rails. Fewer garments per metre run can be
accommodated on a circular rail than on a straight rail owing to the curvature:
this has been allowed for in the figures shown.

Cabinets

Various types of glass fronted cabinets are used for storage
and display. Typical metric dimensions are 1600mm long ×
560mm deep × 1400mm high. Wall mounted cabinets are
1600mm long × 530mm deep × 300mm, 600mm, or 1200mm
high. See **13.20** and **13.22**. Freestanding gondolas vary from
a display platform of about 230mm high × 1520mm ×
530mm or 1070mm wide to units of similar base size but
1180mm high. Counters have to be between 840mm and
910mm high.

Shelving

Common sizes for shelves as at present manufactured (to
the nearest mm) are as follows. Figures in brackets are
obvious metric roundings: 1600mm long × 200, 248 (250),
299 (300), 350, 400mm deep in wood or glass or 533mm (500
or 550) in wood only. Standards should not be more than
910mm apart and brackets to fit them come in sizes projecting 120, 178 (180), 200, 210, 240, 300, 350, 400, 500, 560 and
660mm. The 120 and 200mm brackets are available only in
lightweight and 560 and 660mm in heavyweight material.

Cubicles

The minimum provision for one trying-on cubicle is an
area of 1200mm × 900mm with curtaining up to 2m high.
Most shops provide 1200mm × 1200mm or 1200mm ×
1800mm when an assistant is expected to help customers.

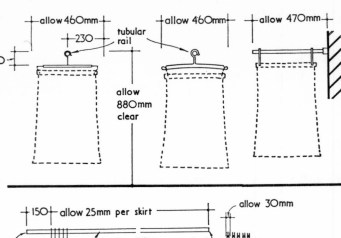

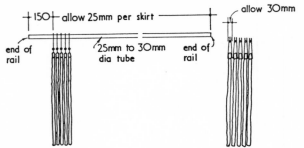

13.15 *Space requirements for skirts hung on hangers or
removable rod fittings*

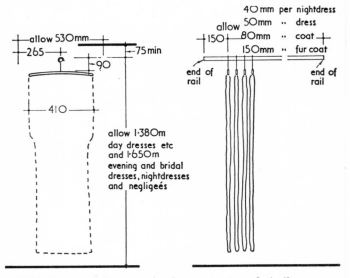

13.16 *Space requirements for dresses, coats and similar
long garments displayed on hangers. Dimensions for short
garments (suits, blouses) similar but overall height to
hanging rail is 915mm. For length of rail allow 40mm per
blouse and 80mm per suit*

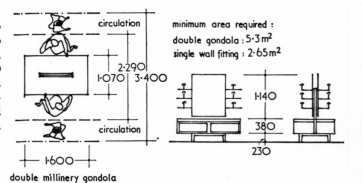

double millinery gondola

13.17 *Millinery gondolas should incorporate a mirror:
basic space requirements should allow for a seated customer.
The same space is needed for a taller base unit*

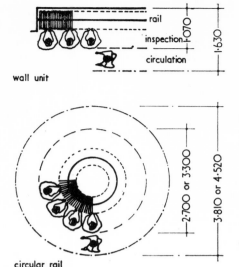

wall unit

circular rail

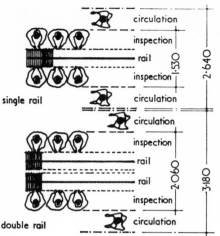

single rail

double rail

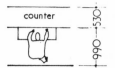

counter

13.19 *Critical dimensions for access to drawer behind counter*

13.18 *Space requirements for racks and cabinets, including circulation around garments hung from circular or straight rails*

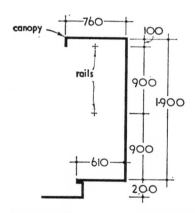

canopy

rails

13.20 *Basic minimum dimensions for cabinets to take one or two tiers of skirts, suits, trousers or jackets. For one tier 1900mm is reduced to 900mm. Cabinets for long garments are similar but with top rail only and height from plinth to top of 1·4m (1·7m for evening dresses etc)*

Shoes

Shoe box sizes vary, but shelving is often arranged to take up to seven boxes in height allowing 90mm in height for women's shoes; 115mm for men's and 300mm in depth. Shoes are displayed on shelving of maximum height 1500mm and depths of 300mm (men) or 230mm to 250mm (women). Some typical dimensions are given in **13.21**.

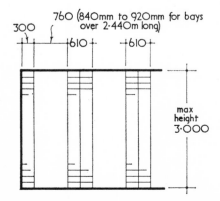

13.21 *Section through stockroom showing recommended dimensions*

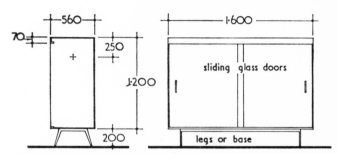

13.22 *Basic dimensions for glass fronted display cabinets*

5 Men's wear

Shelving for bolts of cloth in tailors shops or departments calls for standards spaced at intervals of 500mm or multiples of 250mm. Shelves are usually 750mm apart and the top shelf may be at a height of 1·5m. Selling areas can be quite small, at minimum size accommodating about 18m run of shelving, a hanging space for 20-30 suits, a minimal counter 1600mm × 530mm, one or two cubicles about 1200mm × 1500mm and customer circulation space of about 9·3m². Trousers hung on a single stand require about 330mm run for 10 pairs occupying about 0·14m² excluding circulation. On a double rail stand, 10 pairs occupy about 0·08m² excluding circulation.

6 Drapery

Materials are stored in rolls of which common sizes are
150 to 200mm diameter × 710mm long
150 to 200mm diameter × 910mm long
360 × 200mm (oval) × 690mm long.
Shelves should have at least 220mm clear height between them and the top shelf should not be more than 1370mm above floor level. A 1600mm counter is the absolute minimum but 3200mm is the normal minimum for measuring and cutting metreage of cloth. There should be at least 1000mm clear between counters and shelving.

Patterns

Pattern books need a table which may be 760mm × 2300mm or more or a lectern ledge 400mm wide and between 750mm and 1000mm above floor level. Customers will require at least 1200mm circulation space round a table or from a lectern ledge.

7 Furniture shops: Floor coverings

In new buildings floors should be designed for loadings of 765kgf/m² (7·5kN/m²) and 1000kgf/m² (10kN/m²) Recommended dimensions and forms of display unit are shown in **13.23** to **13.27**.

Table III Standard sizes of floor coverings

Item	Size (mm) as currently manufactured (to nearest mm)
Carpet squares	Ranging from 2285 × 1830 to 4570 × 3660 with size increases of approximately 500mm (at present half a yard or 457mm) in each direction
Body and stair carpet	Roll widths 457, 572, 686, 914, 1372
Broadloom carpet	Roll widths 1830, 2285, 2745, 3200, 3660, 4115 and 4570 and occasionally wider
Indian and Oriental rungs	914, 610, 1525 × 1067, 1372 × 686 and 1525 × 915
Rugs	1220 × 610, 1372 × 686, 1525 × 915
Sisal matting Haircord Underflet	Range of sizes as for body carpet and broadloom
Rush mats	Irregular sizes may be circular or oval but usually not larger than 2500 diameter
Sheet linoleum, vinyl, rubber etc	In rolls 2m wide. A space 3·7m × 2·4m is required for cutting
Linoleum, vinyl, rubber cork etc tiles	230 × 230 and 305 × 305

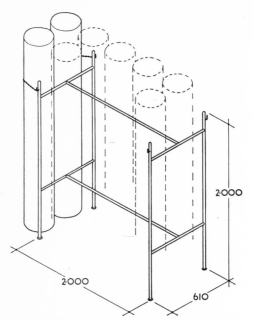

13.23 *Methods of displaying rolls of linoleum on end, suitably anchored*

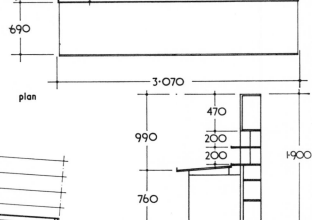

plan

section

13.25 *Method of displaying 230mm × 230mm floor tiles*

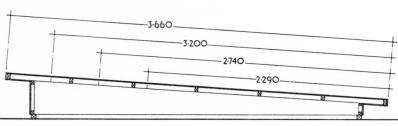

13.24 *Sloping carpet display platform*

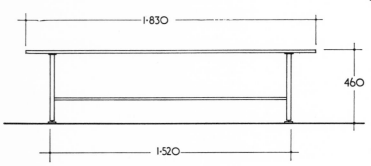

side elevation

13.26 *Display platform for small rugs*

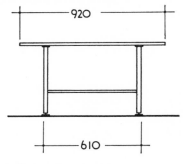

end elevation

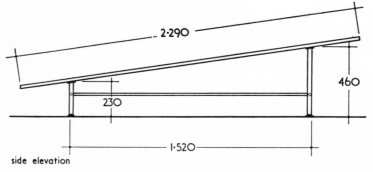

side elevation

13.27 *Sloping display platform for large rugs*

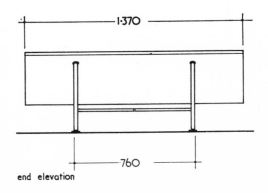

end elevation

8 Workrooms

Typical plans are shown on this **page.**

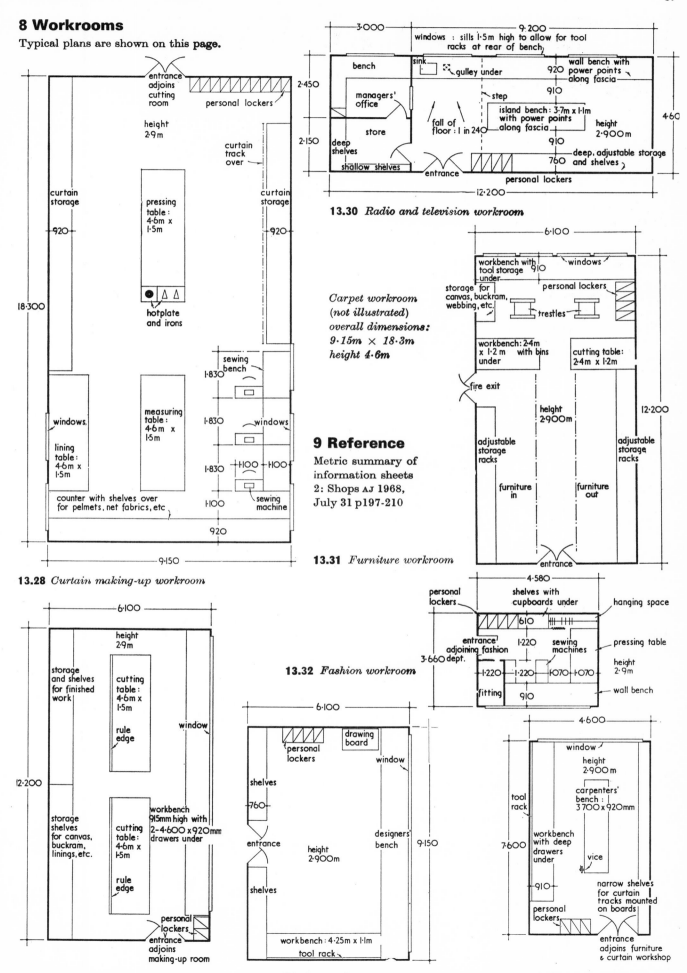

13.28 *Curtain making-up workroom*

13.29 *Curtain cutting workroom*

13.30 *Radio and television workroom*

Carpet workroom
(*not illustrated*)
overall dimensions:
9·15m × 18·3m
height 4·6m

9 Reference

Metric summary of
information sheets
2: Shops AJ 1968,
July 31 p197-210

13.31 *Furniture workroom*

13.32 *Fashion workroom*

13.33 *Display workroom*

13.34 *Carpentry workroom*

14 Garages and service stations

Contents
1 Filling stations
2 Lubrication servicing
3 Vehicle washing
4 General repair workshops
5 Body repairs, coach building and paint shop
6 Stores
7 Showrooms
8 References and sources

1 Filling stations

Pump islands for two, three and four pumps are shown in
14.1 with circulation allowances in **14.2** to **14.4**.
Sizes of petrol tanks are given in table i. Details of petrol
tanks are shown in **14.10** on p89.
Typical filling stations and service stations are shown in
14.5.

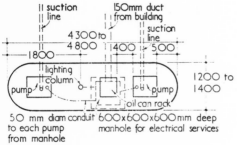

two-pump island

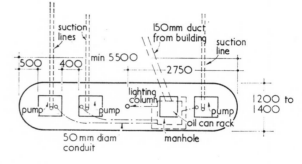

three-pump island

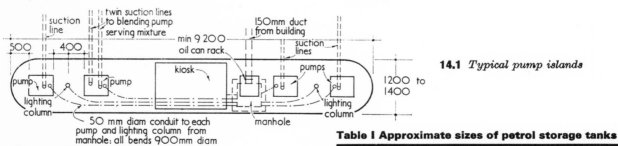

four-pump island

14.1 *Typical pump islands*

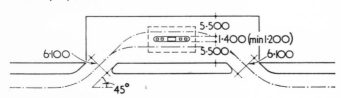

14.2 *Single-island station: circulation requirements*

Table I Approximate sizes of petrol storage tanks

Nominal capacity		Diameter	Overall length
galls	litres	(mm)	(mm)
500	2270	1370	1750
600	2730	1370	2060
750	3410	1370	2600
1000	4550	1370	3350
1250	5680	1530	3430
1500	6820	1830	2900
2000	9100	1830	3800
2500	11370	2000	4040
3000	13640	2130	4120
5000	22730	2290	5950

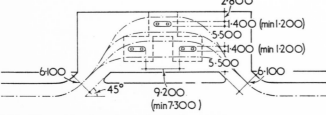

14.3 *Three-island station: circulation requirements*

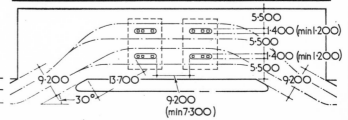

14.4 *Four-island station: circulation requirements*

2 Lubrication servicing

Type of service offered and number of units provided determine basic size of the department. The oil company or consultant will provide advice on the number of stages required in line installations:

Approximate areas required are:

Individual unit: private vehicles   23·2 to 32·5m²
               commercial   37·2 to 50·2m²
Moving line: per stage   23·2 to 32·5m²

Commercial vehicles vary greatly in size, and unless exact requirements are known space allowances should be generous.

Consideration should also be given to the following factors:
Vehicle circulation Carriageways should be at least 3050mm wide with curves of radius not less than 6100mm to outside kerb for private cars. Curves of 7600mm are preferable. For commercial vehicles curves of 9150mm to 10700mm should be allowed. Where space is valuable, turntables can be used and these can be up to 4900mm diam for private cars and 6850mm for commercial vehicles.

Note: Sizes given in this section are those needed to carry out particular activities and are not necessarily related to any controlling dimensions as recommended in BS 4330.

Table II Sizes of lubrication bays

Type of bay	Width (mm)		Length (mm)		Height (mm)	
	Mini-mum	Recom-mended	Mini-mum	Recom-mended	Mini-mum	Recom-mended
Private vehicles						
Individual unit	3650	4600	6700	7900	3650	4100
Moving line (each stage)	3350	4000	6100	6700	3650	4100
Commercial vehicle						
Individual unit	4270	4600	8530	11 000	4570	5500

Typical service pits are illustrated in **14.6** and **14.7**. Average lift dimensions are given in table III, but the manufacturers should be consulted during early planning stages.

Table III Average car lift dimensions

Type of lift	Capacity (kg)	Length (mm)	Width (mm)	Height of lift (mm)
Two-post	3550 to 16250	4570 to 5200 (wheelbase)	1900 to 2590 (wheelbase)	1670 to 1850
Four-post	2280 to 5080	5070 to 7340	3100 to 3650	1570 to 1680
Six-post	8130 to 12190	7670	3280	1470
Centre post*	1030 to 10160	4570 to 6700	740 to 2340	1530 to 1650

* Require excavated foundation to accommodate the ram when lowered; usually measures 2440 × 1220 × 1220mm deep

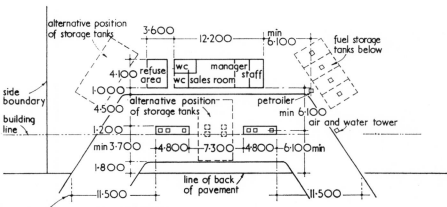

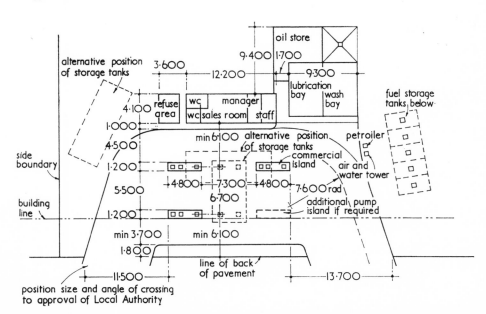

14.5 *Typical filling and service stations. Two-island filling station above; three-island service station below*

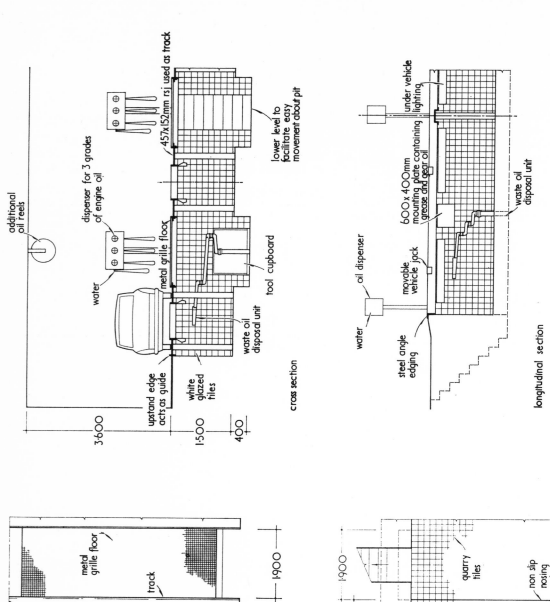

additional oil reels

dispenser for 3 grades of engine oil

water

metal grille floor

457×152mm rsj used as track

lower level to facilitate easy movement about pit

tool cupboard

waste oil disposal unit

white glazed tiles

upstand edge acts as guide

3·600

1·500

400

cross section

water

oil dispenser

movable vehicle jack

steel angle edging

600 x 400mm mounting plate containing grease and gear oil

under vehicle lighting

waste oil disposal unit

longitudinal section

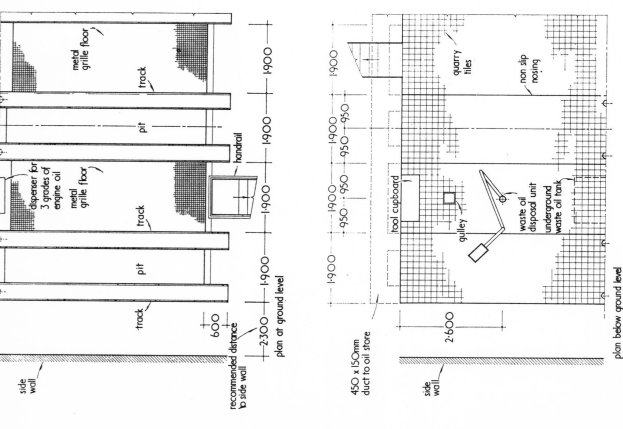

side wall

metal grille floor

track

pit

dispenser for 3 grades of engine oil

metal grille floor

track

pit

track

handrail

1·900

1·900

1·900

1·900

600

2·300 1·900

recommended distance to side wall

plan at ground level

450 x 150mm duct to oil store

side wall

quarry tiles

non slip nosing

tool cupboard

gulley

waste oil disposal unit

underground waste oil tank

1·900

1·900 950

1·900 950

1·900 950

1·900 950

2·600

plan below ground level

14.6 *Service pits (private vehicles)*

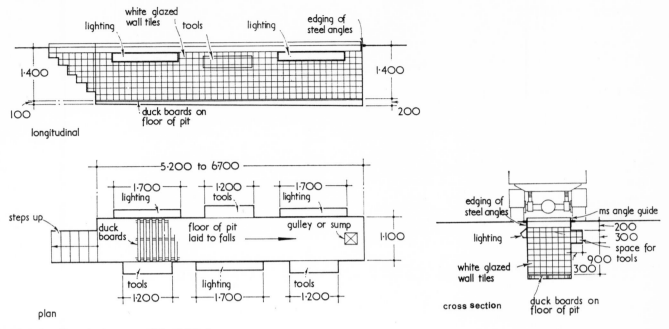

longitudinal

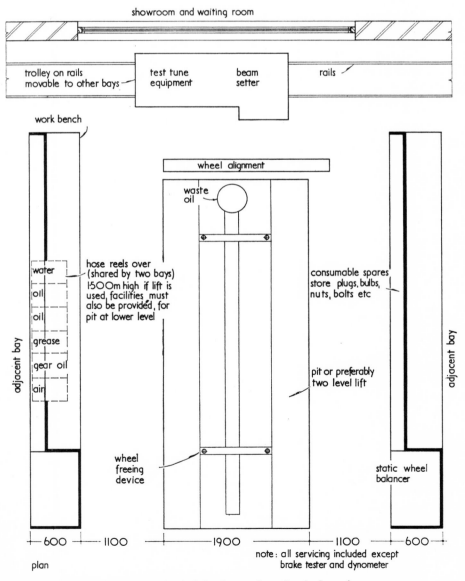

plan

cross section

14.7 *Service pits (commercial vehicles)*

showroom and waiting room

trolley on rails movable to other bays

test tune equipment

beam setter

rails

work bench

wheel alignment

waste oil

hose reels over (shared by two bays) 1·500m high if lift is used, facilities must also be provided, for pit at lower level

water

oil

oil

grease

gear oil

air

adjacent bay

consumable spares store plugs, bulbs, nuts, bolts etc

pit or preferably two level lift

adjacent bay

wheel freeing device

static wheel balancer

+ 600 + 1100 + 1900 + 1100 + 600 +

plan

note: all servicing included except brake tester and dynometer

14.8 *'One-station' service bay for lubrication and mechanical services*

3 Vehicle washing

Minimum space needed for manual wash: 6100 × 3650mm
Minimum space needed for gantry wash: 4600 × 15 250mm
A typical small washing bay with spray arch is shown in **14.8**.

4 General repair workshops

Minimum areas for qualification to be entered in the register
of the motor industry are shown in table IV.

Table IV Minimum areas for workshops

(a) Basic minimum	139·4m²	with accommodation for four cars
(b) For registration as commercial vehicle dealer	185·8m²	
(c) For registration as joint car and commercial vehicle dealer	325·2m²	

Access gangway widths	
Private vehicles	minimum 4570mm (6100mm recommended)
Commercial vehicles	minimum 6100mm (7600mm recommended)

Table V Size of repair bays

Type of bay	Width (mm)		Length (mm)	
	Minimum	Recommended	Minimum	Recommended
Private vehicles				
General repairs	3050	3350	6100	7320
Brake testing	4270	4880	7320	7950
Beam testing	3050	3350	6100	7320
Tuning and testing	4270	4880	10360	10970
Wheel alignment:				
pit		3810		7920
flush		2450		7920
Tyre service	3050	3350	6100	7320
Electrical maintenance	3050	3350	6100	7320
Commercial vehicles				
General repairs	3650	4270	7620	9150

Access doors	
Private vehicles	Minimum 4270mm wide × 3050mm high
Commercial vehicles	Minimum 4570mm wide × 4570mm high

Other workshop spaces	
Diesel testing	Minimum 3000 × 2450mm usually 6100 × 3100mm
Electrical, radio and battery shop	Minimum 9·3m²

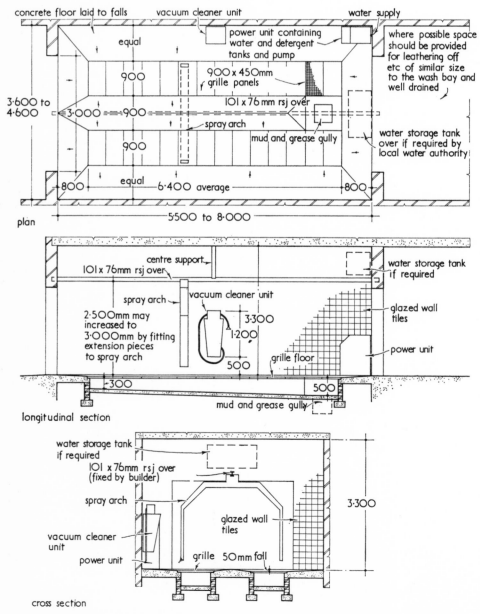

14.9 *Typical small washing bay with spray arch*

5 Body repairs, coach building and paint shop

Table VI Areas required for repairs and painting

Type	Width (mm)	Length (mm)
Private vehicles		
Body repair	7600	12200
Individual unit:		
paint shop	7600	9150
spray booth	3650	7300
Production line:		
preparation area	3950	8250
spray booth	3650	7300
finishing area	3650	7300
Commercial vehicles		
Body repair	9150	13700
Paint shop	9150	10050
Spray booth	4600	9750

6 Stores

The stores area should be about 15 per cent of the area of the whole building.

Access gangways widths:

    main routes 1500mm

    secondary routes 900mm

Maximum length of storage unit 5500 to 6100mm

Maximum height of storage unit 2150mm

7 Showrooms

Area per vehicle 23m² to 28m²

Height:

    private vehicles 3050mm

    commercial vehicles 4450mm

Length as great as possible to allow maximum display

Width:

    private vehicles minimum 6100mm

    commercial vehicles minimum 12 200mm

8 References and sources

AJ information sheet 1251 Fuel sales and forecourt services. AJ, 1964, April 1 CI/SfB 220

AJ information sheet 1252 Lubrication servicing. AJ, 1964, April 1 CI/SfB 220

AJ information sheet 1253 Vehicle washing. AJ, 1964, April 8 CI/SfB 220

AJ information sheet 1254 Workshops. AJ, 1964, April 8 CI/SfB 220

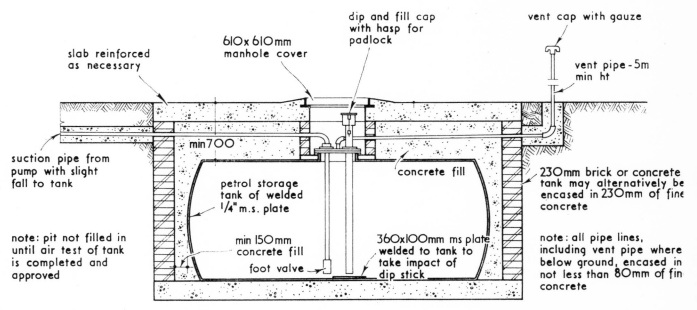

typical underground petrol storage tank

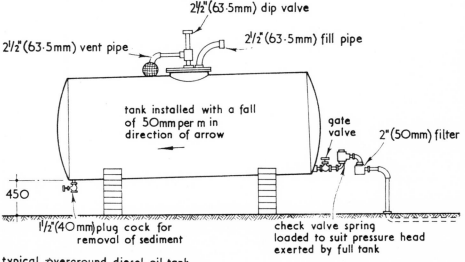

typical overground diesel oil tank

Notes on underground tank:

1 Area above tank and around suction and vent pipe may alternatively be filled with non-corrosive hardcore.

2 Vent pipe need extend only 3·7 m above GL if adequate for surrounding development.

3 Fall to suction pipe (and corresponding section of vent pipe) to be about 40 mm in 3 m.

4 Where tank is under forecourt heavy duty MH cover and frame is required. raised 25 mm. Elsewhere rc slab may not be necessary and light duty MH cover could be used, raised 50 mm.

5 Some authorities may have requirements differing from those shown here

14.10 *Petrol and diesel oil storage tanks*

15 **Industrial buildings**

Contents

1 Controlling dimensions
2 Loading bays
3 Doors for storage buildings
4 Storage buildings: pallets
5 References and sources

1 Controlling dimensions

Recommendations for horizontal and vertical controlling dimensions for industrial buildings are contained in *Dimensional co-ordination for building* documents DC4, 5, 6 and 7 and in BS 4330:1968.

These recommendations are illustrated below.

Horizontal controlling dimensions are shown in **15.1** and table I. Vertical controlling dimensions are shown in **15.2**. A summary of vertical dimensions is given in tables II and III.

Typical heights for use with forklift trucks and stacker cranes are shown in **15.3**, **15.4** and **15.5**. These heights are not necessarily related to the vertical controlling dimensions recommended in the BS document but are related to the types of goods stored. They should, of course, be suitably adjusted to conform to the recommended controlling dimensions.

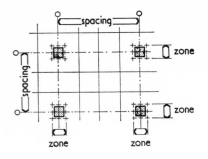

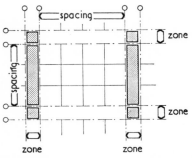

15.1 *Horizontal controlling dimensions. Diagram to define meaning of spacing and zone (to be read in conjunction with table I opposite)*

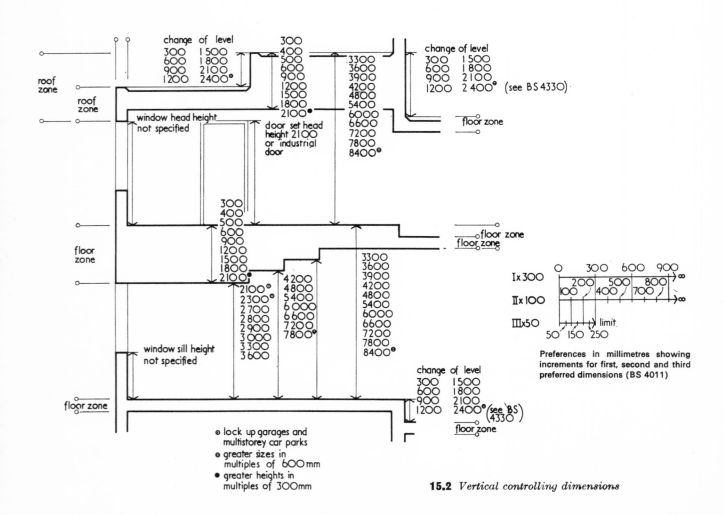

Preferences in millimetres showing increments for first, second and third preferred dimensions (BS 4011)

● lock up garages and multistorey car parks
⊖ greater sizes in multiples of 600mm
● greater heights in multiples of 300mm

15.2 *Vertical controlling dimensions*

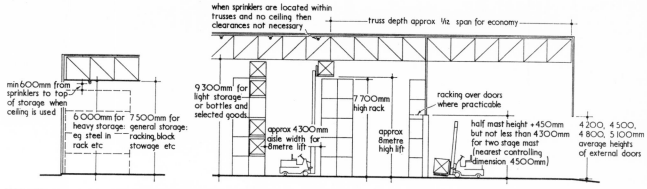

15.3 *Typical section through warehouse for forklift truck handling*

Table I Horizontal controlling dimensions between boundaries or centre lines of loadbearing walls and columns (to be read in conjunction with 15.1)

Spacing (mm)	Zone (mm)
3 000	100
4 500	200
6 000	300
7 500	400
9 000	500
10 500	600
12 000	
13 500	
15 000	
16 500	
18 000	

Note that the economic structural grids for most storage and production requirements are, under normal conditions: maximum 18 000mm; minimum 9000mm.

Table II Typical internal clear heights for production areas (see also 15.2)

Height (mm)	Use
4200	Normal minimum
6600	Normal maximum
5400	Recommended for general purpose factories (and for main entrance and exit doors and others through internal fire walls). This dimension conforms to the current recommended controlling dimension, but note that the previous recommended dimension (as contained in AJ information sheet 1379) was 18ft (5500mm)

Table III Typical internal clear heights for storage areas (see also 15.2)

Minimum clear Internal height* (mm)	Type of storage
7200	Probable minimum for any industrial storage building
7800	When high lift trucks are used
9000 to 11 400	May be needed with modern methods of block storage, racking and post-pallet stacking, and for certain categories like bottling stores
18 500	Fully automatic, computer controlled warehouses and usually where stacker cranes are to be used

* Clearances for structural members, sprinklers, lighting must be added to obtain overall height of buildings

2 Loading bays

Table IV Bay widths for loading and unloading of vehicles

No of vehicles at one time	Recommended width of bay (m)	Minimum width of bay (m)	Minimum depth of bay (within building) (m)
Side loading within building			
1	11	10	
2	18	16	18 to 19
3	24	22	
More than 3	Add 6·5 per vehicle	Add 6 per vehicle	
End loading			
Per vehicle	4	3·75	14

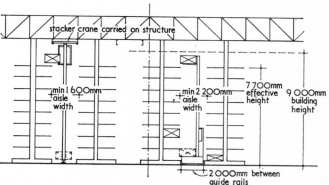

15.4 *Typical section through warehouse for stacker crane handling*

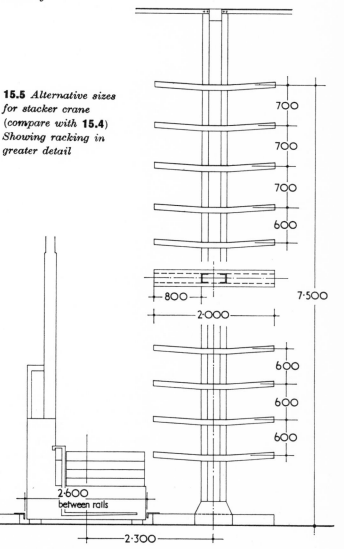

15.5 *Alternative sizes for stacker crane (compare with 15.4) Showing racking in greater detail*

3 Doors for storage buildings

Door width and height must be related to equipment.

Width

3m is usually adequate, but door openings with access to loading and unloading bays may be up to 30m or more.

Height

Should be same throughout building. Minimum is 4200mm; if vehicles enter the building, minimum is 4800mm.

4 Storage buildings: pallets

Dimensions are set out in table v.

Table V Dimensions of standard pallets

Plan dimensions (mm)	Height from ground to underside of deck (mm)
800 × 1000	130 maximum
800 × 1200	
900 × 1200	
1000 × 1000	
1000 × 1200	
1200 × 1600	140 maximum
1200 × 1800	

Clearances are normally 100mm between sides of pallets, 50mm between back and front of adjacent pallets, and 100mm from back of last pallet to wall. There should be 100mm clearance between the side of a pallet and the wall **15.6** and **15.7**.

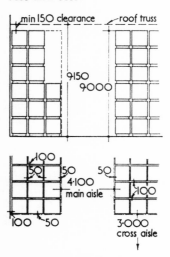

15.6 *Clearance between pallets in stacks with aisle for forklift truck*

Block Stowage

The most economical method of stowage is to form goods into blocks of about 1200mm × 1200mm × 1500mm high, weighing up to 1000kg for direct stacking one on top of the other **15.8**. Current trends suggest that these sizes may be increased to 2400mm × 2400mm × 3000mm high.

5 References and sources

AJ information sheet 1401 Industrial storage buildings: Mechanical handling equipment. AJ, 1966, August 3 CI/SfB 28

AJ information sheet 1402 Industrial storage buildings: Storage systems. AJ, 1966, August 3 CI/SfB 28 (76)

AJ information sheet 1379 Industrial production buildings: Critical design factors. AJ, 1966, April 6 CI/SfB 27

MINISTRY OF PUBLIC BUILDING AND WORKS DC4 Dimensional co-ordination for building: Recommended vertical dimensions for educational, health, housing, office and single-storey general purpose industrial buildings. 1967, HMSO CI/SfB (F4j)

MINISTRY OF PUBLIC BUILDING AND WORKS DC5 Dimensional co-ordination for building: Recommended horizontal dimensions for educational, health, housing, office and single-storey general purpose industrial buildings. CI/SfB (F4j)

MINISTRY OF PUBLIC BUILDING AND WORKS DC6 Dimensional co-ordination for building: Guidance on the application of recommended vertical and horizontal dimensions for educational, health, housing, office and single-storey general purpose industrial buildings. 1967, HMSO CI/SfB (4Fj)

MINISTRY OF PUBLIC BUILDING AND WORKS DC7 Dimensional co-ordination for building: Recommended vertical controlling dimensions for educational, health, housing and office buildings, and guidance on their application. 1967, HMSO CI/SfB (F4j)

BRITISH STANDARDS INSTITUTION BS 4330: 1968 (metric units) Recommendations for the co-ordination of dimensions in building: Controlling dimensions CI/SfB (F4j)

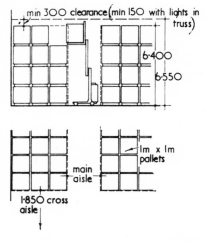

15.7 *Arrangement of stacks with aisle for reach truck*

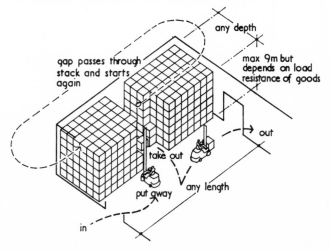

15.8 *Diagram of block stowage technique*

16 **Farm buildings**

Contents

1 Controlling dimensions
2 Farm animals
3 Farm machinery
4 Dairy cattle housing
5 Beef cattle and calf housing
6 Sheep housing
7 Pig housing
8 Crop storage

1 Controlling dimensions

Recommendations for horizontal and vertical controlling dimensions for farm buildings are contained in BS 4330:1968 (metric units). Recommendations for horizontal controlling dimensions are illustrated in **16.1** and table I, and for vertical controlling dimensions in **16.2**.

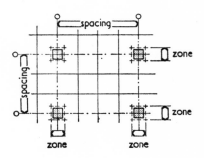

Table I Horizontal controlling dimensions (to be read in conjunction with 16.1). Dimensions are between boundaries or centre lines of loadbearing walls and columns.

pacing (mm)	Zone (mm)
2400	200
3000	300
3600	400
4500	600
6000	
7500	
9000	

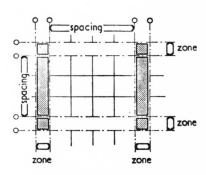

16.1 *Horizontal controlling dimensions. Diagram to define meaning of space and zone (to be read in conjunction with table I)*

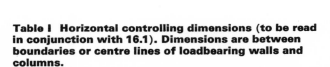

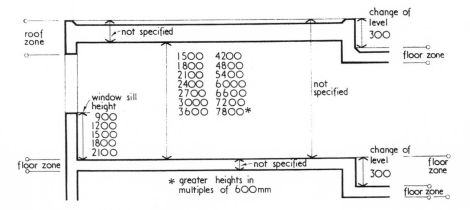

16.2 *Vertical controlling dimensions*

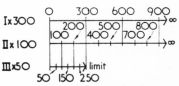

Preferences in millimetres showing increments for first, second and third preferred dimension (BS 4011)

2 Farm animals

Average sizes and weights of animals are shown in **16.3**. Width of animal given is normal trough space allowed (ie about ⅔ of overall width). Length given is normal standing (not fully extended).

baconer (full grown) 100kg
1400 x 300 x 650mm high

sow and litter
2500 x 1000mm

calf (3months) .100kg
1900 x 380 x 1100 mm high

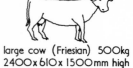

large cow (Friesian) 500kg
2400 x 610 x 1500mm high

bull (small) or steer (large) 1000kg
2600 x 500 x 1800mm high

large ewe (downland) 75kg
1150 x 400 x 750mm high

hen 2kg
400 x 200 x 350mm high

16.3 *Farm animals average sizes and weights*

3 Farm machinery

Average sizes and weights of tractors and other machinery are given in **16.4**.

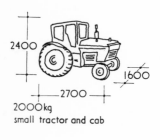

2400

1600

2700

2000kg
small tractor and cab

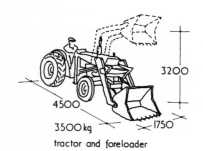

3200

4500

3500kg   1750

tractor and foreloader

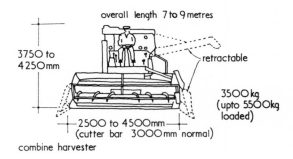

overall length 7 to 9 metres

3750 to
4250mm

retractable

3500 kg
(upto 5500kg
loaded)

2500 to 4500mm
(cutter bar 3000mm normal)

combine harvester

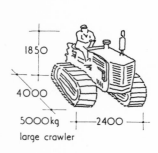

1850

4000

5000kg   2400

large crawler

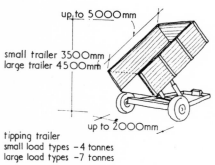

up to 5000mm

small trailer 3500mm
large trailer 4500mm

up to 2000mm

tipping trailer
small load types – 4 tonnes
large load types – 7 tonnes

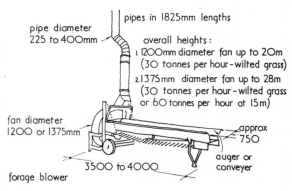

pipes in 1825mm lengths

pipe diameter
225 to 400mm

overall heights :
1. 1200mm diameter fan up to 20m
   (30 tonnes per hour-wilted grass)
2. 1375mm diameter fan up to 28m
   (30 tonnes per hour - wilted grass
   or 60 tonnes per hour at 15m)

fan diameter
1200 or 1375mm

approx
750

auger or
conveyer

3500 to 4000

forage blower

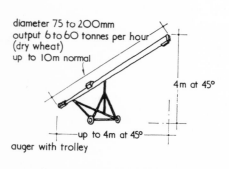

diameter 75 to 200mm
output 6 to 60 tonnes per hour
(dry wheat)
up to 10m normal

4m at 45°

up to 4m at 45°

auger with trolley

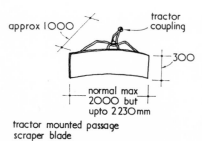

approx 1000

tractor
coupling

300

normal max
2000 but
upto 2230mm

tractor mounted passage
scraper blade

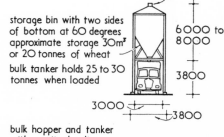

inlet

storage bin with two sides
of bottom at 60 degrees
approximate storage 30m³
or 20 tonnes of wheat

bulk tanker holds 25 to 30
tonnes when loaded

6000 to
8000

3800

3000   3800

bulk hopper and tanker
with gravity loading

16.4 *Farm machinery: average sizes and weights*

4 Dairy cattle housing

Buildings suitable for 120 cow unit including parlour complex are shown in **16.5** to **16.9**.

Allow approximate building area up to 27m wide × 55m long plus 10m turn area at one end plus 4m road.

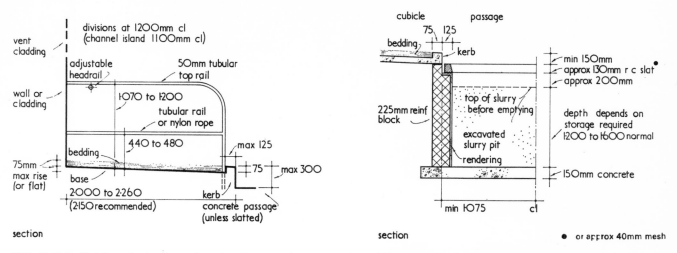

16.5 *Cubicle division: basic dimensions for Friesian Cows*

16.6 *Cubicle division: enlarged detail of passage if slatted*

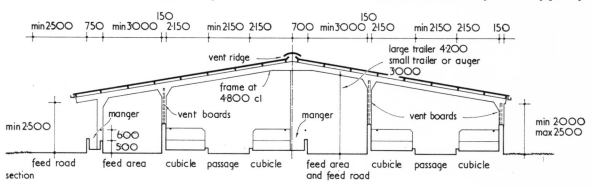

16.7 *Cubicle house: alternative sections showing perimeter feeding to left of centre line; centre feeding to right*

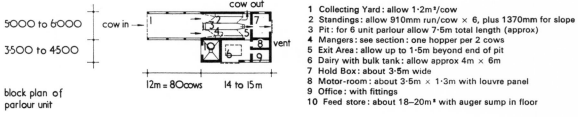

1 Collecting Yard: allow 1·2m²/cow
2 Standings: allow 910mm run/cow × 6, plus 1370mm for slope
3 Pit: for 6 unit parlour allow 7·5m total length (approx)
4 Mangers: see section: one hopper per 2 cows
5 Exit Area: allow up to 1·5m beyond end of pit
6 Dairy with bulk tank: allow approx 4m × 6m
7 Hold Box: about 3·5m wide
8 Motor-room: about 3·5m × 1·3m with louvre panel
9 Office: with fittings
10 Feed store: about 18–20m² with auger sump in floor

16.8

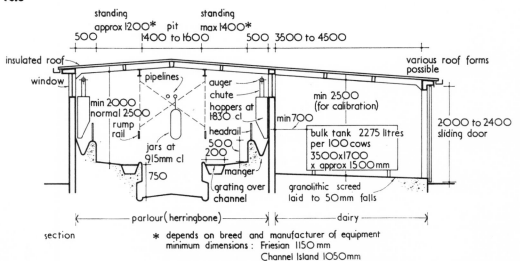

16.9 *Parlour Unit showing overall layout (**16.8**) and typical section: This is a 6 unit/12 standing parlour suitable for operation by one man.*

5 Beef cattle and calf housing

Strawed and slotted yards for beef cattle are shown in
16.10 and **16.11**. Typical calf house is illustrated in **16.12**.

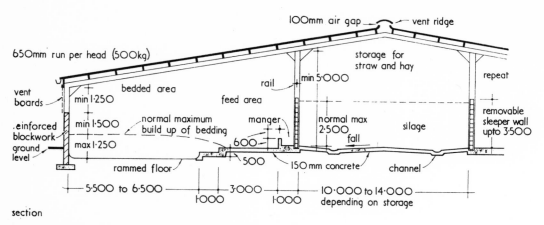

16.10 *Strawed yard for beef cattle with easy feeding*

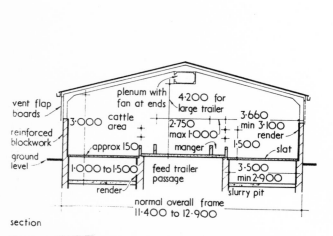

16.11 *Slatted yards for beef (self-unloading trailers). Note:
fully slatted yards not approved by Brambell committee*

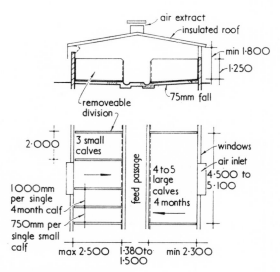

16.12 *Typical calf house*

6 Sheep housing

A typical section of sheep housing is shown in **16.13**. A
dipping tank suitable for large breeds is shown in **16.14**.

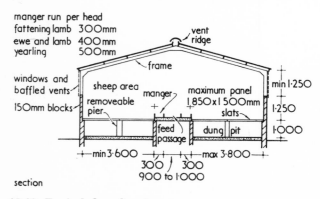

16.13 *Typical sheep housing*

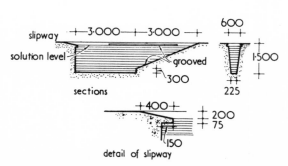

16.14 *Dipping tank (large breeds: 800 ewes) Allow 2.25
litres solution per head*

7 Pig housing

Three types of fattening house are shown in **16.15** to **16.17**
and two types of farrowing house in **16.18** and **16.19**.

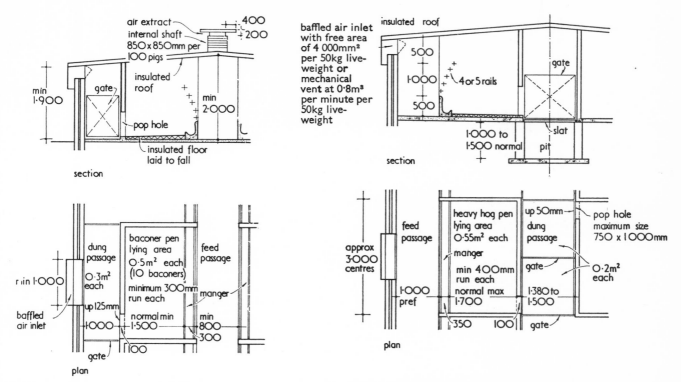

16.15 *Fattening house: side dung passage*

16.16 *Fattening house: centre slatted dung passage*

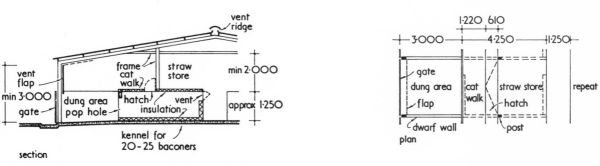

16.17 *Fattening house: Strawed system (with floor feeding)*

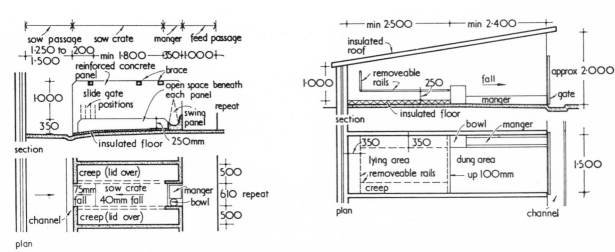

16.18 *Farrowing house: crate (permanent)*

16.19 *Farrowing house: Soleri (open front)*

8 Crop storage

A number of methods of storing a variety of crops is shown in **16.20** to **16.26**.

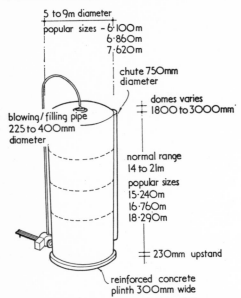

16.20 *Tower silo (wilted grass: 40-50% dry matter)*
Note: wet grain normally in towers of under 12m height × 6m diam approx.

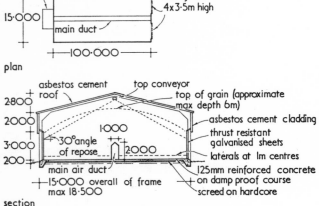

16.21 *Food storage: Grain. Plan shows no lateral system and allows approx 1200 tonnes storage*

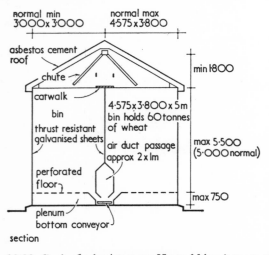

16.22 *Grain drying/storage: Nest of bins (square or rectangular) with roof*

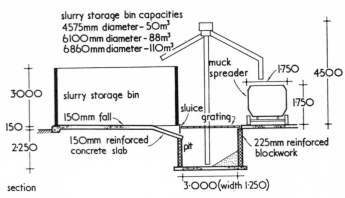

16.23 *Slurry storage above ground*

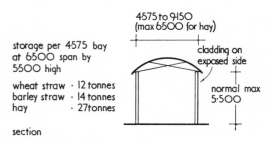

16.24 *Bale storage: Dutch barn*

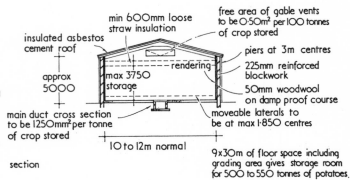

16.25 *Floor storage: Potatoes*

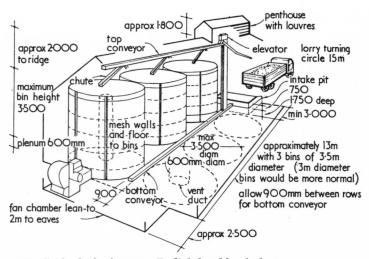

16.26 *Grain drying/storage. Radial flow bins in barn*

17 **Hospitals**

Contents

1 Controlling dimensions
2 Anthropometric data
3 Equipment sizes
4 Room sizes and layouts
5 References and sources

1 Controlling dimensions

Recommendations for horizontal and vertical dimensions for health buildings are contained in *Dimensional co-ordination for building* documents DC4, 5, 6 and 7 and BS4330:1968. These recommendations are illustrated below. Horizontal controlling dimensions are shown in **17.1** and table **1**.

Vertical controlling dimensions are shown in **17.2.**

Table 1 Horizontal controlling dimensions for health buildings between boundaries of zones for loadbearing walls and columns (to be read in conjunction with 17.1)

Spacing (mm)	Zone (mm)
2 700	100
3 000	200
4 200	300
4 500	
5 400	
5 700	
6 000	
6 300	
8 700	
9 000	
9 300	
9 600	
11 700	
12 000	
12 300	
12 600	
12 900	
13 200	
17 400	
18 600	
19 800	

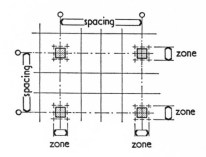

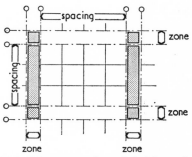

17.1 *Diagram to show definitions of spacing and zone for horizontal controlling dimensions (to be read in conjunction with table 1). BS 4330, Table A6 indicates that only bottom drawing is applicable to health buildings*

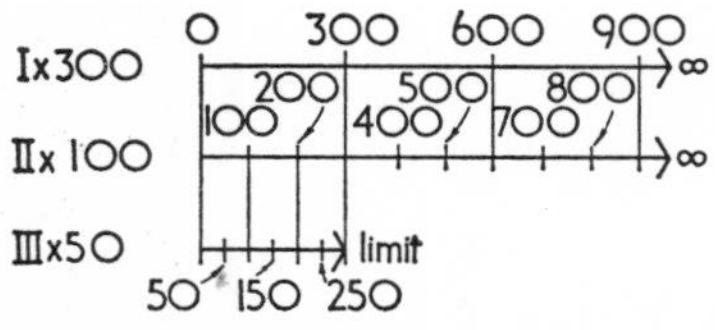

Preferences in millimetres showing increments for first, second and third preferred dimensions (BS 4011)

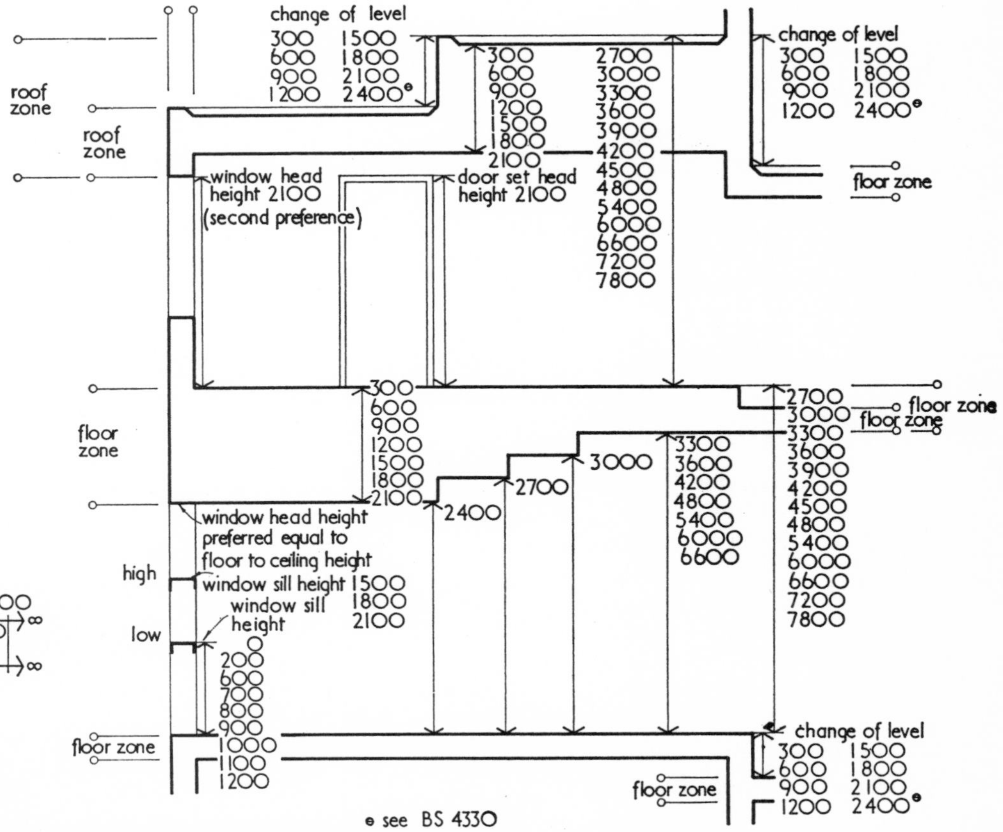

e see BS 4330

17.2 *Vertical controlling dimensions*

2 Anthropometric data

Illustrations **17.3** to **17.19** show some typical space requirements for activities within hospital buildings. See also section **8** of this handbook for general anthropometric data.

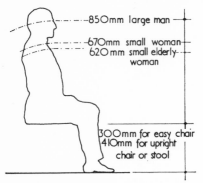

17.3 *Sitting eye levels for large man, small woman and small elderly woman*

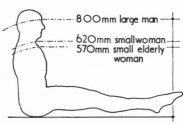

17.4 *Compressed mattress height showing eye levels for large man, small woman and small elderly woman*

17.5 *Using lavatory basin with elbow action valves (small woman and large man)*

17.6 *Lifting supplies onto racks or shelves (small man, where activity involves men only)*

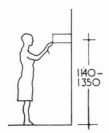

17.7 *Using towel dispenser (lower dimension for small woman in wheelchair; higher dimension for large man)*

17.8 *Carrying a small child*

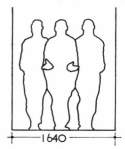

17.9 *Walking helped by two people (large man)*

17.10 *One person walking helped by a handrail. Height of handrail is a compromise between opposing criteria: the large man and the small elderly woman)*

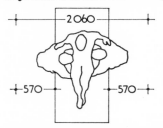

17.11 *Lifting patient in bed (large man)*

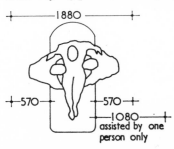

17.12 *Lifting patient in bath (large man)*

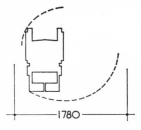

17.13 *Minimum turning space for independent chairbound person: one wheel stationary*

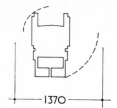

17.14 *Minimum turning space for independent chairbound person: equal and opposite motion of wheels*

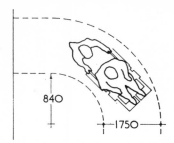

17.15 *Minimum turning circle for assisted chairbound person*

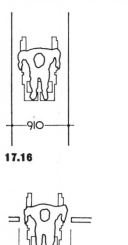

17.16   **17.17**

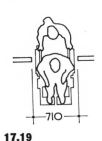

17.18   **17.19**

17.16 *Width for circulation (large man)*

17.17 *Width for circulation (large man helped by large man)*

17.18 *Minimum clear width of openings (large man)*

17.19 *Minimum clear width of openings (large man helped by large man)*

3 Equipment sizes

A selection of some common items of equipment is shown in **17.20** to **17.26**. Sizes given are only approximate (to the nearest 5mm).

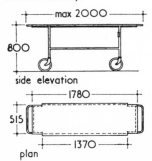

17.20 *Patient trolley*

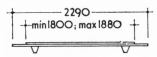

17.21 *Stretcher*

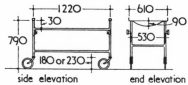

17.22 *Stretcher trolley*

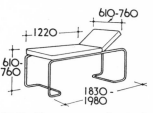

17.23 *Treatment and examination couch*

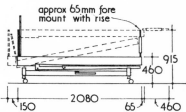

17.24 *Prototype hospital bed, overall width 970mm*

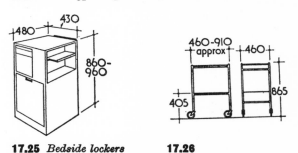

17.25 *Bedside lockers*   **17.26**

17.26 *Front and side elevation of typical instrument trolley (Dressing-trolleys are similar size but of different design and overall height to top of handrail is 940mm)*

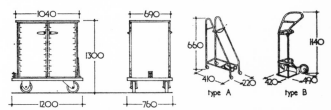

17.27 *Helimatic trolley for insulated trays*   **17.28** *Gas cylinder trolleys*

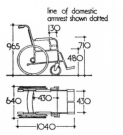

17.29 *Standard self-propelled wheelchair*

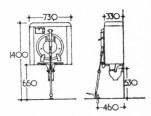

17.30 *Surface mounted bedpan washer*

4 Room sizes and layouts

The drawings **17.31** to **17.35** are taken from *Health Service Design Note No. 3* and show user requirements of spaces common to many departments.

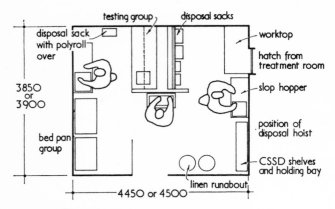

17.31 *Dirty utility room for wards and accident room (17·1m² to 17·6m²)*

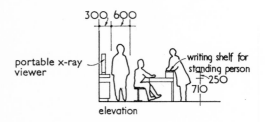

elevation

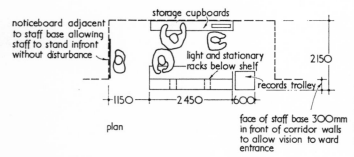

plan

17.32 *Staff base*

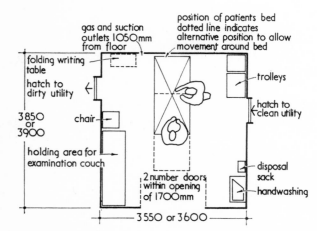

17.33 *Treatment room 13·7m² to 14·0m²*

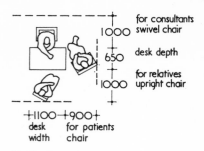

17.34 *Consulting-room*

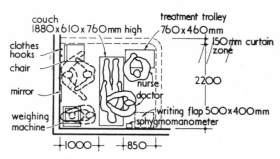

17.35 *Curtained area for weighing and for diagnostic and therapeutic procedures*

5 References and sources

AJ information sheet 1436 Dimensions of common hospital equipment. AJ, 1966, December 28 CI/SfB 41 (7–) (F4j)

MINISTRY OF HEALTH INTER-BOARD STUDY GROUP User requirements: Section A Studies of spaces common to many departments. London, 1966, The Ministry. *Not generally available but may be consulted in* DHSS *library.*

MINISTRY OF PUBLIC BUILDING AND WORKS DC4 Dimensional co-ordination for building: Recommended vertical dimensions for educational, health, housing, office and single-storey general purpose industrial buildings. 1967, HMSO CI/SfB (F4j)

MINISTRY OF PUBLIC BUILDING AND WORKS DC5 Dimensional co-ordination for building: Recommended horizontal dimensions for educational, health, housing, office and single-storey general purpose industrial buildings. 1967, HMSO CI/SfB (F4j)

MINISTRY OF PUBLIC BUILDING AND WORKS DC6 Dimensional co-ordination for building: Guidance on the application of recommended vertical and horizontal dimensions for educational, health, housing, office and single-storey general purpose industrial buildings. 1967, HMSO CI/SfB (F4j)

MINISTRY OF PUBLIC BUILDING AND WORKS DC7 Dimensional co-ordination for building: Recommended vertical controlling dimensions for educational, health, housing and office buildings, and guidance on their application. 1967, HMSO CI/SfB (F4j)

BRITISH STANDARDS INSTITUTION BS 4330:1968 (metric units) Recommendations for the co-ordination of dimensions in building: Controlling dimensions CI/SfB (F4j).

18 **Restaurants and bars**

Contents

1 Space allowances

Table 1 shows space allowances for various types of eating and drinking space. The figures in brackets are minimum allowances, eg for temporary or emergency use where lower standards of comfort may be acceptable. Space requirements per diner are shown in **18.1** and **18.2**.

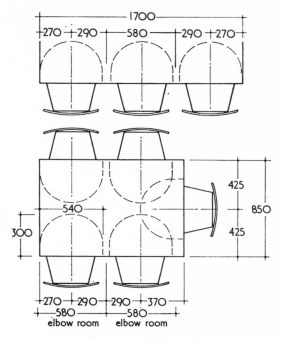

18.1 *Area required by individual diner*

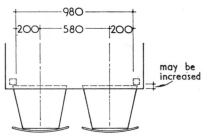

allowing comfort for diners but chairs cannot be pushed inside table legs

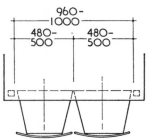

allowing stowing of chairs when not in use

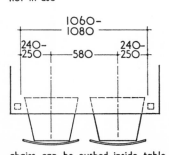

chairs can be pushed inside table legs when diner seated in comfort

Obstruction by table legs

number of seats	table size: drinking mm	table size: eating mm
1	450 to 600	600 to 700
2	600 square	750 square
4	750 square	900 x 950
	—	1500 x 750
6	—	1400 x 950
	—	1700 x 750
8	—	1750 x 900
	—	2300 x 750

number of seats	table size: drinking mm	table size: eating mm
1	450 to 600	750
2	600	850
4	900	1050
6	1150	1200
8	1400	1500

18.3 *Selection of recommended table sizes related to number of persons*

Table I   Space allowances

Type of eating and drinking space	Area per diner (m²)
Commercial restaurants:	
table service	1·0 to 1·3 (0·9)
counter service	1·4 to 1·9 (1·0)
cafeteria service	1·4 to 1·7 (1·1)
Banqueting-rooms (long tables)	0·9 to 1·0 (0·65)
Canteens (industrial and office):	
cafeteria service, tables for four to six	1·1 to 1·4
cafeteria service, tables for eight or over	0·74 to 0·9
School dining-rooms:	
primary schools : counter service	0·74
family service	0·83
secondary schools generally	0·9
colleges of further education	1·1

2 Table sizes

A selection of table sizes (extracted from *Bord*) is given in table II and **18.4**.

Tables for drinking only may be slightly smaller **18.3**.

number of seats

	1	2	3	4	5	6	7	8
A								
B								
C								
D								
E								
F								
G								
H								
I								
J								
round								

18.4 *Tables for varying numbers of people (to be read in conjunction with table II)*

Table II   Table sizes for one to twelve diners (to be read in conjunction with 18.4)

No of persons	Row	400	450	500	550	600	650	700	750	800	850	900	950	1000	1050	1100	Round Diam (mm)
1	D	950	900	850	750	700	650										
	H				750	700	650										
	Round																750
2	A								750								
	E								750								
	D	1350	1250	1150	1100	1100	1100	1100									
	G				950	900	850	800	750								
	I	1300	1250	1050	950	900	850	800									
	Round																850
3	B	1800	1700	1700	1700	1700	1700	1700	950	850							
	D								1300	1100	1100	1100	1100	1100	1100	1100	
	E				1350	1350	1350	1350	950	900	850						
	F				1250	1200	1100	1050									
	Round																950
4	A								1150	1100	1100	1100	1100	1100	1100	1100	
	C								1100	1050	1000	950	900				
	Round																1050
5	B								1350	1350	1300	1300	1250	1200	1100	1100	
	Round																1100
6	A								1700	1700	1700	1700	1700	1700	1700	1700	
	C								1550	1550	1500	1450	1400	1250	1250	1200	
	Round																1200
7	B								1850	1800	1750	1750	1700	1700	1700	1700	
	Round																1300
8	A								2300	2300	2300	2300	2300	2300	2300	2300	
	C								1950	1900	1850	1750	1700	1700	1700	1700	
	J															1600	
	Round																1500
9	B								2400	2400	2350	2300	2300	2300	2300	2300	
10	A								2850	2850	2850	2850	2850	2850	2850	2850	
	C								2550	2500	2400	2350	2300	2300	2300	2300	
11	B								3000	2950	2950	2900	2850	2850	2850	2850	
12	A								3450	3450	3450	3450	3450	3450	3450	3450	
	C								3100	3050	3000	2900	2850	2850	2850	2850	

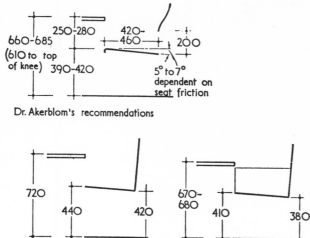

Dr. Akerblom's recommendations

(a) chair without arms
Erik Berglund's recommendations

(b) chair with arms

chairs and tables

18.5 *Heights of chairs and tables*

length of booth, governed by waiter's reach
1140 to 1220mm (sometimes decreased
with reduced comfort): aisle end of
table often rounded at corners and
set back 25 to 50mm

any extra
space allowance
should be
added to this
dimension

760

1900 – 1930
(1550 with 600mm wide table)

18.6 *Banquettes and tables (wall booths)*

Heights of chairs and tables

Recommendations by Dr B. V. Akerblom (in his study *Standing and sitting posture*) and by Erik Berglund (*in Bord*) are given in **18.5**.

However, the present standard height of most tables is about 760mm with chair seat height of 455mm to 460mm.

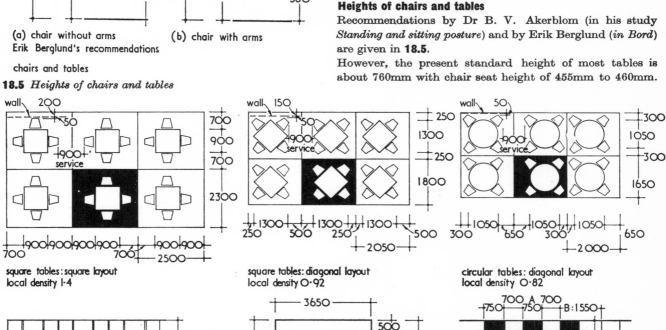

square tables: square layout
local density 1·4

square tables: diagonal layout
local density 0·92

circular tables: diagonal layout
local density 0·82

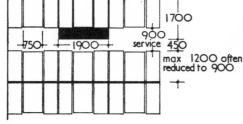

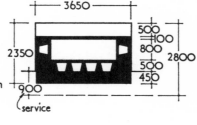

banquette booth seating
local density 0·8

large booth in recess
local density 0·86 if seating 10 people
or 1·1 if only two people sit on bench seat

counter service
local density 1·26
dimensions A and B increased where
two waiters employed

18.7 *Layout arrangements*

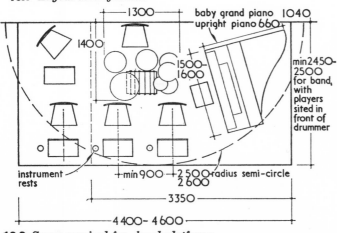

18.8 *Space required for band platforms*

3 Table layouts

Various arrangements with an indication of local densities are shown in **18.7**. No allowance is made for main circulation, entry, queuing, payment and exit.

4 Band platforms

The size and shape of a band platform depends on number of musicians, type of instrument and use of the platform by other artists. Suitable arrangements are shown in **18.8**.

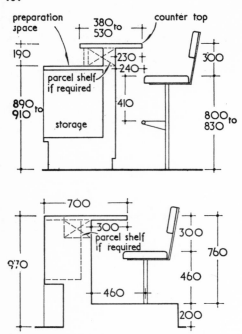

18.9 *Counters and stools*

5 Bars and counters

Typical bar and counter sections are shown in **18.9** to **18.11**.

6 References and sources

AJ information sheet 1257 Eating and drinking spaces: Space requirements and layout. AJ, 1964, April 29 CI/SfB 510

AJ information sheet 1258 Eating and drinking spaces: Dining-room equipment. AJ, 1964, May 13 CI/SfB 510 (72)

BERGLUND, E. Bord (tables) för måltider och arbete i hemmet. Stockholm, 1957, Svenska Slöjdföreningen (Swedish Society of Industrial Design) CI/SfB (72)

KAROLENSKA INSTITUTET Standing and sitting posture. B. V. Akerblom. Stockholm, 1948, Nordiska CI/SfB (E2d)

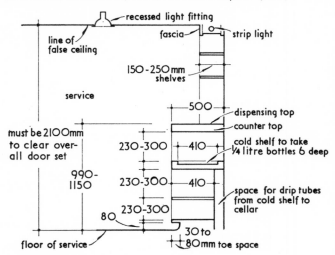

section through typical back counter

18.10 *Counter and back bar fitting*

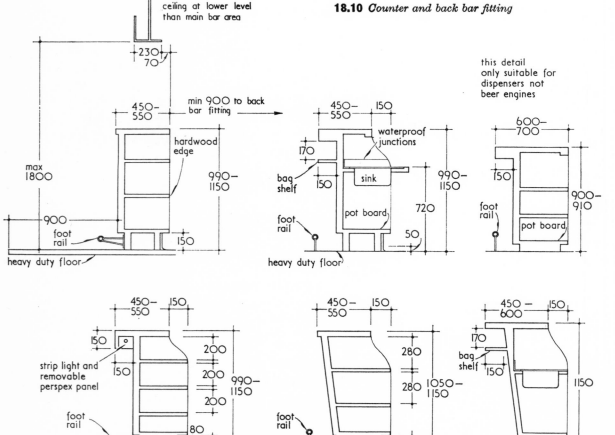

18.11 *Typical counter sections*

19 **Public cloakrooms**

Contents

1 Sizes of items commonly stored

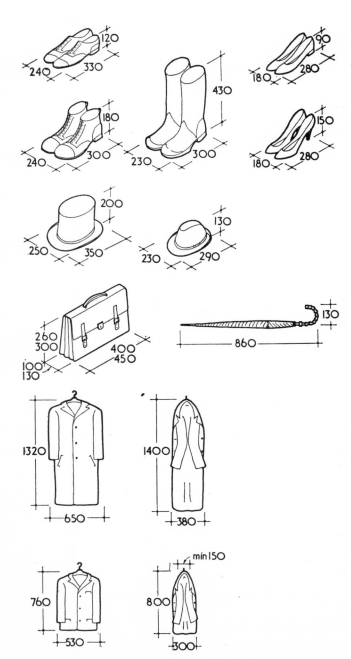

19.1 *Sizes of items commonly stored in cloakrooms*

2 Attended storage

Typical arrangements and space requirements per user are shown in **19.2** and **19.3**.

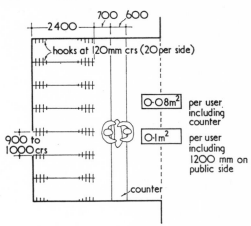

19.2 *Fixed rows of hooks*

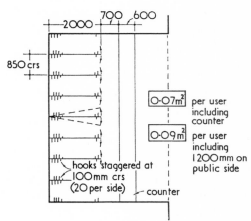

19.3 *Hinged rows of hooks*

3 Unattended storage

The space allowances per user in **19.4** to **19.8** are based upon hangers or hooks at 150mm in rows 3600mm long with 1050mm clear circulation space at ends of rows.

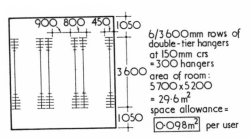

19.4 *Method of calculating space required by each user*

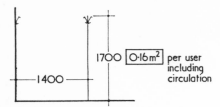

19.5 *Hooks in line*

1700 ☐0·16m² per user including circulation

1400

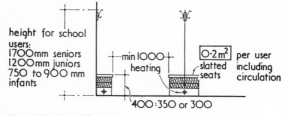

height for school users:
1700mm seniors
1200mm juniors
750 to 900 mm infants

min 1000 heating

☐0·2m² per user including circulation

slatted seats

400 : 350 or 300

19.6 *Hooks with seating*

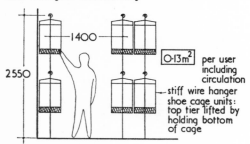

1700

1000 to 1200

protection

heating

shoe cages

☐0·26m² per user, including circulation

19.7 *Hangers with seating*

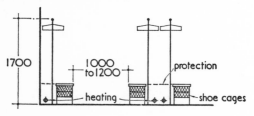

1400

2550

☐0·13m² per user including circulation

stiff wire hanger
shoe cage units:
top tier lifted by
holding bottom
of cage

19.8 *Double tier hangers*

4 Mobile storage

These are proprietary units and measurements **shown in 19.9** are approximate.

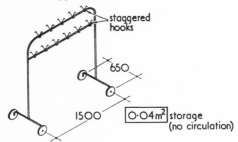

staggered hooks

650

1500   ☐0·04m² storage (no circulation)

19.9 *Mobile coat rack*

5 Lockers

Lockers may be *full height* with a hat shelf and space to hang a coat and store shoes or parcels; or *half height* to take a jacket; or *quarter height* to take either parcels or folded clothes—**19.10** to **19.11**. Combination units such as **19.12** are also available.

Note: Many of the units shown in this section are proprietary systems and metric measurements are only approximate. Manufacturers should be consulted after preliminary planning stages.

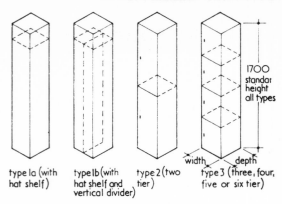

type Ia (with hat shelf)

type Ib (with hat shelf and vertical divider)

type 2 (two tier)

type 3 (three, four, five or six tier)

1700 standar height all types

width   depth

		reference size (mm)					
		A	B	C	D	E	F
type Ia	width	300	300	300	400	400	500
	depth	300	400	500	400	500	500
type Ib	width	–	–	–	400	400	500
	depth	–	–	–	400	500	500
type 2	width	300	300	300	400	400	500
	depth	300	400	500	400	500	500
type 3	width	300	300	300	400	400	500
	depth	300	400	500	400	500	500

19.10 *Locker dimensions according to B.S.*

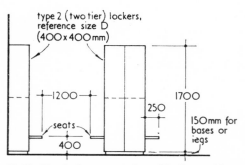

type 2 (two tier) lockers,
reference size D
(400 x 400 mm)

1200

1700

250

150mm for bases or legs

seats

400

19.11 *Typical cross-section of lockers with seats (1200mm dimension not taken from B.S.)*

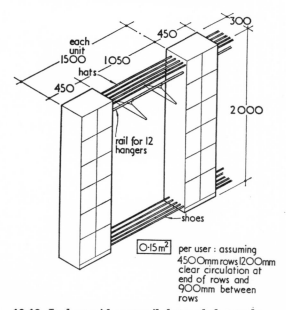

each unit
1500

450

1050

450

hats

300

2000

rail for 12 hangers

shoes

☐0·15m² per user : assuming 450mm rows 1200mm clear circulation at end of rows and 900mm between rows

19.12 *Lockers with coat rail, hat and shoe racks*

6 References and sources

AJ information sheet 1209 Cloakrooms. AJ, 1963, July 17
CI/SfB 96

20 Theatres

Contents

1 Auditorium seating

Minimum dimensions are shown below and in **20.1** to **20.7**. Table I shows the distance of seats from gangways. It should be noted that the width of individual seats is taken as 500mm. The standard dimension at present is 1ft 8in (508mm). While manufacturers' dimensions cannot be anticipated, 500mm appears to be a sensible metric size. Dimensions given in this section may, therefore, be used for preliminary planning purposes but exact sizes of seats must be checked with manufacturers.

Minimum dimensions

A Back-to-back distance between rows of seats with backs: 750mm (minimum).
B Back-to-back distance between rows of seats without backs: 600mm (minimum).
C Width of seats with arms 500mm (minimum).
D Width of seat without arms 450mm (minimum).
E Unobstructed vertical space between rows (seatway) 300mm. See table I.
F For normal maximum distance of seat from gangway see table I. But rows with more than twenty-two seats could be possible, provided that the audience was not imperilled.
G Minimum width of gangway 1000mm.

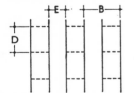

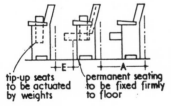
tip-up seats to be actuated by weights
permanent seating to be fixed firmly to floor

20.1 *Seating with backs*

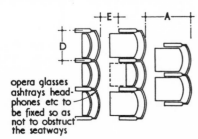
opera glasses ashtrays head-phones etc to be fixed so as not to obstruct the seatways

20.2 *Plan of seating without backs*

20.3 *Plan of seating with backs and arms*

Table I Distance of seats from gangways

Minimum seatway (measured between perpendiculars) E (mm)	Maximum distance of seat from gangway (500mm seats) F (mm)	Maximum number of 500mm wide seats per row	
		Gangway both sides	Gangway one side
300	3000	14	7
330	3500	16	8
360	4000	18	9
390	4500	20	10
420	5000	22	11

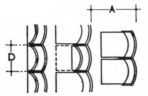

20.4 *Plan of seating without arms*

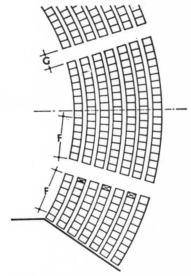

20.5 *Part plan of auditorium*

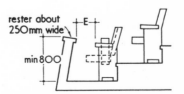
rester about 250mm wide
min 800

20.6 *Section through balcony front*

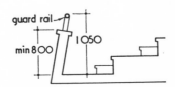
guard rail
1050
min 800

20.7 *Balcony front opposite gangway*

2 Exits

Minimum requirements as to number and width of individual exits are given in table II, but more than the minimum number may be required, depending on such factors as maximum travel distance to an exit and layout of seating and gangways. Principal regulations applying to design of exit corridors and staircases are illustrated in **20.8** and **20.9**.

Table II Exit requirements

Number of people accommodated on each tier or floor	Minimum number of exits*	Minimum width (mm)
200	2	1050†
300	2	1200
400	2	1350
500	2	1500
750	3	1500
1000	4	1500

*Plus one additional exit of not less than 1500mm for each extra 250 persons or part thereof
†Would not normally apply to exit corridors or staircases serving auditorium of a theatre

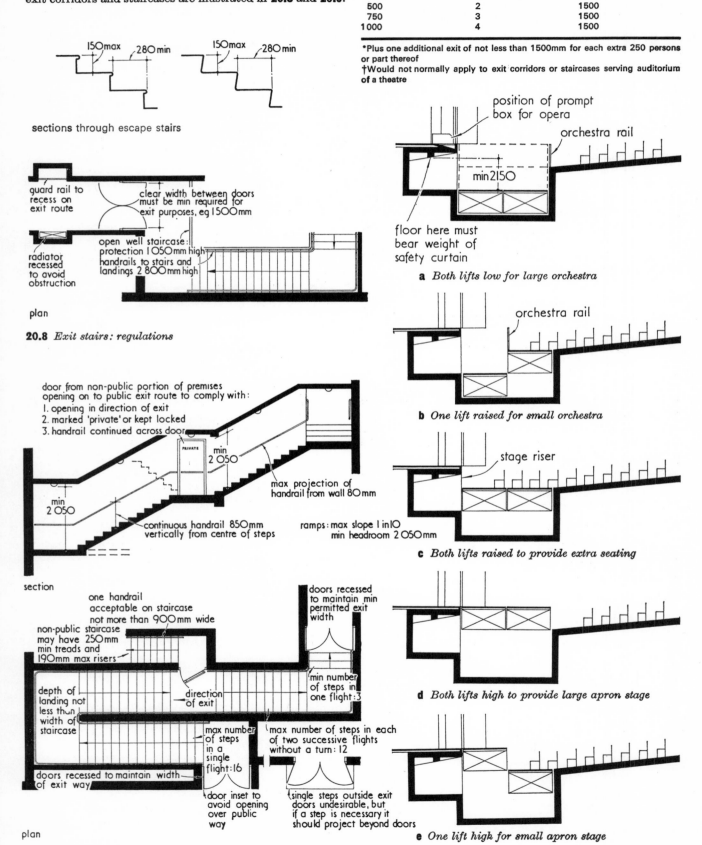

sections through escape stairs

20.8 *Exit stairs: regulations*

20.9 *Exit stairs: regulations*

plan

a *Both lifts low for large orchestra*

b *One lift raised for small orchestra*

ramps: max slope 1 in 10
min headroom 2 050mm

c *Both lifts raised to provide extra seating*

d *Both lifts high to provide large apron stage*

e *One lift high for small apron stage*

20.10 *Use of adjustable apron stage*

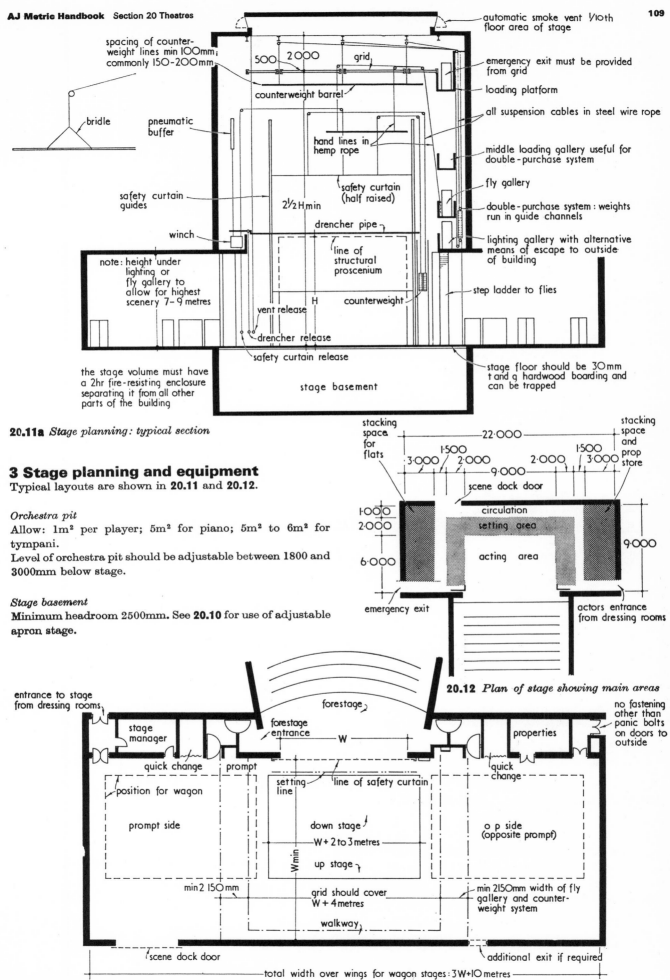

20.11a *Stage planning: typical section*

3 Stage planning and equipment
Typical layouts are shown in **20.11** and **20.12**.

Orchestra pit
Allow: 1m² per player; 5m² for piano; 5m² to 6m² for tympani.
Level of orchestra pit should be adjustable between 1800 and 3000mm below stage.

Stage basement
Minimum headroom 2500mm. See **20.10** for use of adjustable apron stage.

20.12 *Plan of stage showing main areas*

20.11b *Stage planning: typical plan*

4 Dressing-rooms

Dressing-room arrangements are illustrated in **20.13** and **20.14**

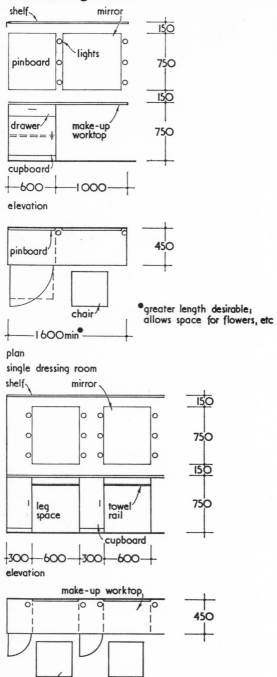

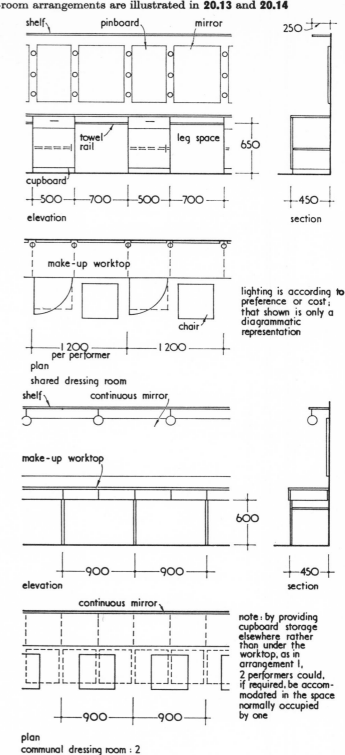

*greater length desirable; allows space for flowers, etc

lighting is according to preference or cost; that shown is only a diagrammatic representation

note: by providing cupboard storage elsewhere rather than under the worktop, as in arrangement 1, 2 performers could, if required, be accommodated in the space normally occupied by one

20.13 *Dressing-room arrangements: space allowances for each performer*

Note also the following dimensions:

Clothes on hangers or on hooks at 600mm centres.

Length of hanging rail: 750mm (normal); 1250mm for shows with numerous changes.

Dressing-room door width: 850mm to 900mm.

Wall space for washbasin: 900mm.

Long mirror: minimum size 1200m × 600m; height 750mm above floor.

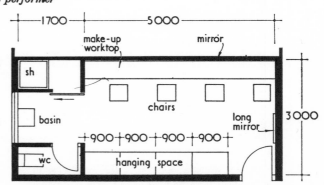

shared dressing room for 4 performers

20.14 *Dressing-room arrangements: complete rooms*

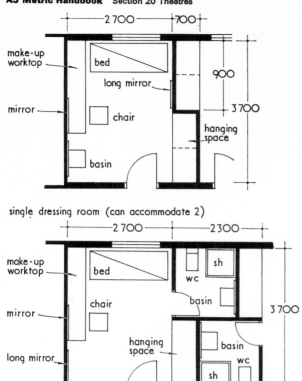

single dressing room (can accommodate 2)

single dressing room with direct access to shower wc and basin (can accommodate 2)

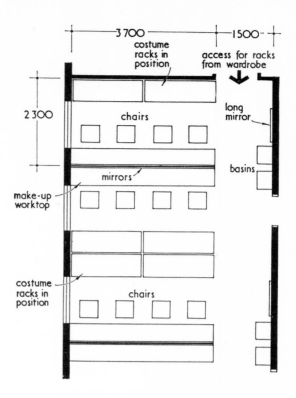

communal dressing room: make-up positions alternating with hanging space

20.4 *(contd) Dressing room arrangements: complete rooms*

5 Overall dimensions

The absolute minimum area of site required for a typical proscenium theatre is 33m × 30m (where no workshops are required); with workshops, restaurant and so on 40m to 52m × 31m to 39m will be required **20.5**.

6 References and sources

MORO, P. Theatre buildings: Shape, size and siting. AJ, 1967, February 1, p321–324 and AJ information sheets 1265, 1270 and 1271 (CI/SfB 524 AJ 1967, February 8, AJ 1964, August 26 and September 2.

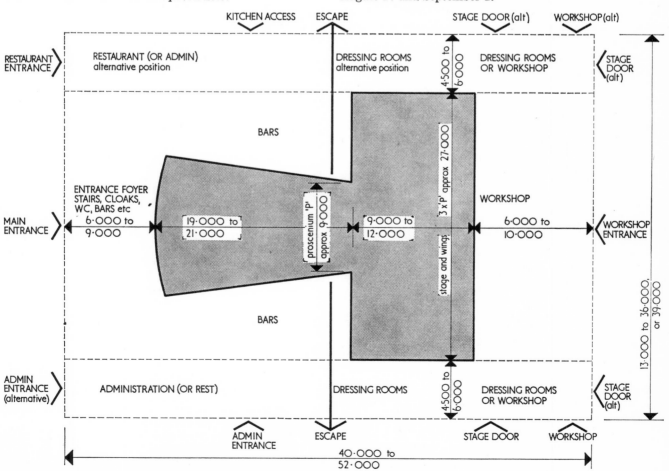

20.5 *Minimum area of site required for typical proscenium theatre*

21 Cinemas

Contents

1 Cinema screens

Screen size and shape are dependent mainly on light source, screen luminance, picture quality, method of projection, screen aspect ratios, viewing conditions, seating capacity and circulation.

The interrelation of these factors does not permit a simple table of sizes to be given, as they depend on ratios and proportions rather than on fixed dimensions.

This information has been published by The Architects' Journal in AJ information sheets 1467, 1469 and 1470, which should be consulted (for details see para 6).

2 Seating

Basic seating layout is shown in **21.1**.

Note also the following:

Minimum spacing back-to-back 750mm. Current comfort standard is 910mm to 920mm back-to-back. Sometimes this is increased to 990mm to 1100mm in balcony or more expensive rear stalls.

The back row requires an additional 75mm to allow for lean-back of seats.

In tip-up seats, minimum distance from front of armrest to back of seat in front is 305mm.

Length of rows: (a) with back-to-back spacing of 760mm,

no seat should be more than 3800mm from the gangway; (b) with back-to-back spacing of 910mm to 920mm, no seat should be more than 4570mm from the gangway.

Allow an extra 25mm to 30mm at each end for arm rests.

3 Gangways and exits

Gangways Minimum width is 970mm in existing premises, 1070mm in new buildings.

Exits

Table I shews minimum number of exits to be provided under statutory regulations.

Table I   Minimum number of exits to be provided under statutory regulations

Number of persons	Number of exits
1 to 60	1
61 to 600	2
610 to 1000	3
1001 to 1400	4
1401 to 1700	5
1701 to 2000	6
2001 to 2250	7
2251 to 2500	8
2501 to 2700	9

Table II   Maximum distances from seat to exit

Structure	Cinema occupancy only (m)	Multiple or dual occupancy (m)
Wholly fire resistant	30	26
30min fire resistance	20	20
Less than 30min fire resistance	15	12

Exit widths are related to use.

Allow forty-five people/min/unit width of 533mm (assume 530mm).

New building: exit doors minimum 2 units (1060mm).

Existing building: exit doors minimum 970mm in width. The audience must be able to leave the auditorium in $2\frac{1}{2}$ minutes.

4 Auditorium layout

The size and position of a cinema screen must be related to the shape and rake of the auditorium floor. For small cinemas and lecture theatres with screens of standard aspect ratio (AJ information sheet 1467) or where 16mm film is exhibited, the limiting factors for satisfactory viewing are shown in **21.2**. The requirements for 35mm projection in **21.3** and for 70mm projection in **21.4**.

5 Projection suites

Typical layouts for projection suites are shown in **21.5** and **21.6**.

6 References and sources

AJ information sheet 1467 Cinemas: Projection systems AJ, 1967, March 22 CI/SfB 525

AJ information sheet 1469 Cinemas: Screens and curtains. AJ, 1967, March 22 CI/SfB 525

AJ information sheet 1470 Cinemas: Auditoria and ancillary accommodation. AJ, 1967, March 22 CI/SfB 5240

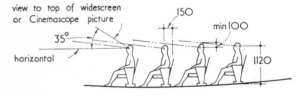

view to top of widescreen or Cinemascope picture

35°

horizontal

front stalls

150

min 100

1120

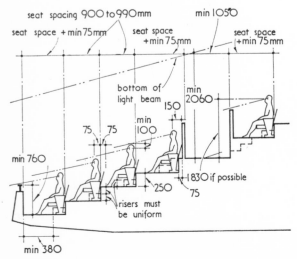

seat spacing 900 to 990mm

min 1050

seat space + min 75mm

seat space + min 75mm

seat space + min 75mm

bottom of light beam

min 2060

150

min 100

75   75

min 760

1830 if possible

250

75

risers must be uniform

min 380

balcony

21.1 *Setting out cinema seating*

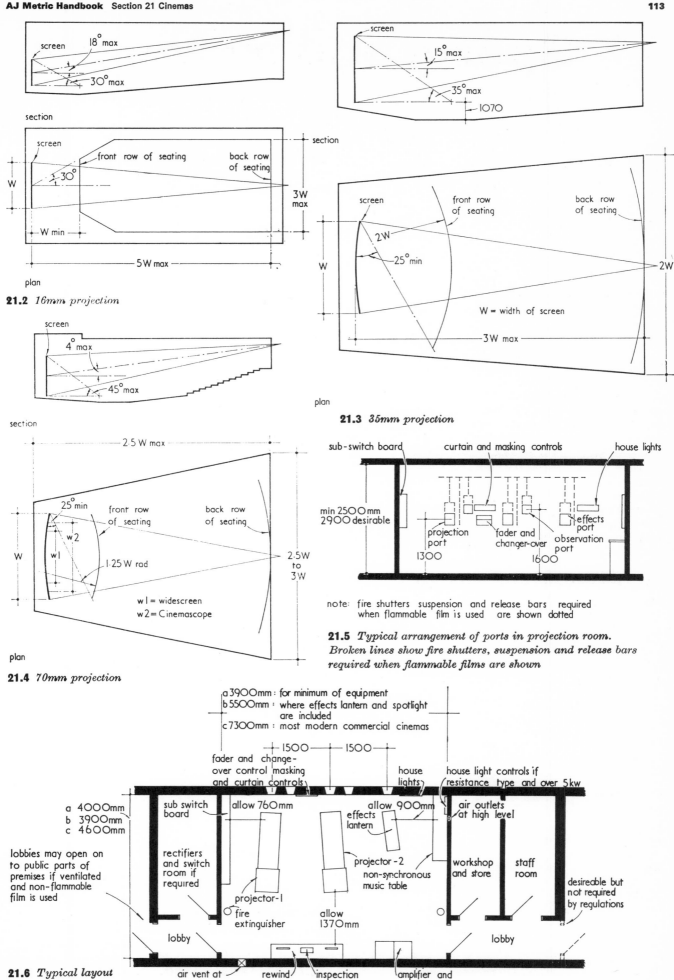

21.2 *16mm projection*

21.3 *35mm projection*

W = width of screen

21.4 *70mm projection*

w1 = widescreen
w2 = Cinemascope

note: fire shutters suspension and release bars required
when flammable film is used are shown dotted

21.5 *Typical arrangement of ports in projection room.
Broken lines show fire shutters, suspension and release bars
required when flammable films are shown*

a 3900mm : for minimum of equipment
b 5500mm : where effects lantern and spotlight
are included
c 7300mm : most modern commercial cinemas

21.6 *Typical layout
for projection suite*

22 **Sports and swimming**

Contents
1 Agreed standards
2 Indoor sports
3 Outdoor sports
4 Other areas
5 Composite spaces
6 Swimming
7 References and sources

1 Agreed standards
Agreement on equivalent metric dimensions for some sports activities in the UK is still some way from approaching finality, so any court and pitch data given here should be checked with the governing bodies concerned.

The issue is complicated by the fact that standard Continental dimensions do not in every case agree with our own, especially where the game concerned began in this country and is rarely played on the Continent. For example badminton dimensions will possibly stay precisely as they are at the present time in this country, and squash and fives courts are rarely seen in Europe. On the other hand, volleyball court sizes could possibly alter to conform to standard Continental practice.

Discussions are taking place between the various bodies involved, and fine adjustments to these data sheets may well be made from time to time. However, the spaces needed around the play area, for run-backs etc, are also indicated and these overall areas will not change.

2 Indoor sports
ACTIVITIES

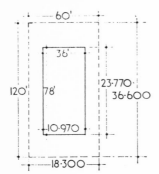

22.1 Tennis
Actual dimensions: 78ft = 23·774; 36ft = 10·973; height: 7·650 to 9·150
Continental dimensions for actual court: 23·770 × 10·970.
For space about court: 36·570 × 18·270
Court markings 50mm wide

Note: Unless stated otherwise dimensions are to the *outside* of court markings. These dimensions relate to tournament standard plans.

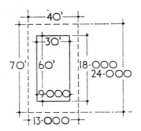

22.2 Volleyball
Actual dimensions: 60ft = 18·288; 30ft = 9·144.
Continental dimensions for actual court: 18·000 × 9·000 (expressed 18m × 9m); height = 7·650
For space about court: 24m × 13m (larger than ours at present). Court markings 50mm wide

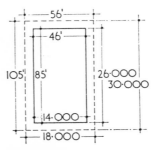

22.3 Basketball
Actual dimensions: 85ft = 25·908; 46ft = 14·021; height: 7·650
Continental dimensions for actual court: 26m × 14m.
For space about court: 30m × 18m
Note: Dimensions are to *inside* of court markings which are 50mm wide

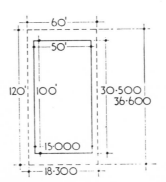

22.4 Netball
Actual dimensions: 100ft = 30·480; 50ft = 15·240.
Continental dimensions—no equivalent as this game is rarely played. Dimensions in metric are suggested above. Check with All England Netball Association.
Court markings are 50mm wide

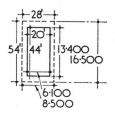

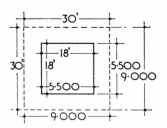

22.5 Badminton

Actual dimensions: 44ft = 13·411; 20ft = 6·096; height:
7·650 to 9·150
Continental dimensions 13·400 × 6·100 (court).
Space about court is very much larger than we usually allow.
Suggested dimensions: 54ft = 16·459 (16·500);
28ft = 8·534 (8·500)
Dimensions are more critical for this game than for any
other, apart from squash. Court markings 38mm wide
Check with Badminton Association, 4 Maderia Avenue,
Bromley, Kent

22.10 Boxing

Actual dimensions: 18ft = 5·486
Space about the ring is purely arbitrary, but in this case it is
assumed that the ring is for practice and that within the
9m × 9m space, allowance has been made for items of
equipment, including a punch ball, two punch bags, wall
bars, and a mirror. This is another activity which would be
included in 12m × 12m practice hall

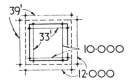

22.6 Judo

Actual dimensions: 33ft = 10·058; 39ft = 11·887
This activity could share a common space, say with boxing,
fencing and so on in a 12m × 12m practice hall

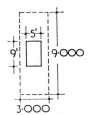

22.7 Table tennis

Four tables can be accommodated in a practice hall 12m ×
12m for recreational play.
Eleven tables can be accommodated in a one-court sports hall
36·600 × 18·300 for coaching and practice. It is usual to
reduce this number to six for high standard competitive play.

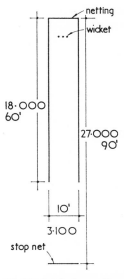

22.11 Cricket practice (indoor)

Height of net over wicket will vary according to height of
room/hall in which wicket is placed
Space behind wicket minimum 1m

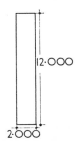

22.8 Fencing piste

No 'space about court' is required for the performers,
but it is necessary to allow for an instructor, and others who
wish to watch, on either side of the piste.
See also notes on Judo 22.6. If electrical scoring equipment
is required, an additional space of 2m × 1m should be
allowed

22.9 Five-a-side football (not illustrated)

This game may be adapted to the size of space in which it is
played. Thus in a one-court sports hall the playing area is
the floor area of the hall 36m × 18m.
Rebound wall surfaces should be allowed for up to at least
3·700 high and preferably up to underside of ceiling
(normally height = 7·650 to 9·150)

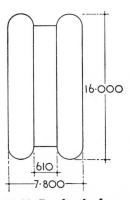

22.12 Rowing basins

The arrangement shown here is for rowing eights practice,
the largest space requirement for this activity. Coxless fours
would occupy approximately the same width but would be
proportionately shorter

22.13 Small bore target shooting (indoor)

*Formerly known as a 25yd range. Width and height
are optional. This activity could possibly occupy a common
space with indoor archery, eg 30·500 × 12·800*

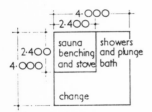

22.14 Sauna

*The size of a sauna is optional and dimensions shown here
are quite arbitrary. They would however be appropriate for a
sauna in a sports hall where it would form part of the
changing accommodation. Two saunas are normally included,
unless the accommodation is to be used by men and women
on different days*

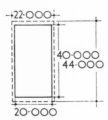

22.15 Hockey (indoor)

*The dimensions shown here are Continental standards, but it
should be noted that they require a sports hall larger than our
one-court hall of 36·600 × 18·300.*

*As with five-a-side football however the game can be adapted
to the size of hall in which it is to be played, and dimensions
of 36m × 18m are suitable for six-a-side hockey. Check with
the Hockey Association, 26 Park Crescent, London W1*

22.16 Handball (not illustrated)

*This game is more frequently seen on the Continent than in
this country. It has exactly the same dimensions as for indoor
hockey—44m × 22m overall—larger than our standard
one-court sports hall (36·600 × 18·300). It is often played in
the open air on multi-purpose, hard surfaced playing areas*

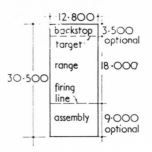

22.17 Archery (indoor)

*Space standards given are for a three-target indoor range
which is a minimum desirable number for practice and
midwinter recreation*
Minimum recommended height = 3·660

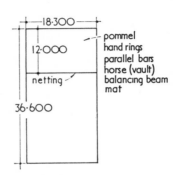

22.18 Olympic gymnastics

*All six items of equipment required can be accommodated in
a space 18·300 × 12·000.*
*For high standard competition however and for users of the
vault who require a long run-up, a one-court sports hall
36·600 × 18·300 is necessary*

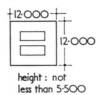

22.19 Trampoline

*Trampolines are normally used in pairs and require
adequate space around for groups being coached and their
instructors. The height factor is critical and should be not less
than 5·500. The arrangement shown here is suggested for a
practice hall 12m × 12m forming part of a sports hall,
which could also be used for boxing, judo and fencing*

22.20 Athletics training (indoor) (not illustrated)

*A one-court sports hall 36·600 × 18·300 is considered to be
the minimum desirable space in which to accommodate
simultaneous training for the long jump*, pole vault*,
high jump*, triple jump, circuit training, shot putt, hammer
and discus and javelin (the last four into nets).*
*A two-court sports hall 36·600 × 32·000 gives considerably
better 'elbow room' between performers and enables indoor
competitions to be held*

* These are best positioned in a bay 6m deep at one end of the hall

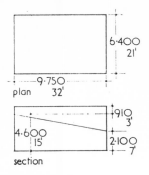

plan

section

22.21 Squash

All dimensions are critical. As this is essentially a UK game, metric dimensions follow closely previous imperial sizes. Dimensions are to outside of court markings which are 50mm wide. Check with Squash Rackets Association, 26 Park Crescent, London W1

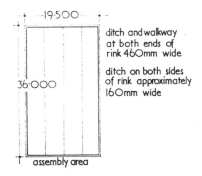

ditch and walkway at both ends of rink 460mm wide

ditch on both sides of rink approximately 160mm wide

assembly area

22.22 Indoor bowls

An assembly area is normally required at one end extending the full width of the hall, and not less than 1·800 wide. Note also that a four-rink hall is the minimum size for recreational play. A six-rink hall is required for competitions and club tournaments in which case the width would be increased to 29·300

22.23 Weight training and weight lifting

This is an activity best housed on its own in order that weights and other pieces of apparatus may be left in place after use.
The dimensions given are arbitrary but would provide a space large enough for twenty to thirty users to train in reasonable comfort. (Experience shows that 'physical culture' users far outnumber those using weights for body strengthening exercises, and that if the latter only are to be provided for, the space shown in this diagram should be reduced by up to a half)

22.24 Fives courts

Dimensions are of UK origin and are transposed closely to existing imperial court sizes 28ft × 18ft

200m or 166m track

22.25 Athletics (indoor)

Indoor track events are normally accommodated on a 200m or 166m track on the Continent and there appears to be little reason for making any alteration to these dimensions in the UK. Field events and the 60yd sprint (54·900) are usually inside the running circuit. Space for spectator seating around is optional. Check with Amateur Athletic Association, 26 Park Crescent, London W1

22.26 Dance and rhythmic movement, climbing practice (on walls), ski training *(not illustrated)*

None of these activities needs specific space allocation. Dance and rhythmic movement is best undertaken in a practice hall say 12m × 12m during normal practice sessions: and in either a one-court or two-court sports hall for public displays or large scale (say, regional) demonstrations. Climbing and ski training may be accommodated either indoors or outside the sports hall. A maximum vertical climb of 7·600 is considered reasonable for training beginners. Artificial ski slopes are approximately 5m wide with an incline of optional length (9m to 12m is frequently seen). Space should also be allowed at the top for an assembly area and bottom for pull-up area

MULTI-PURPOSE SPACES

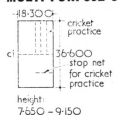

cricket practice

stop net for cricket practice

height: 7·650 – 9·150

22.27 One-court sports hall

Called a one-court hall because it can accommodate to tournament standard one tennis court, one basketball court, one netball and one volleyball court, and four to five badminton courts. (Netting positions normally seen in practice are shown dotted.) It is worth noting that the French equivalent is 44m × 24m × 9m, and the German 42m × 26m × 7m

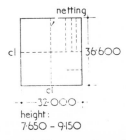

netting

height: 7·650 – 9·150

22.28 Two-court sports hall

Doubles the number of courts listed for the one-court sports hall. Top class competition is normally staged in centre of hall with seating (usually of the bleacher variety) arranged around as required

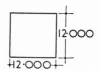

22.29 Practice hall

This size of hall can accommodate for training purposes: boxing, judo, trampoline, fencing, four table tennis tables and dance and rhythmic movement classes. Height should be 5·500 if a trampoline is to be accommodated, otherwise 3·650 would be acceptable

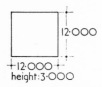

22.30 Training room (in a sports hall)

This space would normally be allocated for stamina training, pulse rate tests and so on

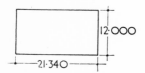

22.31 Gymnasium

These are standard dimensions in UK schools

3 Outdoor sports

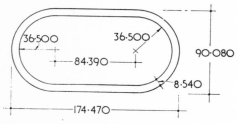

22.32 Athletics (outdoors)

These are standard Continental sizes for a 400m track. A final check for UK interpretation should be made with Amateur Athletic Association, 26 Park Crescent, London w1, or National Playing Fields Association, 57b Catherine Place, London sw1

FIELD EVENTS

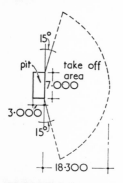

22.33 High jump

Continental dimensions are shown. UK dimensions are: pit size 5m × 4m, radius 18m

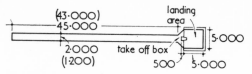

22.34 Pole vault

Standard Continental dimensions are shown with UK dimensions in brackets

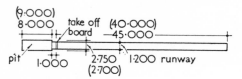

22.35 Long jump

Continental dimensions are shown with UK dimensions in brackets

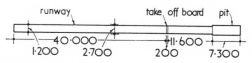

22.36 Triple jump

UK dimensions are transposed into metric

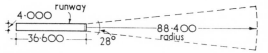

22.37 Javelin

UK dimensions are transposed into metric

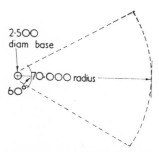

22.38 Discus and Hammer

UK dimensions are transposed into metric
Discus base shown
Hammer base diameter is 2·135

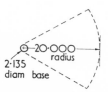

22.39 Shot

UK dimensions are transposed into metric
Continental base diameter is 2·000

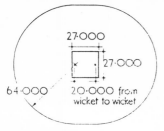

22.40 Cricket
Existing dimensions are:
table = 90ft × 90ft (27·432 actual)
wicket to wicket = 22yd (20·117 actual)
crease to boundary = 70yd (64·008 actual)
It is normal practice for county standard club boundaries
to be 50yd (45·720 actual)

22.41 Tennis *(not illustrated)*
As indoor dimensions, ie overall space of 36·000 × 18·300.
Three courts placed side by side are considered the
minimum economic provision for public use

22.42 Netball and basketball *(not illustrated)*
As indoor dimensions. These two activities frequently share
a common hard surfaced outdoor space

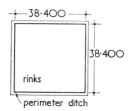

22.43 Bowls
Overall size may vary between 36·600 to 40·200.
38·400 is normally recommended

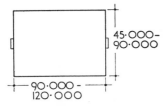

22.44 Soccer
The normal dimensions of a soccer pitch are 64m × 100m
and allowing for space about the pitch, 66m × 104m.
These are standard Continental dimensions

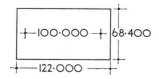

22.45 Rugby
Dimensions shown are standard on the Continent

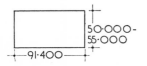

22.46 Hockey
An overall space of 59·000 × 99·400 allows for circulation
around the pitch and is standard on the Continent.

4 Other areas
Where a hard porous outdoor training space may be needed, an area of 64m wide × 100m long is frequently provided. An eighteen-hole golf course will require an area of 70 to 75 hectares

5 Composite spaces
A number of the foregoing areas can be grouped together for economy, subject to compatible programming, as follows.

Fencing, indoor cricket practice, small bore target shooting, and indoor archery, can share a common space with minimum dimensions of 30·500 × 12·800 × 3·660. This is usually included as part of the ancillary accommodation in a sports hall, and would normally also include a store or armoury in addition to the above.

A cricket pitch as shown in **22.40** can also accommodate 2 hockey pitches.

A hard porous training area 100m × 64m (equivalent to a small soccer pitch), can be shared by hockey, five-a-side football, netball and basketball, tennis, and athletics practice.

A hard porous surface (hard-bound as compared with the loose-bound surface of the training area), of 54·900 × 36·600 is frequently shared by tennis (3 courts), netball and basketball (3 courts each), and five-a-side football.

6 Swimming
Most common types of pools are listed in table I which gives the dimensions and capacities deriving from each size of tank. Note that the standard width of 12·5m will accommodate five racing lanes of 2·5m each; this is the latest international standard for top class pools. In other than top class pools, six lanes may be used within the same width. A lane width of 2·1m may be adopted in certain pools (eg some local authority pools) resulting in a pool width of 12·6m (for 6 lanes) or 10·5m (for 5 lanes).

Consult the Amateur Swimming Association, 64 Cannon Street, London, EC4 for any revisions to information on this and the following three pages.

7 References and sources
G. A. Perrin with help from the West German Technical Advisory Bureau, Cologne, and the National Playing Fields Association.

AJ information sheet 1505 Indoor swimming baths: Design data for types of pool. AJ, 1967, July 26 CI/SfB 541

Table I   Design data for pools only pool halls (see also illustrations 22.47 to 22.49)

POOL TYPE		POOL TANKS							HALL
	Size and depth (m)	Water surface area (m²)	Water volume (m²)	Net area of tank walls below water line (m²)	Net area of tank walls above water line (approximately 0·3m high) (m²)	Total area of tank walls (m²)	Area of tank bottom (m²)	Total area inside tank (m²)	Size (m)
1 Learner	12·5 × 9·0 × 0·75	112·5	84·375	32·25	12·9	45·15	112·5	157·65	16·5 × 13·0
	12·5 × 9·0 × 0·9	112·5	101·25	38·7	12·9	51·6	112·5	164·1	
	12·5 × 7·5 × 0·75 (or 0·9)	93·75	70·31	30·0	12·0	42·0	93·75	135·75	16·5 × 11·5
2 Swimming	25·0 × 12·5 × 0·9 (constant)	312·5	281·25	67·5	22·5	90·0	312·5	402·5	
	25·0 × 12·5 × 1·0 (constant)	312·5	312·5	75·0	22·5	97·5	312·5	410·0	
	25·0 × 12·5 × 0·9 to 2·0 (constant slope)	312·5	453·625	108·75	22·5	131·25	312·5	443·75	16·5 × 31·0
	25·0 × 12·5 × 0·9 to 2·0 (4m at 0·9; 4·5m at 2·0; slope 1 in 15 over 16·5m)	312·5	456·56	109·30	22·5	131·8	312·5	444·8	
	25·0 × 12·5 × 2·0 (constant)	312·5	625·0	150·0	22·5	172·5	312·5	485·0	
	33·33 × 12·5 × 2·0 (constant)	416·625	833·25	183·32	(27,498) 27·5	(210·718) 210·72	416·625	627·343	39·33 × 16·5
3 Deep water diving	3a 7·5 × 9·0 × 3·0 (two 1m springboards)	67·5	202·5	99·0	9·9	108·9	67·5	176·4	11·5 × 15·0 × 5·6
	3b 8·5 × 10·5 × 3·5 (one 1m and one 3m springboard)	89·25	312·375	133·0	11·4	144·4	89·25	233·65	12·5 × 16·5 × 7·6
	3c 10·0 × 10·5 × 3·5 (two 1m and one 3m springboards)	105·0	367·5	143·5	12·3	155·8	105·0	260·8	14·0 × 16·5 × 7·6
	3d 11·0 × 12·0 × 3·8 (one 1m and one 3m springboard and one 5m platform)	132·0	501·0	174·8	13·8	188·6	132·0	320·6	15·0 × 18·0 × 8·8

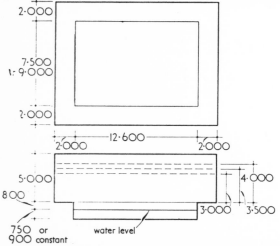

22.47 *Basic dimensions of learners' pool: plan (above) and section (below). See also table I, pool type 1*

Water polo (district and county competitions) requires minimum playing area of 20m x 8m x 1m deep.

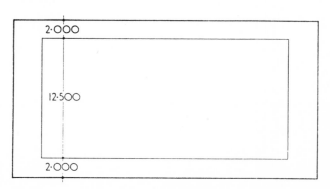

22.48a *Basic dimensions of swimming pools: plan. See also table I, pool type 2. Walkway dimensions at ends: 3·350 (deep end) and 4·270 (shallow end).*

Area (m²)	Volume (m³) at heights:					Internal wall area (m²) at heights:					Pool surround area (m²)	CAPACITIES					
	3m	3.5m	4m	5m	6m	3m	3.5m	4m	5m	6m		Divers	Swimmers at one per 4m³ in water over 1.5m deep	Bathers at one per m³ in water less than 1.5m deep	Total in the water	Plus users on surround	Gross number of users
215.5	646.5	754.25	862	1077.5	—	177	206.5	236.0	295.0	—	103.0	—	—	112	112	37	149
189.75	569.25	644.125	759.0	948.75	—	168	196	224.0	280.0	—	96.0	—	—	93	93	31	124
511.5	—	—	2046	2557.5	3069	—	—	380.0	4750.0	570.0	199.0	—	—	312	312	104	416
												—	—	312	312	104	416
												—	36	170	206	69	275
												—	38	163	201	67	268
												—	—	78	78	26	104
645.0	—	—	2580	3225	3870	—	—	446.64	558.3	669.96	228.375	—	104	—	104	34	138
	5.6m	7.6m	8.8m	—	—	5.6m	7.6m	8.8m	—	—							
172.5	966.0					296.8					105.0	2	—	—	2	9	11
206.25		1567.5					440.8				117.0	2	—	—	2	9	11
231.0		1755.6					463.6				126.0	—	—	—	3	9	12
270.0			2376.0					580.8			138.0	—	—	—	3	9	12

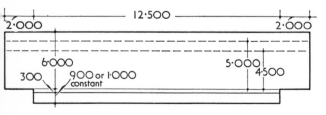

22.48b *Basic dimensions of swimming pools: cross-section*
See also table I, pool type 2

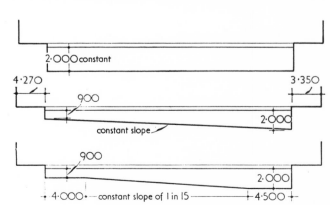

22.48c *Long sections illustrating the three pool profiles:*
constant; constant slope; constant slope of 1 in 15 over 16.5m
If a 1m springboard is provided, depth at deep end is increased to 3.000

See also table I pool type 2

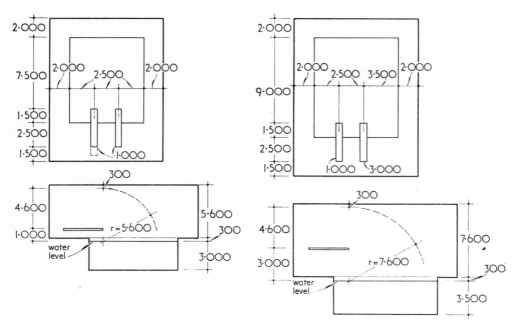

22.49a *Deep water diving pool with two 1m springboards: plan (above), section (below). See also table 1, pool type 3a*

22.49b *Deep water diving pool with springboards at 1m or 3m: plan (above) section (below). See also table 1, pool type 3b*

Note: *Preferred position for single 1m springboard in a 25m pool is on the long side about 5m from the end. Minor amendments have since been made to the diving pool dimensions. Consult ASA (see page 119).*

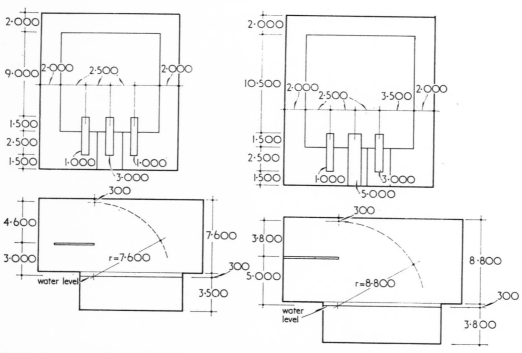

22.49c *Deep water diving pool with two 1m and one 3m springboards: plan (above), section (below). See also table 1, pool type 3c*

22.49d *Deep water diving pool with springboards at 1m and 3m and fixed platform at 5m: plan (above), section (below). See also table 1, pool type 3d*

23 **Educational buildings**

Contents
1 Controlling dimensions
2 Furniture and equipment
3 Area allowances
4 General requirements
5 References and sources

1 Controlling dimensions

Recommendations for horizontal and vertical controlling dimensions for educational buildings are contained in *Dimensional co-ordination in building documents* DC4, DC5, DC6 and DC7 and in BS 4330:1968 (metric units). These recommendations are illustrated below.

Horizontal controlling dimensions are shown in **23.1** and table I.

Vertical controlling dimensions are shown in **23.2**.

Co-ordination of components

See *Building Bulletin 42* for basic data in metric co-ordinated terms

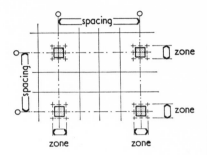

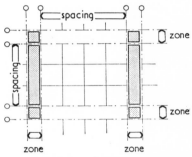

23.1 *Horizontal controlling dimensions. Diagram to define meaning of spacing and zone. (To be read in conjunction with table I.) For public sector buildings only the top drawing is applicable (See table A6 in BS 4330).*

Table I Horizontal controlling dimensions for educational buildings (see also 23.1). Dimensions are between centre lines of loadbearing walls and columns

Spacing (mm)	Zone (mm)
1 800	100
2 400	150 (see BS 4330)
2 700	2000
3 000	50 (see BS 4330)
3 600	200
4 200	300
4 500	600
4 800	
5 400	
6 000	
6 300	
6 600	
7 200	
8 100	
8 400	
9 000	
9 600	
9 900	
10 800	
11 700	
12 000	
12 600	
14 400	
16 200	
18 000	

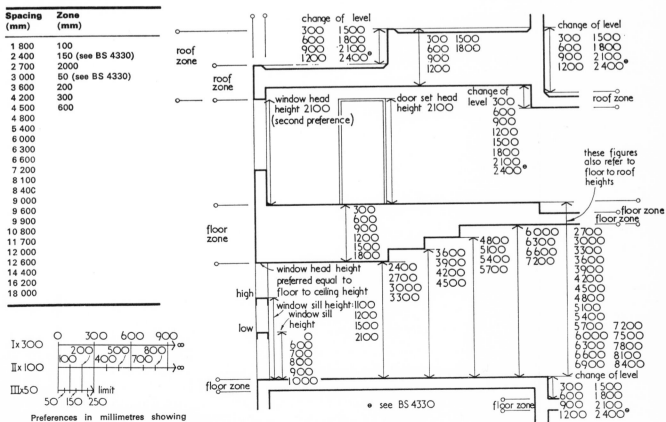

Preferences in millimetres showing increments for first, second and third preferred dimensions (BS 4011)

23.2 *Vertical controlling dimensions*

2 Furniture and equipment

BS 3030:1959 already gives metric dimensions for pupils' tables and chairs (part 3) and dining-tables and chairs (part 5).

They are not rationalised metric but merely metric conversions. These table heights are listed in Table II with the nearest preferred dimension (from BS 4011) given in the last column. They are also shown in graphic form, with other relevant dimensions, in **23.5**.

Also shown on **23.5** are heights of standing working planes. Average heights and reach (translated from DES Building Bulletin 41) are shown in **23.6**.

Guiding dimensions for individual work spaces (translated from DES Building Bulletin 25) are shown in **23.4** and a study carrel (Building Bulletin 41) in **23.3**.

See also DES Bulletin 38. Appendix 3 lists metric equivalents of standing and reaching dimensions.

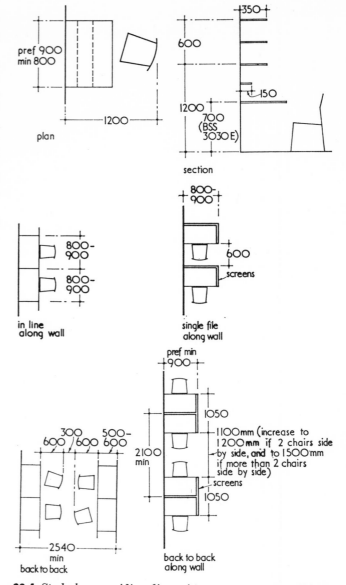

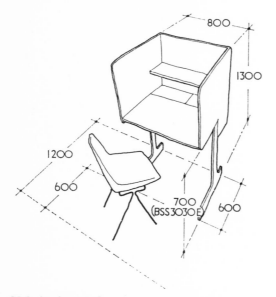

23.3 *Study carrel*

23.4 *Study bays: guiding dimensions for individual work spaces*

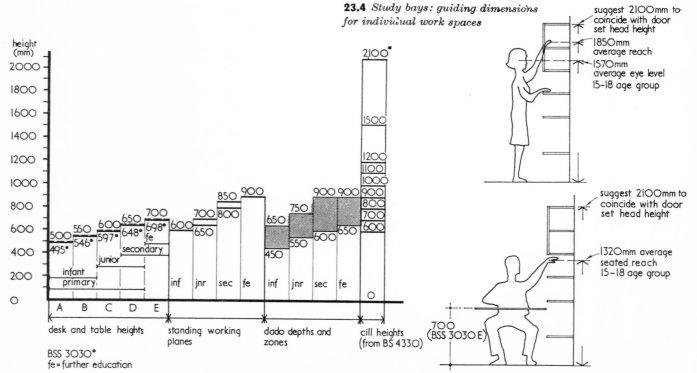

23.5 *Heights of desks, tables, dados, working planes and sills*

23.6 *Average heights and reach (sixth form)*

Primary school furniture

A selection of some useful sizes of primary school furniture (translated from Building Bulletin 36) is illustrated in **23.7** to **23.21**. Letters in the captions refer to classification employed in BS **3030** (in rational metric sizes) and shown also in tables II and III and **23.3**. Sizes shown are typical only: other sizes are available.

23.7 *Square table: sizes:*
Nursery 900 × 900 × 430mm
BS 3030:
A   900 × 900 × 495mm (500)
A   1100 × 1100 × 495mm (500)
B   1100 × 1100 × 546mm (550)

23.8 *Oblong table: sizes:*
Nursery 900 × 450 × 430mm
BS 3030:
A   900 × 450 × 495mm (500)
A   1100 × 550 × 495mm (500)
B   1100 × 550 × 546mm (550)
C   1100 × 550 × 597mm (600)

23.12 *Round table: sizes:*
Nursery 850 diam × 430mm
BS: 3030
A   850 diam × 495mm (500)
B   850 diam × 546mm (550)
C   1270 diam × 597mm (600)

23.13 *Square pedestal table: sizes:*
BS 3030:
B   1100 × 1100 × 546mm (550)
C   1100 × 1100 × 597mm (600)

23.17 *Teacher's table: size:*
BS 3030:
D   1050 × 700 × 648mm (650)

23.18 *Workbench: sizes:*
1200 × 600 × 558mm (550)
1200 × 600 × 609mm (600)
1200 × 600 × 660mm (650)
1200 × 600 × 711mm (700)

23.9 *Book storage and display trolley:*
size: 900 × 450 × 900mm each

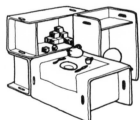

23.14 *Rostrum boxes: 550 × 400 × 280mm*

23.19 *Oblong and trapezoidal rostra:*
900 × 450 × 225mm

23.10 *Trolley locker: sizes:*
900 × 450 × 600mm
900 × 450 × 660mm (650)
900 × 450 × 710mm (700)

23.15 *Easel: 800 × 1200mm high*

23.20 *Coat trolley: 1370 (1400) × 1060 (1000) × 635 (650)mm*

23.11 *Mobile bin: 600 × 450 × 500m*

23.16 *Folding rostrum: 900 × 900 × 450mm*

23.21 *Staff locker and writing unit:*
2400 × 342 (350)mm; writing flap
overall width 500mm; height 1270
(1200)mm

Table II Table heights (including dining-tables)

	Metric conversion (mm)	Nearest preferred dimension (mm)
A	495	500
B	546	550
C	597	600
D	648	650
E	698	700

Table III Minimum length of top surface per pupil

	Metric conversion (mm)	Nearest preferred dimension (mm)
A	457	450
B	508	500
C	559	550
D	559	550
E	559	550

Table IV Heights of standing working planes

	Metric conversion (mm)	Nearest preferred dimension (mm)
Further education	889	900
Secondary	864	850
	838	850
	813	800
Junior	711	700
	660	650
Infant	610	600

3 Area allowances

The following notes are extracted from Statutory Instrument 1959 No 890 *Standards of School premises* (HMSO, 1s 6d). and from DES Administrative memorandum 14/68 Dec. 1968. The Department of Education and Science is in process of revising the Building Code. As an interim measure the schedule of metric equivalents or analogues below should be used.

Only dimensional standards are included (para and part numbers refer to the Standards)

3.1 Primary schools (part II)

AREA OF SITES FOR PRIMARY SCHOOLS (reg 3)
(figures are rounded using an equivalent 4000m² = 1 acre)

Not more than 25 pupils	2000m² = 0·20 hectares
26 to 50 pupils	2500m² = 0·25 hectares
51 to 80 pupils	3000m² = 0·30 hectares
For every additional 40 pupils, or part, add	500m² = 0·05 hectares

HARDPLAY AREAS (reg 3)

No of pupils	Area (m²)	Derived from rectangles (m)
Infant schools		
Up to 100	300	25 × 12
101–280	612	34 × 18
281–400	1032	34 × 18 + 28 × 15
401–480	1224	34 × 33

HARDPLAY AREAS *continued*

Infant and junior schools		
Up to 25	300	25 × 12
26–80	420	28 × 15
81–170	612	34 × 18
171–280	1032	34 × 18 + 28 × 15
281–440	1734	34 × 33 + 34 × 18
441–480	2244	34 × 33 (2)
Junior schools		
Up to 140	612	34 × 18
141–280	1122	34 × 18 + 28 × 15
281–360	1734	34 × 33 + 34 × 18
361–480	2244	34 × 33 (2)

PLAYING FIELD ACCOMMODATION (reg 4)

No of junior pupils	Appropriate area (using an equivalent 4000m² = 1 acre)
Under 50	2000m² = 0·20 hectares
51–120	4000m² = 0·40 hectares
121–200	6000m² = 0·60 hectares
201–280	9000m² = 0·90 hectares
More than 280	12000m² = 1·20 hectares

TEACHING ACCOMMODATION (reg 5)

Number of pupils	Area allowances (m²)
Up to 25 pupils	3·7m² per pupil
26–75 pupils	93m², plus 2·1m² for each pupil in excess of 25
76–119 pupils	227m², plus 2m² for each pupil in excess of 75
120 or more	312m², plus 63m² for every 40 pupils in excess of 120, interpolating as necessary for numbers which are not multiples of 40.

Every teaching space (reg 5(2)) *shall comprise an area of not less than:*

For first 10 pupils or part	18·6
For each of next 20 pupils	1·4
Thereafter	0·4

Accommodation for meals in primary schools (reg 14(1))
Per diner (two sittings) 0·5
Per diner (one sitting) 1·0

3.2 Secondary schools (part III)

AREA OF SITES FOR SECONDARY SCHOOLS (reg 15(1))
(figures rounded using an equivalent 4000m² = 1 acre)

Not more than 150 pupils	6000m² = 0·6 hectares
151–210 pupils	7000m² = 0·7 hectares
211–300 pupils	8000m² = 0·8 hectares
301–360 pupils	9000m² = 0·9 hectares
361–420 pupils	10000m² = 1·0 hectares
421–450 pupils	12000m² = 1·2 hectares
For every additional 50 pupils or part add	1000m² = 0·1 hectares

HARDPLAY AREAS (reg 15(3))

No of pupils	Area (m²)
Not more than 420	1850
421–600	3180
For every additional 150 or part add	465
For a boys' school or a girls' school with not more than 180 pupils:	970

PLAYING FIELDS (reg 16) (figures rounded using an equivalent 4000m² = 1 acre)

Boys' schools (pupils aged 11–15 years)

Not more than 150 pupils	18000m² = 1·8 hectares
151–300 pupils	30000m² = 3·0 hectares
Every additional 150 pupils or part add	6000m² = 0·6 hectares

Girls' schools (pupils aged 11–15 years)

Not more than 150 pupils	16000m² = 1·6 hectares
151–300 pupils	26000m² = 2·6 hectares
301–450 pupils	34000m² = 3·4 hectares
Every additional 300 pupils or part add	4000m² = 0·4 hectares

Playing fields for pupils (boys or girls) of 16+ years

Every 120 pupils or part	5000m² = 0·5 hectares

AREAS PER STUDENT

Number of pupils under 16 for which the school is designed	Area per pupil aged:			
	Under 11	11 & 12	13 & 14	15 and over
Not more than 150		3·72	4·65	5·20
151–300	2·14			
301–450	2·04	3·62	4·55	5·11
451–520	1·95			
521–700	1·86	3·53	4·46	5·02
710–800		3·48	4·41	4·97
801–900		3·39	4·32	4·88
901–1050		3·25	4·18	4·74
1051–1200		3·21	4·13	4·69
1201–1350		3·16	4·09	4·65
1351–1500		3·10	4·02	4·58
1501–1650		3·07	3·99	4·55
1651–1800		3·04	3·97	4·52
1801–1950		2·99	3·92	4·48
1951–2100		2·97	3·90	4·46

In addition:

Each classroom shall be:

for first 10 pupils or part: min	18·6m²
for each additional pupil:	1·4m²
Accommodation for school meals	
Area per diner (2 sittings)	0·5m²
(1 sitting)	1·0m²

3.3 Nursery schools (part IV)

AREA OF SITES FOR NURSERY SCHOOLS (reg 28) (figures rounded using an equivalent 4000m² = 1 acre)

Up to 40 children	1000m² = 0·1 hectares
every additional 20 or part	500m² = 0·05 hectares

GARDEN PLAYING SPACE (reg 29)

Every class requires	9·3m²
of which	3·7m² is to be paved

3.4 Special schools (part V)

AREA OF SITES FOR SPECIAL SCHOOLS (reg 36) (figures rounded using an equivalent 4000m² = 1 acre)

Not more than 25 pupils	2000m² = 0·20 hectares
26–50 pupils	2500m² = 0·25 hectares
51–80 pupils	3000m² = 0·30 hectares
Every additional 40 pupils or part add	500m² = 0·05 hectares

Schools for senior pupils, increase site as follows:

Not more than 50 pupils	500m² = 0·05 hectares
51–120 pupils	1000m² = 0·10 hectares
121–160 pupils	1500m² = 0·15 hectares
more than 160 pupils	2000m² = 0·20 hectares

HARDPLAY AREA FOR SPECIAL SCHOOLS (reg 36(2))
(figures rounded to nearest 5m²)

Juniors and infants

Up to 105 pupils	612m²
More than 105 pupils	910m²

All age schools and schools for seniors

Up to 50 pupils	910m²
51–120 pupils	1122m²
More than 120 pupils	1850m²

PLAYING FIELD ACCOMMODATION (reg 37)

Not more than 120 pupils	5000m² = 0·5 hectares
121–200 pupils	10000m² = 1·0 hectares

TEACHING ACCOMMODATION (reg 38)

First 100 pupils or part (total)	520m²
Each subsequent 20 pupils or part add	56m²
Each class shall be	28m²
For first 10 pupils or part	
Each additional pupil	2·2m²

3.5 Boarding accommodation (part VI)

DORMITORIES AND CUBICLES (reg 42)

Minimum distance between beds	900mm
Floor area per space first 2 beds	5m²
Each additional bed add	4·2m²
Separate bed cubicles	5m²
Single bedrooms	6m²

DAYROOM SPACE (para 45)

Minimum per pupil	2·3m²

SICK ROOM (para 46)

Minimum distance between beds	1800mm
Minimum area per bed	7·4m²

4 General requirements

Only regulations incorporating a unit of measurement requiring a metric analogue are included.

4.1 Lighting (reg 52)

In all teaching accommodation and kitchens the lowest level of maintained illumination and the minimum daylight factor on the appropriate plane in the area of normal use, shall be 110 lux and 2 per cent respectively.

In all teaching accommodation and kitchens no luminous part of any lighting unit, or mirrored image thereof, having a maximum brightness greater than 5100 candela per sq metre (cd/m²) or an average brightness greater than 3400 cd/m² shall be visible to any occupant in a normal position within an angle at the eye of 135 degrees from the perpendicular from the eye to the floor.

4.2 Ventilation and heating (reg 53)

Every room in every school building and in all boarding accommodation shall be provided with means of ventilation capable of securing that the amount of fresh air available to each person for whom the room is designed shall be not less than the amount specified in this regulation as appropriate to the cubic space available per person:

Cubic space per person in m³	Appropriate number of air changes per hour
Not more than	
5m³	6
5m³–5·7m³	5
5·7m³–7m³	4
7m³–8·5m³	3
8·5m³	1½

The heating system in every school and in all boarding accommodation shall be such as to secure that, when the outside temperature is 0°C and when the heating system is heating air at the rates specified in this paragraph, the temperature, at a height of not more than 900mm from the floor, shall be the temperature specified in this regulation as appropriate to the type of room or other space, or as near as may be thereto:

Type of room or space	Number of air changes per hour to be heated by the heating system	Temperature (Celcius)
Convalescent sitting	3	18·5°
Medical inspection rooms	3	18·5°
Changing rooms, bathrooms and shower rooms	3	18·5°
Teaching rooms	2	17·0°
Nursery playrooms	2	17·0°
Common rooms	2	17·0°
Staff rooms	2	17·0°
Sanatorium and sickrooms	3	14·5°
Halls	1½	14·0°
Dining rooms	2	14·0°
Gymnasiums	2	14·0°
Cloakrooms	2	13·0°
Corridors	1½	13·0°
Dormitories	2	11·0°

4.3 Washing and sanitary accommodation (reg 57)

The surfaces of the floors of washing and sanitary accommodation, including bath and shower compartments, and of the walls thereof to a height of not less than 1·8m shall be finished with a material which resists the penetration of water and which can be easily cleaned.

4.4 Building Bulletin 7

Fire and the Design of Schools Administrative Memorandum 14/68 provides for the metric equivalent of all imperial dimensions as do the National Building Regulations current proposals.

This will inevitably cause considerable difficulties, eg 4ft = 1·22m NOT 1·20m, 6in = 152·4mm NOT 150mm.

It is essential that the dimensions are rationalised and not merely equated at the earliest date.

5 References and sources

MINISTRY OF PUBLIC BUILDING AND WORKS DC4, DC5, DC6, DC7 Dimensional co-ordination for building CI/SfB (F4j) *For full details see section 31.*

BRITISH STANDARDS INSTITUTION BS 4330:1968 (metric units). Recommendations for the co-ordination of dimensions in building: Controlling dimensions CI/SB (F4j)

BRITISH STANDARDS INSTITUTION BS 3030 School furniture. Part 1: 1959 Materials, workmanship and finish. Part 2: 1959 Performance tests. Part 3: 1959 Pupils' classroom chairs and tables. Part 4: 1959 Chalkboards. Part 5: 1960 Pupils' dining tables and chairs. London, The Institution CI/SfB 710 (72)

BRITISH STANDARDS INSTITUTION BS 4011:1966 Recommendations for the co-ordination of dimensions in building. Basic sizes for building components and assemblies. London, 1966, The Institution CI/SfB (F4j)

DEPARTMENT OF EDUCATION AND SCIENCE Building Bulletins 25 Secondary school design. Sixth form and staff. 1965; 36 Eveline Lowe Primary School. London. 1967; 41 Sixth-form centre: Rosebery County School for Girls, Epsom, Surrey. 1967. Building Bulletin 38, Standing & Reaching. Building Bulletin 42, The Co-ordination of Components. All HMSO CI/SfB 71

Statutory Instrument 890 Standards for school premises regulations. 1959, HMSO CI/SfB 71 (Ajk)

Administrative Memoranda 14/68 and Amendment 1 (18. 7. 69)

Amendments to the Building Code 20 Dec 1968 (for Code copy holders only)

24 Libraries

Contents
1 Books
2 Layouts and critical dimensions
3 Reading table heights
4 Structural modules
5 References and sources

1 Books
Traditional book sizes are given in **24.1**. Note that A4 size (297mm × 210mm) is just accommodated within top limits allowed for quarto.

The average number of books of various types which can be accommodated is given in table I.

Illustrations **24.2** to **24.4** show recommended shelf heights.

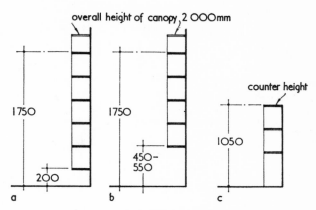

24.2 *Recommended shelf heights for lending library: (a) full height; (b) alternative full height; (c) counter top height*

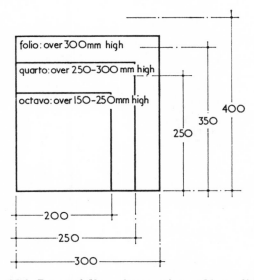

24.1 *Range of dimensions coming within traditional book sizes*

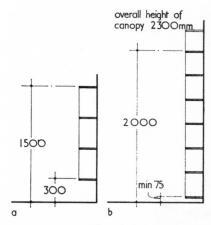

24.3 *Recommended shelf heights for: (a) children's library; (b) bookstack area*

Table I Books per 300mm run of shelf

Type of book	No per 300mm run of shelf	No per 900mm run of shelf	Recommended shelf depth (mm)
Children's books	10 to 12	30 to 36	200 to 300
Loan and fiction stocks in public libraries	8	24 to 25	200
Literature, history, politics, economics	7	21	200
Scientific and technical	6	18	250
Medical	5	15	250
Law	4	12	200
Averages	**7**	**21**	

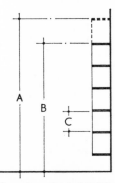

A - senior school: 2000mm
B - elementary and junior: 1700mm
C - average 250mm clear height

24.4 *Recommended shelf heights for school libraries*

2 Layouts and critical dimensions

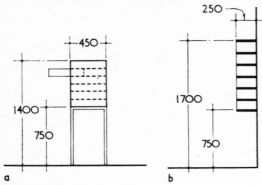

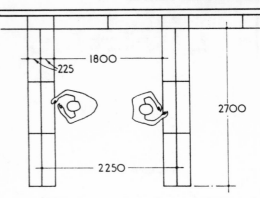

24.5 *Recommended heights for: (a) card catalogue cabinet; (b) sheaf catalogue binder shelves*

24.8 *Recommended minimum plan dimensions in open access bookshelf area with shelving arranged in alcoves*

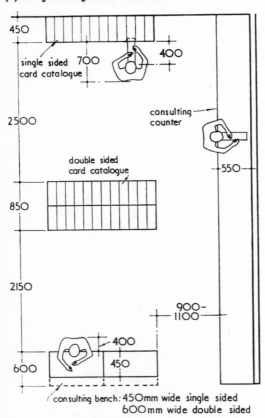

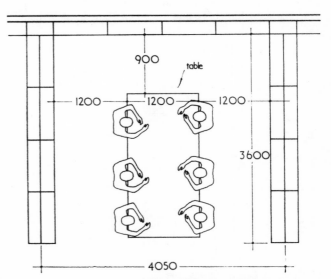

24.6 *Recommended minimum plan dimensions in card catalogue area*

24.9 *Recommended minimum plan dimensions in open access bookshelf area with shelving arranged in alcoves containing reading tables*

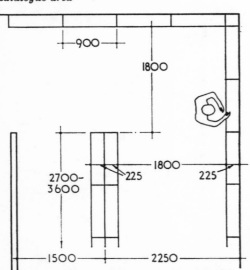

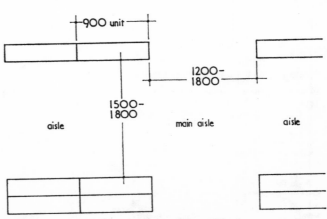

24.7 *Recommended minimum plan dimensions in open access bookshelf area*

24.10 *Recommended minimum aisle widths in open access bookstack area*

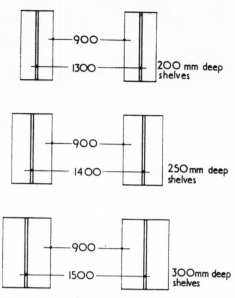

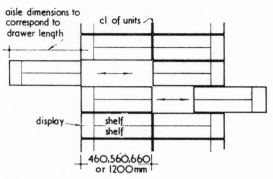

24.11 *Recommended aisle widths between rows of bookshelf units in closed access bookstack area for 200mm, 250mm and 300mm deep shelves respectively*

24.14 *Drawer type bookshelf units*

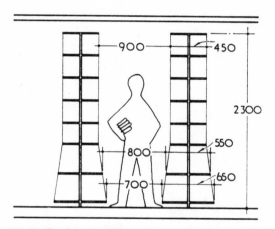

24.12 *Section through shelves in closed access bookstack area showing how aisle width can be reduced to a minimum by arranging variable bookshelf depths in each unit*

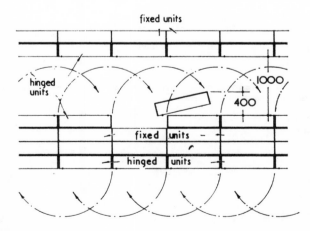

24.13 *Various proprietary systems of compact storage are available such as the hinged bookshelf unit shewn above where 66 per cent increased capacity is possible compared with ordinary bookstacks. The drawer unit in* **24.14** *can achieve up to 100 per cent space saving. Other types (not illustrated) are rolling bookshelf units which roll either in the direction of the shelf length or in a direction at right angles to the shelf length*

24.15 *Dimensions of six-person reading tables*

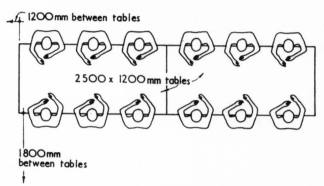

4.16 *Dimensions of eight-person reading table*

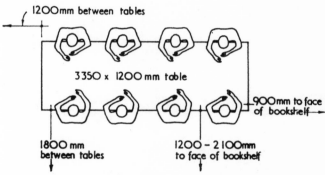

24.17 *Dimensions for single-sided table for four persons*

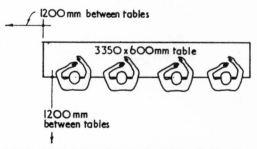

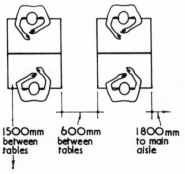

24.18 *Layout dimensions for one-person tables*

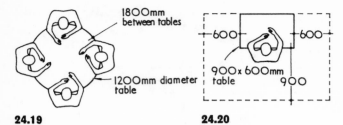

24.19

24.20

24.19 *Dimensions for round reading table*

24.20 *Recommended minimum space requirements for one-person reading table*

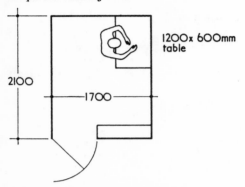

24.21 *Recommended plan dimensions for one-person enclosed carrel*

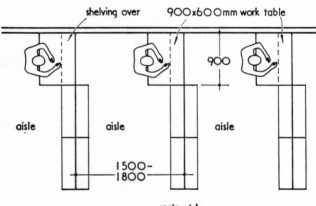

24.22 *Suggested arrangement for open carrels in bookshelf areas*

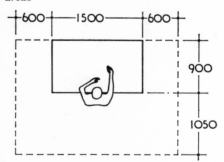

24.23 *Recommended minimum space requirements for library staff*

5 References and sources

AJ information sheet 1318 Library planning 2: Space standards. AJ, 1965, February 24 CI/SfB 76 (E6)
AJ information sheet 1319 Library furniture and equipment 1: Book storage. AJ, 1965, March 3 CI/SfB 76 (76)
AJ information sheet 1320 Library/furniture and equipment 2: General. AJ, 1965, March 3 CI/SfB 76 (7–)
AJ information sheet 1593 Library planning: structural modules. AJ, 1968, February 28 CI/SfB 76 (2–) (F4j)

3 Reading table heights

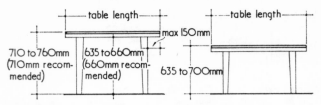

24.24 *Reading table heights for adults (left) and children (right)*

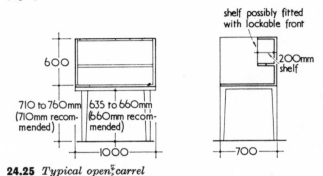

24.25 *Typical open carrel*

4 Structural modules

The basic layout of bookstacks in a stack area is given in **24.26**. The structural grid should be a multiple of the bookstack centres and should allow for at least three possible stack centres. Table II shews dimensional relationships between stack centres and structural grids conforming to spacing sizes recommended in BS 4330.

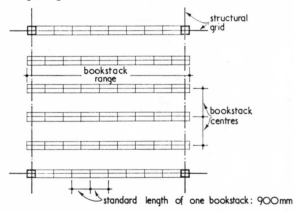

24.26 *Principal relationships between bookstacks and structural grids*

Table II   Dimensional relationships between stack centres and structural grids of various spans

Span of structural grid (mm)	Stack centres		
	Closed access (mm)	Open access (mm)	Periodicals display (mm)
5400	1080	1350	1800
6000	1200	1500	2000
6900	1150	1380 1725	1725
7200	1200	1400	1800
7500	1070 (approx) 1250	1500	1875
7800	1114 (approx)	1300 1560	1950
8400	1200	1400 1680	1680 2100

25 **Hotels**

Contents

1 Controlling dimensions

Recommendations for horizontal and vertical controlling dimensions for hotel buildings are contained in BS 4330:1968 (metric units). Recommendations for horizontal controlling dimensions are illustrated in **25.1** and table I, and for vertical controlling dimensions in **25.2**.

2 Space standards for hotel bedrooms

The areas of bedrooms vary considerably between different classes of hotel. Table II lists absolute minimum sizes; table III gives recommended minimum sizes.

Table I Horizontal controlling dimensions (to be read in conjunction with 25.1). Dimensions are between boundaries or centre lines of loadbearing walls and columns.

Spacing (mm)	Zone (mm)
1 800	100
2 100	200
2 400	300
2 700	400
3 000	500
3 300	600
3 600	
3 900	
4 200	
4 500	
4 800	
5 100	
5 400	
5 700	
6 000	
6 300	
6 600	
6 900	
7 200	
7 500	
9 000	
10 500	
12 000	

25.1 *Horizontal controlling dimensions. Diagram to define meaning of space and zone (to be read in conjunction with table I)*

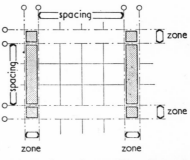

Preferences in millimetres showing first, second and third preferred dimensions. This refers also to horizontal dimensions (BS 4011)

25.2 *Vertical controlling dimensions*

Table II Absolute minimum sizes of hotel bedrooms

Category of room	Dimensions (mm)	Area (m²)
Single	3000 × 2450	7·35
Combination (includes divan)	3350 × 2900	9·7
Double	3350 × 3350	11·2
Private bathroom (wc lb and bath)	1600 × 2000	3·2

Table III Recommended minimum sizes of hotel bedrooms

Category of room	Dimensions (mm)	Area (m²)
Single	3800 × 2600	9·9
Combination (includes divan)	3800 × 3400	12·9
Double	4400 × 3800	16·7
Private bathroom (wc, lb and bath)	1600 × 2200	3·5

Note that the combination rooms may be used as either single or double rooms.

3 Bedroom furniture and equipment

The drawings illustrate various items of hotel bedroom furniture and their space requirements 25.4.

4 Bedroom suites

Standard layouts are illustrated in 25.3. The sizes shewn fall between the dimensions given in Tables II and III, but the drawings are intended mainly to show typical arrangements rather than particular dimensions.

5 References and sources

AJ information sheet 1328 Space standards for hotel bedrooms. AJ, 1965, May 12 CI/SfB 850 (E8)
AJ information sheet 1329 Furniture and equipment for hotel bedrooms and bathrooms. AJ, 1965, May 19 CI/SfB 850 (7–)
BRITISH STANDARDS INSTITUTION BS 4330:1968 (metric units) Recommendations for the co-ordination of dimensions in building: Controlling dimensions CI/SfB (F4j)

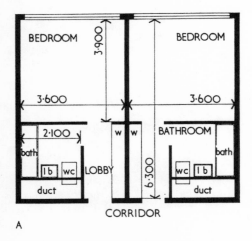

A

Notes on the drawings
Bathrooms

A saving of space is achieved by the use of internal bathrooms. These require mechanical extract ventilation which can be easily planned in the service ducts. Sufficient intake ventilation should be provided by means of a grille or a space under the door.

Flushing cisterns are best accommodated in the service ducts to prevent tampering as well as to save space.

Lobbies

The lobby can be used as a dressing annexe to the bedroom. With such an arrangement no door is required between the two spaces (see fig E).

For full metric information on hotel planning, see *Principles of Hotel Design*, Architectural Press, £2·50.

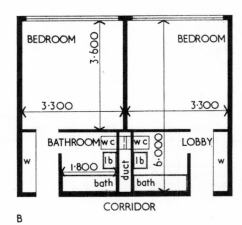

B

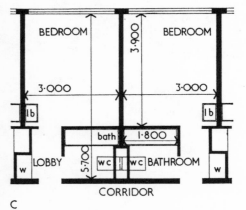

C
INTERNAL   BATHROOMS

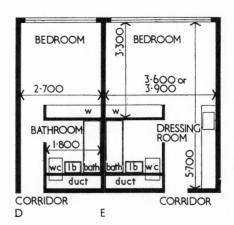

D   E

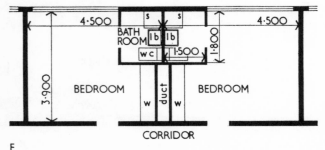

F
EXTERNAL   BATHROOMS

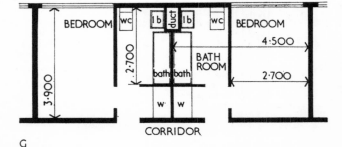

G

25.3 *Some typical plan arrangements of bedrooms and bathrooms*

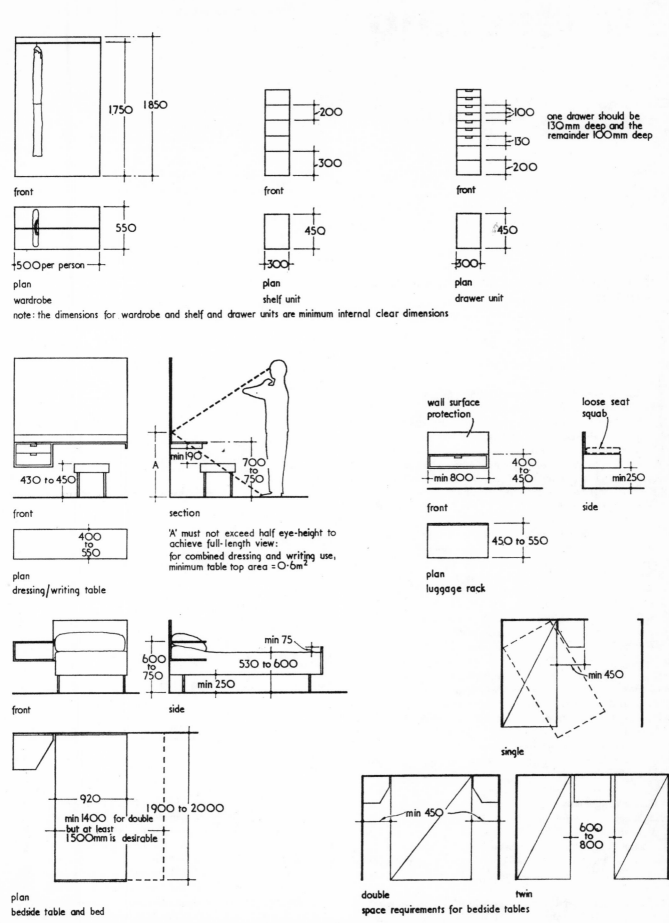

25.4 *Space requirements for various items of hotel bedroom furniture*

26 **Housing**

Contents

1 MOHLG requirements for housing in the public sector

The requirements which follow are taken from MOHLG Circular 1/68.

MANDATORY MINIMUM STANDARDS FOR LOCAL AUTHORITY HOUSING
General: Appendix 1 to Circulars 36/67 (England) and 28/67 (Wales)—metric version

The housing standards set out in this appendix are the principal standards recommended as minima by the Parker Morris Committee appointed as a subcommittee by the minister's Central Housing Advisory Committee and recommended in its report *Homes for today and tomorrow* published by HMSO in 1961. Not every metric measurement shown in this appendix is a precise equivalent of the corresponding imperial measurement.

The new standards are based on minimum floor areas according to the intended size of the household and, in accordance with the committee's recommendation, compliance with minimum room sizes is no longer required. However, it is essential that designers have full regard to user requirements and respect the requirements and variety of home activities as reflected in the amounts of furnishing normally required and its convenient arrangements. For this purpose designers are referred to Design Bulletin 6 *Space in the home** and also to additional information set out in E below. The new standards relate to all schemes of public authority housing other than (*a*) flatlets for old people and (*b*) single persons' accommodation provided with common-rooms and other communal facilities. The standards do, however, relate to single-person dwellings where communal facilities are not provided. For self-contained dwellings for old people only the space standards apply.

A Plan arrangement
Mandatory from 1 January 1969
1 A dwelling shall have (i) an entrance hall or lobby with space for hanging outdoor clothes and (ii) for three-person and larger houses and three-person and larger dwellings served by a lift or ramp a space for a pram (1400mm × 700 mm).
2 Except in one-person or two-person dwellings access from bedroom to the bathroom and a wc shall be arranged without having to pass through another room.
3 The kitchen in a dwelling for two or more persons must

* Sections of Design Bulletin 6 are included in this handbook. See p140, 141

provide a space where casual meals may be taken by a minimum of two persons—see also E1.
4 In addition to kitchen storage, the sink and space for a cooker, a minimum of two further spaces shall be provided in convenient positions to accommodate a refrigerator and a washing machine. The latter may be in the kitchen or in a convenient position elsewhere. These spaces may be provided under work top surfaces.
5 Most house layouts now provide for public access to both sides of the house, but where public access to a house of three or more persons is from one side only, a way through the house from front to back shall be provided and this must not be through the living-room. In such cases the dustbin compartment shall be on the front.
6 Access to dwellings shall not involve a climb through more than two storeys to the front entrance doors.

B Space
Mandatory from 1 January 1969
Standards A home for occupation for the number of people shown in the table below shall be designed to provide areas of net space and general storage space not less than those set out in the table and fulfilling the conditions in the notes following the table. N = net space (note 1); S = general storage space (note 2).

| | | Number of people (ie bedspaces) per dwelling | | | | | | |
		1 (m²)	2 (m²)	3 (m²)	4 (m²)	5 (m²)	6 (m²)	7 (m²)
Houses One-storey	N	30	44·5	57	67	75·5	84	
	S	3	4	4	4·5	4·5	4·5	
Two-storey: semi-detached or end	N				72	82	92·5	108
	S				4·5	4·5	4·5	6·5
intermediate terrace	N				74·5	85	92·5	108
	S				4·5	4·5	4·5	6·5
Three-storey (excluding garage, if built-in)	N					94	98	112
	S					4·5	4·5	6·5
Flats	N	30	44·5	57	70*	79	86·5	
	S	2·5	3	3	3·5	3·5	3·5	
Maisonettes	N				72	82	92·5	108
	S				3·5	3·5	3·5	3·5

*(67 if balcony access) **Tolerance:** Where dwellings are designed on a planning grid and not otherwise a maximum minus tolerance of 1½ per cent shall be permitted on the net space

Note 1 Net space is the area on one or more floors enclosed by the walls of a dwelling measured to unfinished faces. It includes the space, on plan, taken up on each floor by any staircase, by partitions and by any chimney breast, flue and heating appliance and the area of any external wc. It excludes the floor area of general storage space (s in table) and dustbin store, fuel store, garage or balcony and any area in rooms with sloping ceilings to the extent that the height of the ceiling does not exceed 1·5m and any porch, lobby or covered way open to the air.
In the case of a single access house, any space within a store required to serve as access (taken as 700mm wide)

from one side of a house to the other shall be provided in addition to the areas in the table.

Note 2 General storage space is the space which shall be provided *exclusive* of any dustbin store, fuel store, pram space located in a store and, in the case of a single access house, any space within a store required to serve as access (taken as 700mm wide) from one side of a house to the other. For *houses* some of the storage space may be on an upper floor but at least 2·5m² shall be at ground level.

Where some of the storage space is provided on an upper floor, it shall be enclosed separately from linen or bedroom cupboards; it shall be accessible from the circulation space or from a room if conveniently accessible in relation to furnishing.

Where there is a garage integral with or adjoining a house, any area in excess of 12·0m² shall count towards the general storage provision.

For *flats and maisonettes* not more than 1·5m² may be provided outside the dwelling and any area in excess of 12·0m² which is provided in a garage integral with or adjoining the dwelling shall count towards this 1·5m².

Fuel storage (*excluded* from the table) where required, shall be a minimum of:

For houses   1·5m² where there is only one appliance
                  2·0m² where there are two appliances or in rural areas

For flats and maisonettes 1·0m² if there is no auxiliary storage.

Note 3 Area per dwelling for form TC2 is the sum of the net space, the general storage space and where applicable:
(*a*) in the case of a single access house, any space within a store required to serve as access (taken as 700mm wide) from one side of a house to the other;
(*b*) fuel store;
(*c*) pram store where provided *outside* the dwelling (where pram space is additional to general storage).

Areas are measured to the unfinished faces of the main containing walls on each floor of the dwelling and include the space, on plan, taken up by private staircases, partitions, internal walls (but not 'party' or similar walls), chimney breasts, flues and heating appliances. The area of the dwelling includes the area of the tenant's storage space whether located within the dwelling or elsewhere.

It excludes any space where the height to the ceiling is less than 1·5m (eg areas in rooms with sloping ceilings, external dustbin enclosures);

any porch, covered way and so on open to the air;

any garage *except* that, where a garage is integral with or adjoining the dwelling, the excess over 12·0m² qualifies as storage space and as part of the area of the dwelling;

all balconies (private, escape and access) and decks;

all public access space (eg tunnel passages, communal entrances, staircases, corridors);

all space for communal facilities or services;

all space for other-than-housing purposes (eg commercial).

C Fittings and equipment

The standard at 1(c) will be mandatory as from 1 January 1969. The date on which the other standards in this section might become mandatory has not yet been fixed.

1 The wc and washbasin provision shall be as set out below:
(*a*) In one-, two- and three-person dwellings, one wc is required, and may be in the bathroom.
(*b*) In four-person two-storey or three-storey houses and two-level maisonettes, and in four-person and five-person flats and single-storey houses, one wc is required in a separate compartment.
(*c*) In two- or three-storey houses and two-level maisonettes

at or above the minimum floor area for five persons, and in flats and single-storey houses *at or above* the minimum floor area for six persons, two wcs are required, one of which may be in the bathroom.

(*d*) Where a separate wc does not adjoin a bathroom, it must contain a washbasin.

2 Linen storage
A cupboard shall be provided giving 0·6m³ of clear storage space in four-person and larger dwellings or 0·4m³ in smaller dwellings.

3 Kitchen fitments
Kitchen fitments comprising enclosed storage space in connection with:
(*a*) preparation and serving food and washing-up;
(*b*) cleaning and laundry operations;
(*c*) food
shall be provided as follows:

Three-person and larger dwellings	2·3m³
One- and two-person dwellings	1·7m³

Part of this provision shall comprise a ventilated 'cool' cupboard and a broom cupboard. The broom cupboard may be provided elsewhere than in the kitchen.

Where standard fitments are used the cubic capacity shall be measured overall for the depth and width, and from the underside of the work top to the top of the plinth for the height.

Work tops shall be provided on both sides of the sink and on both sides of the cooker position. Kitchen fitments shall be arranged to provide a work sequence comprising work top/cooker/work top/sink/work top (or the same in reverse order) unbroken by a door or other traffic way.

4 Electric socket outlets shall be provided as follows:

Working area of kitchen	4
Dining area	1
Living area	3
Bedroom	2
Hall or landing	1
Bedsitting-room in family dwellings	3
Bedsitting-room in one-person dwellings	5
Integral or attached garage	1
Walk-in general store (in house only)	1

D Space heating
Mandatory from 1 January 1969
The minimum standard shall be an installation with appliances capable of maintaining the kitchen and circulation spaces at 13°c, and the living and dining areas at 18°c when the outside temperature is − 1°c.

E Furniture
Mandatory from 1 January 1969
All dwelling plans must show the furniture drawn on and should be designed to accommodate furniture as set out below:

1 Kitchen
Small table unless one is built-in
2 Meals space
Dining-table and chairs
3 Living space
Two or three easy chairs
Settee
Tv set
Small tables
Reasonable quantity of other possessions, such as radiogram, bookcase

4 *Single bedrooms*
Bed or divan (2000mm × 900mm)
Bedside table
Chest of drawers
Wardrobe or space for cupboard to be built-in

5 *Main bedrooms*
Double bed (2000mm × 1500mm)—and where possible two single beds* (2000mm × 900mm) as an alternative
Bedside tables
Chest of drawers
Double wardrobe or space for cupboard to be built-in
Dressing-table

6 *Other double bedrooms*
Two single beds (2000mm × 900mm) each
Bedside tables
Chest of drawers
Double wardrobe or space for cupboard to be built-in†
Small dressing-table
Note Spaces for wardrobes, or space for cupboards to be built-in later should be on the basis of 600mm run of hanging space per person. The space provided for a cupboard depth should be not less than 550mm internally.

F Play space
The date on which this might become mandatory has not yet been fixed
Play space must be provided on schemes of 200 persons per hectare and above on the basis of 1·5m² to 2·0m² per bed space, with a minimum of 1·0m² in exceptionally favourable circumstances, such as where an estate has existing playgrounds easily accessible in the immediate vicinity.

The housing cost yardstick—appendix II to circular 36/67
Metrication of densities
The cost table in appendix II to Circular 36/37 may still be used for calculating the housing cost yardstick figure applicable to any given metric scheme. The density in persons per hectare should first be converted to persons per acre by multiplying by 0·405 after which the table may be used in the normal way.

Example
In order to apply the yardstick table, find the equivalent in persons per acre of 375 persons per hectare:
375 × 0·405 = 151·875 ie 152 persons per acre.

Car spaces per acre
The same factor (0·405) may also be used to convert car spaces per hectare to car spaces per acre. This figure will be needed in order to establish the 'equivalent higher density' applicable to schemes incorporating more than thirty car spaces per acre—see paragraph 16 of appendix II to Circular 36/67.

Notation: In general, building dimensions such as ceiling heights and main areas are expressed in metres and furniture sizes or furniture spaces in millimetres.
This is the end of the extracts from Circular 1/68.

2 Critical dimensions
Table I is freely translated from standards adopted by the Architects Department of Wates Ltd in designing residential accommodation, which experience has shown to be desirable minima. The metric dimensions are not necessarily exact equivalents of the imperial figures.

Table I Critical dimensions: recommended minima

Item		Recommended minimum dimension (mm)	Notes
Doorsets (Door and frame)	Front doors Other doors generally Bathroom, wc or cupboards	900 800 600	Preferably 900mm
Entrance	Lobby	1500	Least dimension
Staircase openings	Single flight	900 width 2700 length	Preferably 1000mm with maximum rise 195mm and minimum going 220mm
	Double flight or stairs plus landing	1800 width 2100 length (dogleg)	
Living Room	Possible 3300mm (min) (depending on access and furniture arrangement)		
Dining Room or Recess	Possible 2700mm (min) (depending on access and furniture arrangement)		
Kitchen	600mm fittings on one side only 600mm fittings on both sides	1800 (min) 2400 (Min)	Could possibly go down to 1700 and 2300 in small and 1 and 2 person dwellings
Bathroom	Bath and W.C. or bath and wash basin side by side Bath basin and W.C. side by side Bath end on or basin sideways W.C. sideways	1500 (Min) 2200 (Min) 2500 (Min) 2400 (Min)	

Bed sizes
The National Bedding Federation Ltd intends to launch new standard sizes in January 1972.
Four basic sizes will be introduced:
Standard beds
Single: 1000 mm × 2000 mm
Double: 1500 mm × 2000 mm
Replacement beds
Single: 800 mm × 1920 mm
Double: 1350 mm × 1920 mm

3 Controlling dimensions
Recommendations for horizontal and vertical controlling dimensions for housing are contained in *Dimensional co-ordination for building* documents DC4, DC5, DC6, and DC7 and in BS 4330:1968. The application of the BS in public sector housing is stated in Design Bulletin 16. These recommendations are illustrated on p139.
Horizontal controlling dimensions are shown in **26.1** and table II.
Vertical controlling dimensions are shown in **26.2**.

---

* Where single beds are shown they may abut or where alongside walls must have a space of 750mm between them. See also note opposite.
† May be provided within easy access outside room.

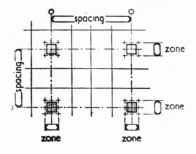

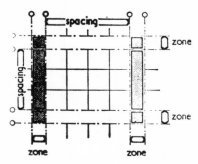

26.1 *Horizontal controlling dimensions. Diagram to define meaning of spacing and zone (to be read in conjunction with table II)*

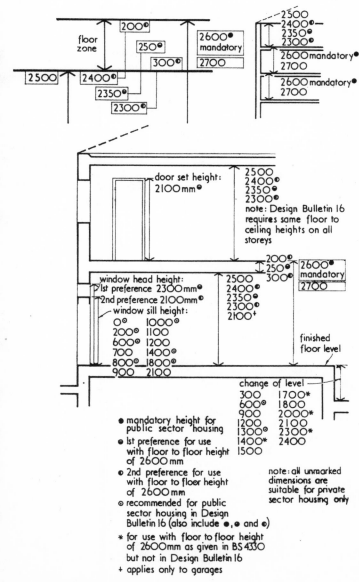

26.2 *Vertical controlling dimensions*

Table II Horizontal controlling dimensions (to be read in conjunction with 26.1). Dimensions are between boundaries of zones for loadbearing walls and columns according to DC publications and MOHLG Design Bulletin 16. Axial spacing not recommended for public sector housing.
Dimensions greater than 7200mm apply to pitched roof construction

Spacing (mm)	Zone (mm)
800	100
900	200
1 200	300
1 500	400
1 800	
2 100	
2 400	
2 700	
3 000	
3 300	
3 600	
3 900	
4 200	
4 500	
4 800	
5 100	
5 400	
5 700	
6 000	
6 300	
6 600	
6 900	
7 200	
7 500	
7 800	
8 100	
8 400	
8 700	
9 000	
9 300	
9 600	
9 900	
10 200	
10 500	
10 800	
11 100	

Preferences in millimetres showing increments for first, second and third preferred dimensions (BS 4011)

4 Space requirements related to activities

Drawings **26.3** to **26.38** are reproduced from Design Bulletin 6 *Space in the home* (metric edition) by permission of the MOHLG and the Controller of HMSO (Crown copyright). The following notes are extracted from *Bord*.

Eating area in the kitchen
A rectangular table is the best shape for a small kitchen, with a minimum width of 750mm.
If the dining area is bounded by walls or work tops on both sides of the table, total width should not be less than 2300mm; or 2600mm if there needs to be free passage behind the chair on one side.
Allow: 700mm between the oven and a chair in use;
800mm between the most commonly used work top and a pushed-in chair.
Recommended table sizes:

Two people	:	750mm × 750mm
Three people	:	750mm × 1000mm
Four people	:	750mm × 1200mm
Five people	:	750mm × 1350mm

Eating area in the living-room or dining-room
The table should be at least 850mm wide, preferably extendable.
Recommended rectangular dining-room table sizes (before extension):

Three to four people	:	850mm × 1050mm
Four people	:	850mm × 1200mm
Five people	:	850mm × 1350mm
Six people	:	850mm × 1500mm
Seven people	:	850mm × 1800mm

FOOD PREPARATION

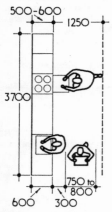

26.3 *At sink: passing with tray: at oven (new* BS *in preparation will supersede depth and length of fitments in* **26.3** *and* **26.4**)

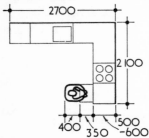

26.4 *Sitting at pull-out work top*

26.5 *Taking things from a sideboard or low cupboard*

26.6 *Maximum vertical reach and maximum shelf height for general use*

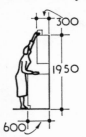

26.7 *Maximum vertical reach over work top*

26.8 *Comfortable vertical reach over work top*

26.9 *Shelf at eye level*

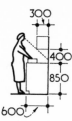

26.10 *Comfortable height of work top for standing position and clearances for cupboards above. Larger dimension of 450mm above top would allow for use of a mixer or similar appliance*

26.11 *Comfortable height of work top for seated position. Seat height 400 mm*

EATING

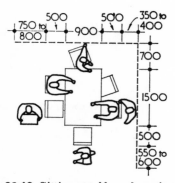

26.12 *Sitting at table and moving around*

26.13 *Sitting at work top with person passing*

REFUSE DISPOSAL

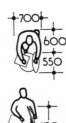

26.14 *Using dustbin*

LEISURE (five-person family)

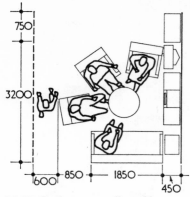

26.15 *Eating at a coffee table*

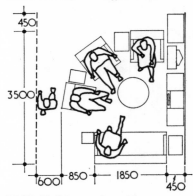

26.16 *Talking and reading*

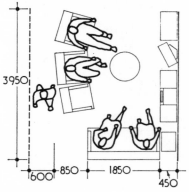

26.17 *Looking at tv*

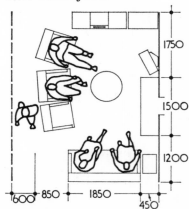

26.18 *Sitting around fireplace while watching tv*

26.19 *Getting up from a table, desk or writing bureau*

SLEEPING

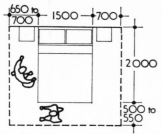

26.20 *Circulation around double bed*

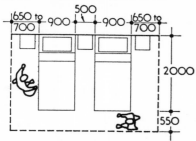

26.21 *Circulation around twin beds*

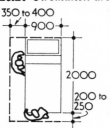

26.22 *Making a bed*

PERSONAL CARE

26.23 *Face washing*

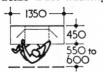

26.24 *At dressing-table*

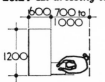

26.25 *Taking clothes from wardrobe drawer*

26.26 *Taking clothes from chest of drawers*

26.27 *Drying after a bath*

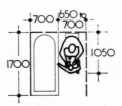

26.28 *Drying a child after a bath*

26.29 *Using wc (low level cistern)*

CIRCULATING

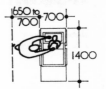

26.30 *Passing between two pieces of furniture one of which is at level lower than table height*

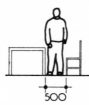

26.31 *Passing between a piece of furniture at or lower than table height and taller piece of furniture or wall*

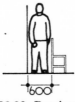

26.32 *Passing between tall piece of furniture and wall*

ENTERING AND LEAVING

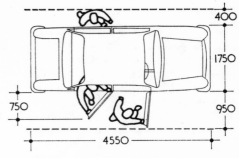

26.34 *Getting in and out of a car*

26.35 *Getting pram ready*

26.36 *Helping on with a coat*

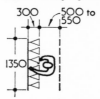

26.37 *Hanging coats on hangers*

26.38 *Hanging coats on hooks*

Note: The dimensions shewn of any furniture or fittings may require revision as new dimensionally co-ordinated metric standards are published by BSI

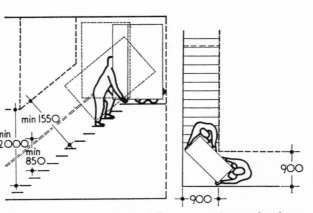

26.33 *Moving a double wardrobe up a staircase showing minimum headroom, handrail height and a going and rise of 215mm and 190mm respectively*

5 Kitchen storage

Shelf and drawer storage

The information in this section is based on an investigation into shopping habits and storage requirements of households, carried out by the Council of Scientific Management in the Home in 1963/64 and on a subsequent study of the space requirements of stored items made by Queen Elizabeth College, the University of London.

Table III lists the area of shelf storage required; table IV lists the amount of drawer storage required by the average family.

Table III Space requirements: Shelves

Item	Storage (m²)	Shelf length (mm)	Shelf depth (mm)	Vertical clearance (mm)
Dry goods*	0·5	2800 / 250	150 / 230	200 / 330
Tinned and bottled goods	0·2	1100 / 300	150 / 150	150 / 300
Drinks	0·3	Length immaterial	150 minimum	350
Bread, cake, biscuits	0·3	900 + bread bin base 300 × 250mm	300	150
Pet foods	0·05	300	150	200
Dairy goods, meat, fish, poultry, frozen foods	0·9			
Table china and glass	1·6	3250 / 450	280 / 280	150 / 200
Cooking china and glass	0·45	1000 / 1600	450 / 280	150 and a small amount of 250
Saucepans, frying pans, and so on	0·7	760 / 1700	280 / 280	150 / 300
Miscellaneous china and glass	0·25	900	280	300
Cookery books	0·05	300	150	230
Empty jars and bottles	0·2 in any form			

* Figures given are for items stored in packets as purchased. If they are stored in canisters the area required could be as much as 0·8m² with 3900mm of 150mm deep shelving and 1100mm of 230mm

Table IV Space requirements: Baskets, drawers

Item	Method of storage	Number required	Storage area (m²)	Dimensions (mm)
Fruit and vegetables	Baskets	2	0·4	480 × 380 × 80
Table cutlery	Drawer	1		450 × 450 × 80
Kitchen cutlery and equipment*	Drawers	1½ / ¾	0·8	450 × 450 × 100 / 450 × 450 × 150
Baking tins	Drawers	1¼		450 × 450 × 150

* Including bread board, chopping board, rolling pin

Other storage requirements

In addition, provide space for the following:
Cleaning materials and equipment
Door height cupboards 760mm wide × 400mm deep
Vacuum cleaner:
upright model approx 350mm × 400mm } plus
cylinder model approx 600mm × 200mm } attachments

Trays
Space of 400mm deep × 150mm wide × 600mm high.

Kitchen linen
At least one drawer 450mm × 450mm × 120mm deep.

Summary

Tables V and VI summarise storage requirements of a kitchen.

Tables V and VI are for families of three, four and five people, but smaller families usually require the same amount of space for most items. Total storage area from tables V and VI is 8·3m². Compare this with the three-to-four bedroom house in table VII.

Table V Area of shelf storage: Summary

Item	Area (m²)
Food, china and glass, saucepans, frying-pans, bread board, chopping board, and so on	4·5
Refrigerator storage	0·9
Broom cupboard	0·7
Bucket and soaps cupboard	0·5
Total area	6·6

Table VI Area of drawer storage: Summary

Item	Area (m²)
Fruit and vegetables, table cutlery, kitchen cutlery and equipment	1·3
Linen	0·4
Total area	1·7

Table VII BS 3705:1964   Recommendations for kitchen provision (approximate metric conversions)

Size of home (number of bedrooms)	Area (m²)		
	Shelves	Drawers	Total
1	4·6	0·9	5·5
2	5·6	1·1	6·7
3 to 4	7·0	1·4	8·4

MOHLG storage requirements

By comparison, MOHLG requirements for housing in the public sector are as follows:
Kitchen fitments comprising enclosed storage space in connection with:
(a) preparation and serving of food, and washing up;
(b) cleaning and laundry operations;
(c) food:
Three-person and larger dwellings   2·3m²
One- and two-person dwellings   1·7m²
Critical dimensions are shown under para 4 of this section, p140.

6 Clothes storage

Storage requirements will vary from family to family. Many of the recommendations in this section are taken from Swedish sources, except for table VIII which summarises the requirements of the MOHLG and the New Scottish Housing Handbook.

If possible, space should be provided for storage of seasonal and rarely used clothes (see table IX).

Wardrobe, shelf and drawer dimensions are given in **26.39** to **26.41**. Shelf or drawer storage for stockable clothes should be provided: allow about 0·3m³ for men and 0·27m³ for women.

Note that in tables IX and X and in **26.39** the Swedish clothing inventory includes more heavy winter garments than in the UK.

7 Linen storage

Suggested dimensions for linen cupboards are shown in table X and alternative arrangements for linen storage in **26.42**.

The MOHLG requires a cupboard giving 0·6m³ clear storage space (four-person or larger dwelling) and 0·4m³ in smaller dwellings.

The New Scottish Housing Handbook recommends the same standards as MOHLG.

Table VIII Recommended sizes for built-in wardrobes

	Homes for today and tomorrow	New Scottish Housing Handbook
Main bedroom	No recommendation	600mm run of hanging space per person, with minimum internal depth of 550mm.
Other bedrooms	600mm of rail per occupier not less than 550mm deep	600mm × 600mm wardrobe cupboard

Table IX Recommended lengths of hanging rail for seasonal and occasional clothes

For seasonal clothes	
Family of two	1000mm
Family of four	1600mm
For clothes not in use	
Family of two	500mm
Family of four	1000mm

Table X Suggested dimensions for linen cupboard (Swedish)

Size of household	No of shelves	Width (mm)	Depth (mm)	Height (mm)
	5	550	300	850
	8	550	300	1500
	8	550	300	2000

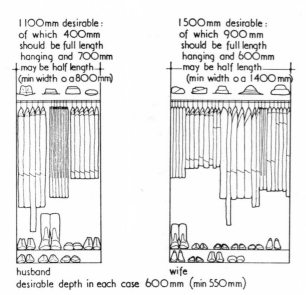

husband　　　　　　wife
desirable depth in each case 600mm (min 550mm)

26.39 *Optimum hanging space for a family of four*

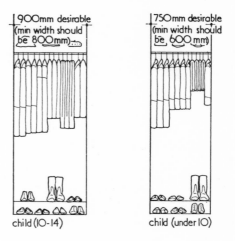

child (10-14)　　　　child (under 10)

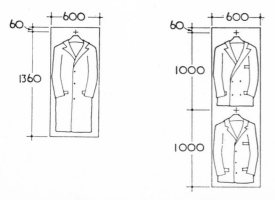

26.40 *Dimensions of wardrobes for full and half-length hanging*

Table XI Dimensions of common items of linen

	Number	Width (mm)	Depth (mm)	Height (mm)
Sheet (linen):	5	610	330	15
2350 × 2540mm—double	6	340	310	30
Sheet (linen):	4	610	460	15
1800 × 2540mm—single	5	610	230	30
Sheet (flannelette):	4	560	510	30
2000 × 2300mm—double	5	560	260	50
Blanket:	4	610	560	80
2300 × 2540mm—double	5	610	280	150
Blanket:	4	610	510	80
2000 × 2540mm	5	610	260	150
Blanket:	4	560	460	80
1800 × 2300mm—single	5	560	230	150
Pillow Case	2	530	220	15
Pillow	1	710	430	150
Hand towel: 600 × 400mm	2	200	310	30
Face towel: 500 × 1000mm	3	260	260	30
Bath towel: 920 × 1370mm	4	350	230	90

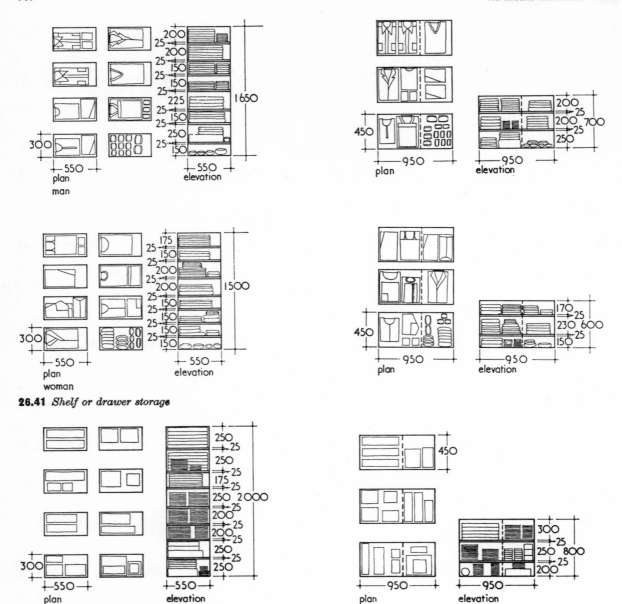

26.41 *Shelf or drawer storage*

26.42 *Alternative arrangements of linen stored by family of five*

8 General storage

Sizes of commonly used equipment which will require storage are shewn in **26.43**.

9 Fuel storage

The recommendations of MOHLG and the Coal Utilisation Council for solid fuel storage are given in table XII.

The space occupied by 1000kg (1 tonne) of solid fuel varies with the bulk density of the particular fuel (table XIII).

Capacities of given floor areas with stacking to an average

height of 1400mm and with a hopper outlet at the base are given in table XIV.

Various examples of domestic fuel stores are shown in **26.44**. These are designed to make the best use of the minimum standards of the MOHLG and have been adapted from the CUC Handbook.

Note that the store should not be more than 46m walking distance from the nearest point accessible by delivering vehicle.

Some domestic fuel stores, with capacities, are shown in **26.45**.

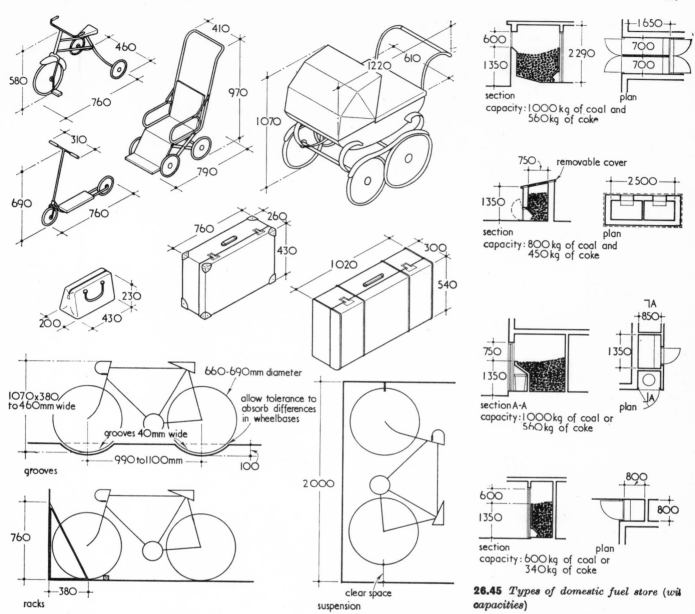

26.43 *Common items of general storage*

section
capacity: 1000 kg of coal and
560 kg of coke

plan

removable cover
section
capacity: 800 kg of coal and
450 kg of coke

plan

section A-A
capacity: 1000 kg of coal or
560 kg of coke

plan

section
capacity: 600 kg of coal or
340 kg of coke

plan

26.45 *Types of domestic fuel store (with capacities)*

clear space
suspension

grooves 40mm wide
grooves
racks

660-690mm diameter
allow tolerance to absorb differences in wheelbases

26.44 *Examples of domestic fuel stores showing minimum dimensions*

doors, if provided, should fold back when open and not obstruct access

minimum clear width of delivery opening

minimum clear width of corridor at opening to fuel store to enable carman to manoevre easily

minimum clear width of all passageways corridors and stair cases used by carmen

plan

recommended height from floor to top of coal boards or sill of delivery opening

minimum clear height of delivery opening

minimum headroom of all passageways corridors and staircases used by carmen

recommended dimensions of hopper

section

section where delivery is through external wall

Table XII   Recommended sizes of solid fuel stores

	MOHLG recommendations	Minimum store area (m²)	CUC recommendations	Minimum store area (m²)
Local authority house	One solid fuel appliance	1·5	Size should relate to size of house but capacity should be at least 1500kg	2·8
	Two solid fuel appliances or in rural areas	2·0		
Private house			At least 2000kg. The space should be divided into two 1000kg compartments	3·7
Flats and maisonettes	If there is no auxiliary storage	1·0	At least one month's supply of winter rate of consumption, preferably six weeks' supply	1·1

Table XIII   Bulk densities of solid fuels

Fuel	Density
Coal Anthracite Phurnacite	1·4m³/1000kg (or 1 tonne)
Gas coke Hard coke	2·1 to 3·1m³/1000kg
Coalite Rexco	2·5m³/1000kg

Note that 1 English ton = 1016kg (approx) ; 1 tonne (metric ton) = 1000kg

Table XIV   Capacities of floor areas to a stacking height of 1400mm

Floor area (m²)	Capacity (kg) for fuel at 1·4m³ per 1000kg	Capacity (kg) for fuel at 2·5m³ per 1000kg
0·2	200	120
0·4	400	230
0·6	600	340
0·8	800	450
1·0	1000	560
1·2	1200	670
1·4	1400	780
1·6	1600	890
1·8	1800	1000
2·0	2000	1120
2·2	2200	1230
2·4	2400	1340
2·6	2600	1450
2·8	2800	1560
3·0	3000	1680

Planning the fuel store

The following points should be noted:

Convenience for householder with no necessity for coalman to enter any part of house or flat on delivery.

Preferable construction of brick or concrete.

Designed to protect contents from weather.

Ventilation at high and low level by airbricks or cutouts in door.

Aperture through which store is filled should have a sill 1320 to 1370mm above floor level, preferably extending the width of the store but not less than 560mm square (sill to rest coal sack on during emptying).

Chutes and hoppers not recommended.

10 References and sources

MINISTRY OF HOUSING AND LOCAL GOVERNMENT AND WELSH OFFICE Circular 1/68 Metrication of housebuilding. 1968, HMSO CI/SfB 81 (F7)

MINISTRY OF HOUSING AND LOCAL GOVERNMENT Circular 36/67 Housing standards, costs and subsidies. 1967, HMSO CI/SfB 81 (Y)

WELSH OFFICE Circular 28/67 Housing standards, costs and subsidies. Cardiff, 1967, The Welsh Office, Cathays Park, Cardiff CI/SfB 81 (Y)

MINISTRY OF HOUSING AND LOCAL GOVERNMENT Design Bulletin 6 Space in the home (metric edition). 1968, HMSO CI/SfB 81 (E6)

MINISTRY OF HOUSING AND LOCAL GOVERNMENT: Design Bulletin 16. Co-ordination of components in housing: Metric dimensional framework. CI/SfB 81 (F4j)

MINISTRY OF HOUSING AND LOCAL GOVERNMENT Homes for today and tomorrow. Report of a subcommittee of the Central Housing Advisory Committee (chairman Sir Parker Morris). 1961, HMSO CI/SfB 81

SCOTTISH DEVELOPMENT DEPARTMENT New Scottish housing handbook. Bulletin 1, HMSO 1968 CI/SfB 812

COAL UTILISATION COUNCIL Fuel stores for houses and flats. London, 1959, The Council CI/SfB 81 (5–)

BRITISH STANDARDS INSTITUTION BS 4330:1968 (metric units) Recommendations for the co-ordination of dimensions in building: Controlling dimensions. CI/SfB (F4j)

BRITISH STANDARDS INSTITUTION BS 3705:1964 Recommendations for provision of space for domestic kitchen equipment. London, 1964, The Institution CI/SfB 93 (73) (E6)

BERGLUND, E. Bord [tables] för måltider och arbete i hemmet. Stockholm, 1957, Svenska Slöjdföreningen (Swedish Society of Industrial Design) CI/SfB (72)

AJ information sheet 1331 Space standards for domestic activities. AJ, 1965, June 2 CI/SfB 81 (E2p)

AJ information sheet 1333 Domestic kitchens: 1 Storage requirements. AJ, 1965, June 9 CI/SfB 81003

AJ information sheet 1209 Cloakrooms. AJ, 1963, July 17 CI/SfB 96

AJ information sheet 1216 Domestic storage. AJ, 1963, September 25 CI/SfB 81 (76)

MINISTRY OF PUBLIC BUILDING AND WORKS DC4 Dimensional co-ordination for building: Recommended vertical dimensions for educational, health, housing, office and single-storey general purpose industrial buildings. 1967, HMSO CI/SfB (F4j)

MINISTRY OF PUBLIC AND BUILDING WORKS DC5 Dimensional co-ordination for building: Recommended horizontal dimensions for educational, health, housing, office and single-storey general purpose industrial buildings. 1967, HMSO CI/SfB (F4j)

MINISTRY OF PUBLIC BUILDING AND WORKS DC6 Dimensional co-ordination for building: Guidance on the application of recommended vertical and horizontal dimensions for educational, health, housing, office and single-storey general purpose industrial buildings. 1967, HMSO CI/SfB (F4j)

MINISTRY OF PUBLIC BUILDING AND WORKS DC 7 Dimensional co-ordination for building: Recommended vertical controlling dimensions for educational, health, housing and office building, and guidance on their application. 1967 HMSO CI/SfB (F4j)

27 **Heating, thermal insulation, condensation**

Contents

Tables

1 Heating and thermal insulation: definitions and terms

1.1 Temperature The distinction between temperature and intervals of temperature may not be generally known. The former was denoted as °F and the latter as degF in the imperial system. This distinction is to be retained in the SI, thus

$$\text{temperature: } °C$$
$$\text{interval of temperature: } degC$$

eg if the inside temperature is $t_1 = 25°C$
and external temperature is $t_e = 10°C$
the temperature difference $\Delta t = 15degC$

The term *degree Celsius* should be used instead of *centigrade* to avoid confusion with the centigrade used in France as an angular measurement, meaning $\frac{1}{10000}$ right-angle. The Celsius scale takes the freezing and boiling points of water and subdivides this interval into 100 degrees. The degree Kelvin as an interval of temperature is the same as a degC, but on the scale of temperatures the absolute zero is the starting point. Thus: $t°K = t - 273 \cdot 15°C$ (or $273 \cdot 15°K = 0°C$) but t degK = t degC.
The absolute (thermodynamic) or Kelvin temperature scale is used in scientific work and is the official SI scale, but for all practical purposes the Celsius scale is used.
See also conversion tables °F to °C on p203.

1.2 Heat Until now there have been special units used for the measurement of heat quantity. The unit was defined as the amount of heat necessary to elevate the temperature of unit mass of water by one degree. Thus
1 Btu elevating 1 lb of water by 1 degF
1 kcal elevating 1 kg of water by 1 degC
1 cal elevating 1 g of water by 1 degC
In SI units heat is considered as any other form of energy and measured in the generally used energy unit **joule (J)**

A joule is the energy or work unit coherently derived from the three basic mks (metre-kilogramme-second) units:

length   m
mass   kg
time   s
velocity   m/s   unit length movement in unit time
acceleration m/s² unit velocity change in unit time
force   kg m/s² = N (newton) causing unit acceleration of unit mass
energy, work kg m²/s² = J (joule)   unit force acting over unit length

For very large and very small quantities the standard metric multiples and submultiples may be used, eg:

MJ (megajoule) = 1 000 000J
kJ (kilojoule) =   1000J
mJ (millijoule) =   0·001J

The unit most likely to be generally used in thermal calculations is the kilojoule   **(kJ)**

1.3 Calorific value is defined as the amount of heat energy released when burning a unit mass of fuel material. Customarily it was expressed in Btu/lb or kcal/kg. The SI unit is   **J/kg**

1.4 Specific heat capacity is the amount of heat energy required per unit mass of a substance per degree rise in temperature.
Both the imperial and the metric units (Btu/lb degF and kcal/kg degC) are replaced by   **J/kg degC**

1.5 Heat flow rate This quantity is used to express heat loss, heat gain, heating or cooling capacity. It was usually given in Btu/h or kcal/h. Now it could be expressed in terms of J/h, but as its dimension is unit energy per unit time, which is the same as the dimension of power, it can be reduced to a second: J/s, which is   **watt (W)**
For large quantities the standard multiples can be used:
MW (megawatt) = 1 000 000 W
kW (kilowatt) =   1000 W
kWh (kilowatthour) and Wh (watthour) are energy units, of the same character as the J.

1.6 Density of heat flow rate is defined as the amount of heat (energy) passing through unit area (of a body or of space) in unit time. It is obtained in calculations as a product of the u-value and the temperature difference, but it can be used to express the density or intensity of all types of energy flow (eg radiation intensity or sound intensity). Both the imperial units (Btu/ft²h or W/ft²) and the metric units previously used (kcal/m²h or cal/cm²h) are now replaced by   **W/m²**

1.7 Conductivity or k-value of a material is defined as the amount of heat energy conducted through unit area of unit thickness in unit time with unit temperature difference between the two faces. The imperial unit was Btu in/ft²h degF. The way of writing this as Btu/in/ft²/h/degF is mathematically meaningless and untidy. The metric unit so far used was kcal/m h degc. This was a simplification of kcal m/m²h degc. A new unit could be J/m h degc, but as the power unit (watt) incorporates a time unit in its denominator (W = J/s) the expression is simplified:

$$\text{W/m degC (or degK)}$$
$$\text{or Wm/m}^2 \text{ degC}$$

(for data see table III p156)

1.8 Transmittance or u value of a construction is defined as the amount of heat energy conducted through unit area of the construction in unit time with unit air-temperature difference between the two sides. The imperial unit was Btu/ft²h degF, and the old metric unit was kcal/m²h degc.

The SI unit is   **W/m² degC**

Conductance has the same dimension, thus the same unit, the difference being that conductance is taken between the two surfaces and not air-to-air, as the transmittance. The transmittance value includes the measure of surface properties, of the air-to-surface and surface-to-air heat transfer, ie the surface or film conductances. This f-value is measured with the same unit (for data see tables II and V p155 and p156).

1.9 Conversion factors

2.1 Design temperatures External design temperature (t_e) can be established as constant for a certain geographical location. In the UK normally −1°c or 0°c is used, as lower temperatures occur infrequently and for short periods, the structures have some degree of thermal inertia and heating systems usually have an overload capacity.

Indoor temperatures (t_1) are given in table I (p155) as functions of room usage or activity to be accommodated. The recommended values largely depend on the state of acclimatisation and the habit-dictated preferences of the occupants. This is demonstrated by the German recommendations included in the table for comparison.

2.2 Transmittance The other quantity to be used in formula 1 is the transmittance or u value of the enclosing construction. Definition of this quantity is given in para **1.8** and the u values of some typical constructions are given in table II p155.

2.3 Computation of transmittance When a form of construction is to be used which is not included in table II, its u value can be computed from its component elements. Transmittance and conductance values are not additive, only the reciprocals: the resistances. The resistance of a layer of any material is the product of its resistivity (1/k) and its thickness. For the purposes of such computations the conductivities (k values) and the reciprocals, the resistivities

Paragraph	Imperial		SI
1.1 Temperature	t °F	$(t - 32) \times \frac{5}{9}$	°c
1.1 Temperature interval	1 degF	$\frac{5}{9}$	degc
1.2 Heat	1 Btu	1055·06	J (joule) or 1·055 06 kJ
	1 Therm	105·506	MJ
1.3 Calorific value	1 Btu/lb	2326	J/kg or 2·326 kJ/kg
1.3a Calorific value, volume basis	1 Btu/gal	232·6	J/litre
or	1 Btu/ft³	37 259	J/m³ or 37·259 kJ/m³
1.4 Specific heat capacity	1 Btu/lb degF	4186·8	J/kg degc
1.4a Volumetric specific heat	1 Btu/ft³ degF	67 066	J/m³ degc or 67·066 kJ/m³ degc
1.5 Heat flow rate	1 Btu/h	0·293	W (watt)
1.6 Density of heat flow rate	1 Btu/ft²h	3·155	W/m²
1.7 Conductivity (k-value)	1 Btu in/ft²h degF	0·144	W/m degc
1.8 Transmittance (u-value)	1 Btu/ft²h degF	5·678	W/m² degc

Some useful conversion scales have been produced by the Regent Street Polytechnic.

For details of the complete set, see note, para **6**.

2 Heat loss

Heating design is based on calculation of heat loss, ie finding the rate at which heat is flowing out of a room or a building. For many straightforward problems steady state conditions can be assumed and this heat flow rate is expressed by the formula $Q_c = A \times U \times _\Delta t$ **(formula 1)** where

Q_c = conductance heat loss in W

A = surface area considered (of room or building) in m²

U = transmittance of the construction in W/m² degc

$_\Delta t$ = air temperature difference ($t_1 - t_e$) in degC

The following paragraphs examine the latter two factors in more detail and para **2.4** gives the calculation method.

(1/k) of some frequently used materials are given in table III p156. By convention, the *-ity* ending implies the property of a material, regardless of its thickness and the ending *-ance* refers to the properties of a construction of given thickness.

The air-to-air resistance (R) is the sum of the following components:

resistances of the several layers of materials: $1/k \times b$ or b/k where b is the thickness in metres;

resistance of the cavity (if any) is 1/C (see table IV p156);

exterior and inside surface (or film) resistances are $1/f_e$ and $1/f_1$ (see table V p156).

Thus

$$R = \frac{1}{f_e} + \frac{1}{C} + \frac{b'}{k'} + \frac{b}{k''} + \ldots + \frac{1}{f_1} \quad \textbf{(formula 2)}$$

The u value is the reciprocal of this sum: $U = \dfrac{1}{R}$ **(formula 3)**

EXAMPLE 1 CALCULATION OF TRANSMITTANCE (U VALUE) OF A COMPOSITE WALL

The wall consists of $4\frac{1}{2}$in outer skin of engineering bricks, a 4in inner skin of dense concrete with $\frac{1}{2}$in gypsum plaster finish on the inside. There is a 2in cavity between the two skins and the wall is facing north with normal exposure.

Stage 1 Note conductivity value of each material from table III p156, the cavity conductance from table IV p156 and the surface conductances from table V p156.

Imperial units			*Metric* (SI) *units*		
Brick	4·5in k =	8·00	Brick	0·120m k =	1·15
Concrete	4in k =	10·00	Concrete	0·100m k =	1·44
Plaster	0·5in k =	3·20	Plaster	0·012m k =	0·46
Cavity conductance			Cavity conductance		
	C =	1·00		C =	5·67
Surface conductances			Surface conductances		
	f_i =	1·43		f_i =	8·12
	f_e =	3·33		f_e =	18·90

Stage 2 Using formula 2, find the overall resistance.

$$R = \frac{1}{f_e} + \frac{1}{C} + \frac{b'}{k'} + \frac{b''}{k''} + \cdots + \frac{1}{f_i}$$

$$R = \frac{1}{3·33} + \frac{1}{1} + \frac{4·5}{8} + \frac{4}{10} + \frac{0·5}{3·2} + \frac{1}{1·43} = 0·3 + 1 +$$

$0·56 + 0·4 + 0·16 + 0·7 = 3·12$ ft²h degF/Btu (imperial)

$$R = \frac{1}{18·9} + \frac{1}{5·67} + \frac{0·12}{1·15} + \frac{0·1}{1·44} + \frac{0·012}{0·46} + \frac{1}{8·12} =$$

$0·053 + 0·176 + 0·104 + 0·07 + 0·026 + 0·123$
$= 0·552$ m² degC/W(SI)

Stage 3 Using formula 3, find the overall transmittance

$$U = \frac{1}{R}$$

$$U = \frac{1}{3·12} = \textbf{0·32 Btu/ft}^2\textbf{h degF}$$

$$U = \frac{1}{0·552} = \textbf{1·81 W/m}^2\textbf{ degC}$$

(In the arithmetic there will be multiplications instead of divisions if the resistivity and resistance values from tables II, IV and V are used in lieu of the conductivity and conductance values.)

When the U value of a construction similar to the proposed one is known, a substitution or addition can be made as follows:

EXAMPLE 2 MODIFICATION OF A GIVEN U VALUE

A roof is to be constructed with a $\frac{1}{2}$in (13mm) wood fibre softboard lining under corrugated asbestos cement cover. In table II p155 the transmittance of a similar roof, but with a $\frac{1}{2}$in (13mm) timber (softwood) boarding is given as

Imperial units	SI *units*
U = 0·38 Btu/ft²h degF	U = 2·16 W/m² degC

From formula 3

$R = \frac{1}{U} = \frac{1}{0·38} = 2·63$	$R = \frac{1}{U} = \frac{1}{2·16} = 0·462$

The resistance of the timber boarding must be deducted. The k value of softwood is obtained from table III p156. (Rt is the resistance of timber board.)

$k = 0·96$ Btu in/ft²h degF	$k = 0·138$ W/m degc
$R_t = \frac{b}{k} = \frac{0·5}{0·96} = 0·521$	$R_t = \frac{b}{k} = \frac{0·013}{0·138} = 0·094$
$R - R_t = 2·63 - 0·521$ $= 2·109$	$R - R_t = 0·462 - 0·094$ $= 0·386$

The resistance of the wood fibre softboard must be added. The k value of the softboard is obtained from table III p156. (R_s is the resistance of softboard.)

Imperial units	SI *units*
k = 0·45 Btu in/ft²h degF	k = 0·065 W/m degc
$R_s = \frac{b}{k} = \frac{0·5}{0·45} = 1·111$	$R_s = \frac{0·013}{0·065} = 0·2$
$R - R_t + R_s = 2·109$ $+ 1·111 = 3·22$	$R - R_t + R_s = 0·368$ $+ 0·2 = 0·568$

Thus from formula 3 the modified U value is

$$U' = \frac{1}{3·22} = \textbf{0·31 Btu/ft}^2\textbf{h degF}$$

$$U' = \frac{1}{0·568} = \textbf{1·76 W/m}^2\textbf{ degC}$$

In computing heat loss from a room or a building, formula 1 must be solved separately for each enclosing element exposed to a different temperature difference and each enclosing element of a different construction (having a different U value). In certain cases it may be more convenient to find the average U value of a wall which consists of areas of different constructions. This is best shown by an example.

EXAMPLE 3 CALCULATION OF AN AVERAGE U VALUE

A typical 12ft (3·6m) × 10ft (3m) high bay of the external wall of an office building consists of the following three elements (U values are obtained from table II p155):

Imperial units

3ft 4in × 9in brick piers, plastered both sides, at 12ft centres	area: A = 33·4ft²
	U = 0·43
8ft 8in × 3ft 4in spandrel wall, 8in thick in-situ clinker concrete	A = 28·8ft²
	U = 0·23
8ft 8in × 6ft 8in window, single glazed	A = 57·8ft²
	U = 0·88
Total area of whole bay	A = 120ft²

SI *units*

1m × 0·234m brick piers plastered both sides, at 3·6m centres	area: A = 3·0m²
	U = 2·44
2·6m × 1m spandrel wall, 0·2m thick in-situ clinker concrete	A = 2·6m²
	U = 1·30
2·6m × 2m window, single glazed	A = 5·2m²
	U = 5·00
Total area of whole bay	A = 10·8m²

$$\text{Average transmittance } U' = \frac{\text{sum of areas} \times U \text{ values}}{\text{total area}}$$

Imperial units

$$U' = \frac{(33·4 \times 0·43) + (28·8 \times 0·23) + (57·8 \times 0·88)}{120}$$

$$= \frac{14 \cdot 4 + 6 \cdot 6 + 51}{120} = \frac{72}{120} = \textbf{0·60 Btu/ft}^2\textbf{h degF}$$

SI *units*

$$U' = \frac{(3 \times 2 \cdot 44) + (2 \cdot 6 \times 1 \cdot 3) + (5 \cdot 2 \times 5)}{10 \cdot 8}$$

$$= \frac{7 \cdot 32 + 3 \cdot 38 + 26}{10 \cdot 8} = \frac{36 \cdot 7}{10 \cdot 8} = \textbf{3·40 W/m}^2\textbf{ degC}$$

2.4 Computation of heat loss The above information is enough for computation of heat loss through an enclosing construction.

EXAMPLE 4 HEAT LOSS THROUGH A WALL
Taking a construction similar to that in example 1, and assuming area 10ft × 10ft (3m × 3m)

$A = 100\text{ft}^2$	$A = 9\text{m}^2$
$U = 0 \cdot 32$	$U = 1 \cdot 81$
$t_e = 32°\text{F}$	$t_e = 0°\text{C}$
$t_1 = 68°\text{F}$	$t_1 = 20°\text{C}$

$_\Delta t = 68 - 32 = 36$ degF　　$_\Delta t = 20 - 0 = 20$ degC
In terms of formula 1: $Q_c = A \times U \times {}_\Delta t$
$Q_c = 100 \times 0 \cdot 32 \times 36$　　$Q_c = 9 \times 1 \cdot 81 \times 20$
$= 1150$ Btu/h　　　　　　　$= 326$W

It must be remembered that the result represents a heat flow rate, in the old form taken per hour, in the new form per second. 326W represents the flow of 326J of heat energy per second. This can be used directly eg for heating design, if the heating capacity is expressed in the same terms—watts (as it is conventionally done for electric heaters). If for some reason the quantity of heat lost in an hour is required, the above rate (given in watts or J/s) must be multiplied by 3600, the number of seconds in an hour, thus:
326 J/s × 3600 = 1 173 600 J/h which can be written as 1173·6 kJ/h.

When computing heat loss from a room or a building, the ventilation heat loss must be added to the above conduction heat loss. Besides the heat loss owing to deliberate, purposeful ventilation, there is always a certain amount of infiltration heat loss. This normally amounts to 0·5 to 1 air change per hour. The figure is non-dimensional, thus the customarily used figures are applicable when doing the calculation in SI units.

It is usual to take 0·02 Btu/ft³ degF as the volumetric specific heat of air, ie the heat necessary to warm unit volume of air by one degree temperature. When working with SI units, the use of 1300 J/m³ degC (1·3 kJ/m³ degC) is recommended, as this corresponds to the specific heat of air at 0°C and at higher temperatures this specific heat is lower.

The conventional formula of ventilation heat loss:
$Q_v = 0 \cdot 02 \times V \times N \times {}_\Delta t$
where
Q_v = ventilation heat loss in Btu/h
V = volume of air in room in ft³
N = number of air changes per hour
$_\Delta t$ = air temperature difference in degF
could now be modified into
$Q_v = 1300 \times V \times N \times {}_\Delta t$
where Q_v is in J/h, V in m³ and $_\Delta t$ in degC.

If the result is to be in watts (J/s), the sum must be divided by 3600.

It may be simpler to use the $\frac{1300}{3600} = 0 \cdot 36$ constant, thus
$Q_v = 0 \cdot 36 \times V \times N \times {}_\Delta t$ (in W)　　　**(formula 4)**

EXAMPLE 5 TOTAL HEAT LOSS FROM A SIMPLE BUILDING
Dimensions and constructions are assumed and U values are taken from table II p155.

Imperial units		*Metric* (SI) *units*	
Plan area	10ft × 10ft	Plan area	3m × 3m
Height	9ft	Height	2·7m
Three sides 11in brick cavity wall	$U = 0 \cdot 30$	Three sides 0·286m brick cavity wall	$U = 1 \cdot 70$
One side timber window wall and door, average	$U = 0 \cdot 89$	One side timber window wall and door, average	$U = 5 \cdot 05$
Roof: 4in reinforced concrete, screed and felt	$U = 0 \cdot 59$	Roof: 0·1m reinforced concrete, screed and felt	$U = 3 \cdot 35$
Floor: concrete on hardcore	$U = 0 \cdot 20$	Floor, concrete on hardcore	$U = 1 \cdot 13$
Ventilation rate		Ventilation rate	

$N = 3$	$N = 3$
$t_1 = 68°\text{F}$	$t_1 = 20°\text{C}$
$t_e = 32°\text{F}$	$t_e = 0°\text{C}$

$_\Delta t = t_1 - t_e = 68 - 32$　　$_\Delta t = t_1 - t_e = 20 - 0$
$= 36$ degF　　　　　　　　　$= 20$ degC

In terms of formula 1, where $Q_c = A \times U \times {}_\Delta t$, here we have the sum of the (A × U) values multiplied by $_\Delta t$ to get the conduction heat loss:

Imperial units
$Q_c = [(9 \times 30 \times 0 \cdot 3) + (9 \times 10 \times 0 \cdot 89) + (10 \times 10 \times 0 \cdot 59) + (10 \times 10 \times 0 \cdot 2)] \times 36 = (81 + 80 + 59 + 20) \times 36 = 240 \times 36 = 8640$ Btu/h

SI *units*
$Q_c = [(2 \cdot 7 \times 9 \times 1 \cdot 7) + (2 \cdot 7 \times 3 \times 5 \cdot 05) + (3 \times 3 \times 3 \cdot 35) + (3 \times 3 \times 1 \cdot 13)] \times 20 = (41 \cdot 3 + 41 + 30 \cdot 2 + 10 \cdot 2) \times 20 = 122 \cdot 7 \times 20 = 2454$W

In terms of formula 4, where $Q_v = 0 \cdot 36 \times V \times N \times {}_\Delta t$ (the constant being 0·02 for imperial units) we find the ventilation heat loss:

$Q_v = 0 \cdot 02 \times 10 \times 10 \times 9$ $\times 3 \times 36 = 1945$ Btu/h	$Q_v = 0 \cdot 36 \times 3 \times 3 \times 2 \cdot 7$ $\times 3 \times 20 = 525$W
total $Q = Q_c + Q_v =$ 8640 + 1945	total $Q = Q_c + Q_v =$ 2454 + 525 = 2979W
= 105 85 Btu/h	**= 2·979 kW**

2.5 Statutory requirements Some statutory regulations prescribe a definite thermal insulation performance for the enclosing constructions of buildings. The most comprehensive such set of performance specifications is contained in *The Building Standards (Scotland) Regulations, 1963.* (The English requirements are similar, but not as detailed.) For residential buildings the following minimum standards are prescribed:

Minimum standards for thermal insulation	ft²h degF/Btu	**Btu/ft²h degF**	m² degc/W	**W/m² degC**
Roof (including ceiling)		max 0·20		**Max 1·13**
$1/f_1 + 1/f_e$ to be taken as	0·85		0·149	
Wall having no window		max 0·30		**Max 1·70**
$1/f_1 + 1/f_e$ to be taken as	1·00		0·176	
Wall including window, average		max 0·42		**Max 2·38**
The following U values are to be used in computation:				
single glazing		1·00		5·67
double glazing		0·50		2·84
When average for windows and openings is more than		0·75		4·25
walls are to be taken as not less than		0·20		1·13
When average for windows and openings is less than		0·75		4·25
walls may be taken as not less than		0·10		0·58
Floor if underside is ventilated		max 0·20		**Max 1·13**
$1/f_1 + 1/f_e$ to be taken as	1·00		0·176	

The Thermal Insulation (Industrial Buildings) Regulations, 1958 prescribe the following minimum standard for the insulation of factory roofs:

Roof of factory buildings		max 0·30		**Max 1·70**
$1/f_1 + 1/f_e$ to be taken as	0·85		0·149	

3 Heating

3.1 Radiators When the heating requirement for a room has been established in terms of kW (kilowatts, as in example 5) the installation can be designed to issue heat at the same rate.

Radiators usually emit around 1·85 Btu/ft²h degF*. This has the same dimension as the transmittance (U value) thus the same SI unit can be used: W/m² degC. The same conversion factor applies (5·67) thus the heat emission of radiators is around 10·5 W/m² degC. The necessary radiator surface can be found by the formula

$$S = \frac{Q}{e \times \Delta t} \qquad \textbf{(formula 5)}$$

where

Q = heat loss in W

S = radiator surface area in m²

e = emission of radiator, average value 10·5 W/m² degC, but catalogues should be consulted for the actual value

Δt = temperature difference between the heating medium (water) in the radiator and the room air, in degC

Eg if the room for which the heat loss has been calculated in example 5 as 2·979 kW, is to be heated by a hot water radiator, and we assume an average water temperature of 80°C, thus Δt is 80 − 20 = 60 degC, radiator area needed is:

$$S = \frac{Q}{e \times \Delta t} = \frac{2979}{10·5 \times 60} = 4·728 m²$$

A suitable size radiator can be chosen from catalogues. Often radiator outputs are shown in catalogues in Btu/h for the given surface area and an assumed temperature difference. This can easily be converted into SI units: 1 Btu/h = 0·293W and the radiator can be chosen without any calculation, on the basis of heating requirement (heat loss).

*Based on the assumption that there is a 100 deg temperature difference between the surface of the radiator and the air in the space.

3.2 Boilers The output rating of boilers, which has so far been given in terms of Btu/h, should now be given in watts (W) or kilowatts (kW). Using the same conversion factor as above (1 Btu/h = 0·293W) any given data can be converted into SI units.

One catalogue lists the following boiler output ratings:

Btu/h	**kW**	Btu/h	**kW**
25 000	7·33	3 000 000	880
50 000	14·65	4 000 000	1170
100 000	29·30	5 000 000	1470
250 000	73·25	6 500 000	1900
500 000	146	8 600 000	2520
1 200 000	351	10 500 000	3080
1 500 000	440	11 500 000	3370
1 800 000	528	14 000 000	4100
2 100 000	600	17 000 000	4980
2 500 000	730	21 000 000	6150

The required boiler can be chosen on the basis of the sum total of the heat loss from all spaces served by the boiler.

$$Q_b = \frac{\Sigma Q}{E_s} \qquad \textbf{(formula 6)}$$

where

Q_b = boiler output rating in kW

ΣQ = sum of all heat losses in kW

E_s = efficiency of the system

This last quantity depends on many factors, such as heat loss from pipework and so on. A consulting engineer's advice should be sought.

Pressure rating of boilers has so far been given in lb/in² (psi). It must be remembered that the lb is used here as pound force (and not as a mass unit), meaning the gravitational force acting on one pound mass.

The SI unit of force is the newton (N) which is 1kg m/s², ie the force which causes unit acceleration (1 m/s²) of unit mass (kg). (See para **1.2.**)

The unit of pressure is the newton per square metre: N/m²: 1lb/in² = 6895 N/m² = 6·895 kN/m²*.

Thus boilers can be classified as follows:

low pressure boilers up to 15 lb/in² = 103 kN/m²
high pressure boilers from 20 lb/in² = 138 kN/m²
high pressure boilers up to 150 lb/in² = 1030 kN/m²

The 'head of water' as a pressure unit, can be retained usefully:

1ft head of water = 0·434 lb/in² = 3   kN/m²
1m head of water = 1·424 lb/in² = 9·8 kN/m²

The specific heat of water, ie the energy required to elevate the temperature of unit mass of water by one degree is 1 Btu/lb degF = 1 kcal/kg degC = 4·187 kJ/kg degC which is, in fact, specific heat capacity.

The following graph gives useful guidance for boiler room sizes as a function of the total heating requirement given in terms of kW.

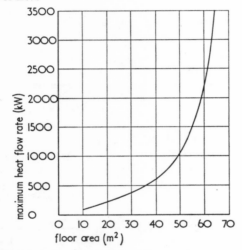

Graph to determine size of boiler room

3.3 Fuels When electricity is used for heating, the appliance can be selected on the basis of formula 6. Here the Q_b will mean the power rating of the appliance in kW.

Eg for the room calculated in example 5, the appliance required will be:

Q = 2·979kW
E = the efficiency of, say, a thermo-fan: 0·98 (98 per cent)

$$Q_b = \frac{Q}{E} = \frac{2·979}{0·98} = 3·039 \text{kW}$$

An appliance with a similar rating (say 3kW) can be selected.

When other types of fuels are used and the boiler has already been chosen, the fuel requirement can be calculated using the formula:

$$F = \frac{Q_b \times 3600}{E_b \times C} \qquad \textbf{(formula 7)}$$

where

F = fuel requirement in kg/h
E_b = boiler efficiency (catalogue or consultant's advice)
C = calorific value of fuel in kJ/kg

The approximate fuel requirement can be found on the basis of the total heat loss (Q) but in this case the overall efficiency factor must be used in the same formula: $E = E_b \times E_s$ (boiler efficiency × system efficiency).

* An alternative to N/m² is the bar (= 10⁵ N/m²). The bar is not an accepted SI unit but is allowable in practice

4 Air-conditioning

As in the previous part of this handbook, so in air-conditioning, the comprehensive treatment of calculation and design methods is beyond the scope of the present notes. Only a few typical examples are dealt with which are affected by the change to the metric (SI) system.

4.1 Heat gain Computation of heat gains is similar to the heat loss calculations, only the direction of the heat flow is the reverse. Solar radiation can be a significant component. The intensity of this was usually given in terms of Btu/ft²h, ie in units of 'density of heat flow rate'. (Btu/h is the heat flow rate and this relative to unit area gives the 'density', see para **1.6**). The SI unit is W/m² and 1 Btu/ft²h = 3·155 W/m².

The *solar constant*, ie the intensity of solar radiation at the outer edges of the earth's atmosphere is 430 Btu/ft²h = 1355 W/m². This is reduced by the optical air mass that the radiation has to penetrate and by the cosine of the angle of incidence. Values can be established as constant for a certain geographical location, in many years' average. Table VI p157 gives computed values of solar radiation incident on vertical planes of different orientations, in England. For windows the values given in table VI can be taken as actually entering through the glass pane. For solid structures or opaque surfaces there is a considerable time-lag for the heat to flow through. Calculation of heat gain is complex and beyond the scope of this section.

4.2 Internal heat gain Besides deliberate heating, the major components of internal heat gain are: lamps, electric motors and human bodies.

Lamps Wattage of all lamps can easily be added up and used in conjunction with other types of heat gain, without any conversion. (Previously this had to be converted into Btu/h, using the conversion factor of 3·412, which is the reciprocal of 0·293, the conversion factor used above.)

Motors Until now the formula $Q = 2544 \times \text{hp}\left(\dfrac{1}{E} - 1\right)$ has been used to get Btu/h, where hp is the horsepower of the motor and E is the efficiency, varying between 0·75 and 0·92. The constant is derived as follows:

1hp = 746W and 1W = 3·412 Btu/h therefore
1hp = 746 × 3·412 = 2544 Btu/h

Working with SI units, if the motor rating in hp is given, the heat produced is

$$Q = 746 \times \text{hp}\left(\frac{1}{E} - 1\right) \qquad \text{(in watts)} \qquad \textbf{(formula 8)}$$

but if the wattage of the motor is known, the formula is simplified

$$Q = W\,(1 - E) \qquad \text{(in watts)} \qquad \textbf{(formula 9)}$$

Human bodies (adult males) produce heat at the following rates:

	Btu/h	**W**
Seated, at rest	390	114
Light work (office)	475	139
Seated, eating	490	144
Walking slowly	550	161
Light bench work	800	234
Medium work, dancing	900	264
Heavy work	1500	440
Exceptionally for short periods	5100	1500

In air-conditioning design the cooling requirement is established as the sum of all types of heat gains, ie:
external heat gain due to hot air and solar radiation;
plus internal heat gain due to lamps, motors, human bodies and any other heat source.

4.3 Conditioners To express the cooling requirement in a building, or to give the cooling capacity of a conditioner or refrigeration plant, the term *ton of refrigeration* has often been used. This is the rate of cooling achieved by 1 ton of $32°F$ ($0°c$) ice when melting into water of the same temperature in twenty-four hours. As the latent heat of fusion of water is 144 Btu/lb and as the ton used in the above expression is the American short ton of 2000lb, 1 ton of

$$\text{refrigeration} = \frac{2000 \times 144}{24} = 12000 \text{ Btu/h.}$$

The dimension is energy/time, the same as for heating capacity, thus the power unit of the si can be used, which is the **watt (W)**
1 ton of refrigeration = $12000 \times 0.293 = 3516W$
for practical purposes = $3.5kW$
It should not cause confusion that the same catalogue may show a kW figure alongside the ton rating. This is not the cooling capacity, but refers to the electric power input of the compressor motor of the cooler. Eg 13.9 ton (= 48.8kW) cooling is produced by a 22.2kW motor.
The contradiction (48.8 > 22.2) is apparent only. Obviously the 48.8kW power cannot be produced by 22.2kW power. The latter, the electric motor, only helps in removing 48.8kJ energy per second.

4.4 Distribution Air velocity in ducts, which has been given customarily in ft/min is now to be measured in metres per second (m/s):
1 ft/min = 0.00508 m/s
thus 100 ft/min = 0.508 m/s
Sensible effects of air movements can be described as follows:
up to 0.25 m/s — unnoticed
0.25–0.50 — pleasant
0.50–1.00 — awareness of air movement
1.00–1.50 — draughty
above 1.50 — annoyingly draughty
Usual air velocities in ducts:
low velocity systems 500 ft/min = 2.5 m/s approx
up to 1500 ft/min = 7.5 m/s approx
high velocity systems 3000 ft/min = 15.0 m/s approx
up to 5000 ft/min = 25.0 m/s approx
Air delivery rate for fans or ducts, measured so far in ft³/min. should now be expressed in m³/s or possibly litres/s:
1 ft³/min = 0.00047 m³/s
thus 1000 ft³/min = 0.47 m³/s
A rough estimate of the duct size required can be obtained from the formula

$$A = \frac{\text{delivery rate}}{v} \qquad \text{(formula 10)}$$

where
A = cross-sectional area of duct in m²
v = air velocity in m/s
and delivery rate is given in m³/s

thus $m^2 = \dfrac{m^3/s}{m/s}$

This formula does not take into account shape factors,

frictional resistances and pressures. For accurate estimate consultants' advice should be sought.

EXAMPLE 6 VENTILATION AND DUCT SIZE
Assume a classroom of 15m × 8m × 3m high, in which three air changes per hour are to be produced:
volume of air in room: $15 \times 8 \times 3 = 360m^3$
air required in an hour: $360 \times 3 = 1080 \text{ m}^3/h$

delivery rate: $\dfrac{1080}{3600} = 0.3 \text{ m}^3/s$

On the basis of this figure a suitable fan can be chosen from catalogues. Selecting a desirable velocity, say 2.5 m/s the approximate duct size can be found from formula 10:

$$A = \frac{0.3}{2.5} = 0.12m^2 \text{ which could be eg } 0.4m \times 0.3m \text{ or}$$

0.6m × 0.2m.
Duct sizes may also be expressed in millimetres, thus:
400mm × 300mm or 600mm × 200mm.

5 Condensation and moisture movement

5.1 Quantity of moisture or vapour, previously measured in grains or pounds, is to be expressed in kilogrammes: **kg**
(probably the submultiples g and μg will be used)

5.2 Vapour pressure or partial pressure of vapour in the air-vapour mixture previously given in atmospheres (Atm), in inches of mercury (in Hg), in pounds-force per square foot (lbf/ft²) or in millibars. All these units are to be replaced by the newton per square metre: **N/m²**

5.3 Permeance (or vapour diffusance) can be defined as the quantity of vapour passing through unit area of a given *construction*, in unit time, when a unit vapour pressure difference exists between the two sides. The following units can be found in existing English literature:
grain/ft²h mbar
grain/ft²h in Hg (often referred to as 'perm')

$$\text{lb/h lbf} \left(\text{simplified by cancellation from } \frac{\text{lb}}{\text{ft}^2\text{h lbf/ft}^2} \right)$$

lb/ft²h Atm
The proposed si unit is μg/Ns
microgramme per newton second)

simplified by cancellation from $\dfrac{\mu g}{m^2 \times N/m^2 \times s}$

The *surface coefficient of vapour transfer* is measured in the same units.

5.4 Vapour resistance is the reciprocal of the above quantity, thus the si unit could be written as $sN/\mu g$, but the accepted rule of the si is that multiple or submultiple prefixes are applied to the whole of the expression only, and not to its parts. Thus the above permeance unit actually means $\mu(g/Ns)$ therefore its reciprocal will be M(Ns/g), normally written as **MNs/g**
(meganewton second per gramme)

5.5 Permeability (or vapour diffusivity) is measured as the amount of vapour passing through unit area of a *material of unit thickness*, when a unit vapour pressure difference exists between the two sides. Units encountered in literature are:

grain in/ft²h mbar

grain in/ft²h inHg (referred to as 'perm-inch')

$$\text{lb ft/h lbf} \left(\text{simplified by cancellation from} \frac{\text{lb} \times \text{ft}}{\text{ft}^2 \times \text{h} \times \text{lbf/ft}^2} \right)$$

lb in/ft²h Atm

The SI unit is: microgramme metre per newton second)

μg m/Ns

5.6 Vapour resistivity is the reciprocal of permeability, and following the same reasoning as for vapour resistance above, the SI unit will be **MNs/g m**

5.7 Heat flow and vapour movement quantities

The above quantities and their handling in calculations are analogous with heat-flow quantities:

	quantity of heat	(J)		quantity of vapour	(kg)
t	temperature	(°C)	vp	vapour pressure	(N/m²)
U	transmittance	(W/m² degc)	P	permeance	(μg/Ns)
f	surface conductance	(W/m² degc)	β	surface coefficient of vapour transfer	(μg/Ns)
R	resistance	(m² degc/W)	R$_v$	vapour resistance	(MNs/g)
k	conductivity	(W/m degc)	δ	permeability	(μgm/Ns)
l/k	resistivity	(m degc/W)	l/δ	vapour resistivity	(MNs/gm)

As the thermal transmittance of a multi-layer construction was:

$$U = \frac{1}{R} \text{ and } R = \frac{1}{f_e} + \frac{b'}{k'} + \frac{b''}{k''} + \ldots \ldots \frac{1}{f_1}$$

so the overall permeance of the same construction is:

$$P = \frac{1}{R_v} \text{ and } R_v = \frac{1}{\beta_e} + \frac{b'}{\delta'} + \frac{b''}{\delta''} + \ldots \ldots \frac{1}{\beta}$$

where b is the thickness of each layer.

5.8 Conversion factors

Permeance	1 grain/ft² h mbar	=	1·935	
	1 grain/ft² h inHg (perm)	=	0·057 15	μg/Ns
	1 lb/h lbf	=	28 321	
	1 lb/ft² h Atm	=	13·383	
Vapour resistance	1 ft² h mbar/grain	=	0·5167	
	1 ft² h inHg/grain	=	17·5	MNs/g
	1 h lbf/lb	=	0·000 039	
	1 ft² h Atm/lb	=	0·0747	
Permeability	1 grain in/ft² h mbar	=	0·049 15	
	1 grain in/ft² h inHg	=	0·00 145	μgm/Ns
	1 lb ft/h lbf	=	8632·24	
	1 lb in/ft² h Atm	=	0·34	
Vapour resistivity	1 ft² h mbar/grain in	=	20·345	
	1 ft² h inHg/grain in	=	689·655	MNs/ gm
	1 h lbf/lb ft	=	0·000 115	
	1 ft² h Atm/lb in	=	2·93	

Necessary data are included in the following tables:
Table VIII Typical permeance and vapour resistance values
Table IX   Some permeability and resistivity values
Table X   Some surface coefficients of vapour transfer and surface resistances of vapour transfer

6 References and sources

INSTITUTION OF HEATING AND VENTILATING ENGINEERS Guide. London, 1965, The Institution CI/SfB (I) (J). *The next edition of the* IHVE *Guide, scheduled for* 1970, *will use* SI *units.*

BUILDING RESEARCH STATION Thermal insulation of buildings: Design data and how to use them. G. D. Nash, J. Comrie, H. F. Broughton. 1955, HMSO CI/SfB (J2)

BUILDING RESEARCH STATION Principles of modern building vol. 1. 1964, HMSO p40-43 CI/SfB (9-) (E1)

BILLINGTON, N. S. Building physics: heat. London, 1967, Pergamon Press CI/SfB (J)

BILLINGTON, N. S. Thermal properties of buildings. London, 1952, Cleaver-Hume CI/SfB (J2)

BUILDING RESEARCH STATION National Building Studies Research Paper 23 Condensation in sheeted roofs. A. W. Pratt. 1962, HMSO, CI/SfB (I6)

BUILDING RESEARCH STATION
Digest 23 (first series) Condensation problems in buildings. 1950, HMSO CI/SfB (I6)
Digest 117 (first series) Condensation and the design of factory roofs. 1958, HMSO CI/SfB 27 (27) (I6)
Digest 132 (first series) Condensation in dwellings. 1960, HMSO CI/SfB 81 (I6)

AMERICAN SOCIETY OF HEATING AND AIR-CONDITIONING ENGINEERS Guide. New York, 1957. The Society p220-238 CI/SfB (I) (J)

BUILDING RESEARCH STATION Digest 91 (second series) Prevention of condensation. 1968, HMSO CI/SfB (I6)

AJ *information sheets*
1232 Office environment. AJ, 1964, February 5 [(82)] CI/SfB 3 (E8)
1275–79 Indoor sports 1–5. AJ, 1964, September 30 [(85)] CI/SfB 562
1280 (revised) Indoor sports 6. AJ, 1967, May 10 [(85)] CI/SfB 562
1281–1285 Indoor sports 7–11. AJ, 1964, September 30 [(85)] CI/SfB 562
1286–1289 Indoor sports spaces. AJ, 1964, October 7 [(85)] CI/SfB 562
1290–93 Social and recreation spaces. AJ, 1964, October 7 [(85)] CI/SfB 562, 53 (1291–1293)
1296 Indoor swimming baths: Internal environment. AJ, 1961, November 11 [(85)] CI/SfB 541 (E8)
1306 Environmental standards for general teaching spaces. AJ, 1964, December 16 [(87)] CI/SfB 70 (E6)
1326 Environmental standards and services for hostel bedrooms. AJ, 1965, April 14 [(88)] CI/SfB 8470 (E8)

See also AJ Handbook **Building environment** sections 4 and 8. This handbook is published in weekly installments in the AJ during the latter half of 1968 and the first half of 1969 For details see p186

Imperial/metric conversion scales for light, heat and mechanics are available from the college registrar, The Polytechnic, College of Architecture and Advanced Building Technology, 309 Regent Street, London W1. Prices vary depending on quantity: single sheets (of which there are four) 1s each

Table I   Recommended room temperatures

	British		German†
	(°F)	(°C)	(°C)
Flats and houses*			
Bedroom (sleeping only)	55–60	13–16	20
Living and dining-room	68–70	20–21	20
Kitchen	60	16	20
Bathroom	60	16	22
Communication spaces. wcs	60	16	15
Hostels, halls of residence			
Bedroom, dormitory (sleeping only)	60	16	20
Study-bedroom, common and reception rooms, sick bay, offices	65	18	20
Lavatories, washrooms	55–60	13–16	15
Bathrooms, showers	65–70	18–21	22
Cloakrooms, entrance lobby, staircases, corridors, linen store	60	16	15
Kitchenettes laundry, drying, dining-room	60	16	20
Office buildings			
Private and general offices	68	20	2
Typing, machine rooms	66	19	20
Conference rooms	68	20	18
Circulation areas	65	18	15
Storage areas	50	10	—
Factories			
Sedentary work	65	18	—
Light work	60	16	—
Heavy work	55	13	—
Educational and recreational buildings			
Teaching rooms	65	18	20
Seminar rooms, offices	65	18	20
Cloakrooms, corridors	55	13	15
Gymnasia	50–55	10–13	15
spectator area	60–65	16–18	—
changing rooms	70	21	22
Billiard, snooker, darts, table games and dance halls	60	16	—
Swimming baths (1 °C above water temperature)	74–80	23–27	—

* These temperatures are higher than the minima laid down by MOHLG for public sector housing. See section 26, p137 part D
† DIN 4701 (1959) Regeln für die Berechnung des Wärmebedarfs von Gebäuden (Rules for the calculation of heating requirements of buildings) as quoted by Reitschel and Raiss in *Der Heiz-und Lüftungstechnik* (Heidelberg, 1962, Springer Verlag)

Table II   Air-to-air transmittance (U value) of various constructions (the lesser this value the better the insulation)

Construction	Btu/ft²h degF	W/m² degC
Walls		
Brick, solid unplastered 4½in (114·3mm)*	0·64	3·64
9in (228·6mm)	0·47	2·67
Brick, solid, plastered both sides 4½in (114·3mm)	0·57	3·24
9in (228·6mm)	0·43	2·44
Concrete, ordinary dense, solid 6in (152·4mm)	0·63	3·58
8in (203·2mm)	0·56	3·18
Stone, medium porous 12 in (304·8mm)	0·50	2·84
18in (457·2mm)	0·40	2·27
Timber tongued and grooved boarding 1in (25·4mm)	0·50	2·84
¼in (6·35mm) asbestos cement sheeting	0·89	5·05
11in (279·4mm) brick wall, fletton facings outer and fletton commons inner skin, inside plastered	0·30	1·70
Do with insulating boards bonded to inside, plastered:		
1in (25·4mm) corkboard	0·15	0·85
½in (12·7mm) fibreboard	0·21	1·19
2in (50·8mm) wood wool slab	0·15	0·85
Do but ⅝in (15·9mm) vermiculite plaster on inside	0·26	1·47
Do but rigid boards on 2in × 1in (50·8 × 25·4mm) battens on inside:		
½in (12·7mm) asbestos board	0·21	1·19
½in (12·7mm) fibreboard	0·17	0·95
⅜in (9·5mm) insulating plasterboard	0·19	1·08
2in (50·8mm) strawboard plastered	0·13	0·74
Do but insulation on battens and plasterboard:		
aluminium foil (½in (12·7mm) corrugated and plain combined)	0·15	0·85
aluminium foil, single sided, paper bonded	0·19	1·08
¾in (19·05mm) glass-, slag- or rock wool quilt	0·13	0·74

* Current sizes of materials and constructions are indicated with exact metric equivalents in brackets

Table II *continued*

Construction	Btu/ft²h degF	W/m² degC
1in (25·4mm) eel grass, glass-, slag- or rock wool quilt	0·12	0·68
Do but inner skin concrete blocks plastered:		
3in (76·2mm) aerated concrete blocks	0·22	1·25
4in (101·6mm) aerated concrete blocks	0·20	1·13
3in (76·2mm) clinker concrete blocks	0·26	1·47
4in (101·6mm) clinker concrete blocks	0·23	1·30
3in (76·2mm) expanded clay concrete blocks	0·24	1·36
4in (101·6mm) expanded clay concrete blocks	0·22	1·25
3in (76·2mm) foamed slag concrete blocks	0·23	1·30
4in (101·6mm) foamed slag concrete blocks	0·21	1·19
10in (254mm) concrete block wall, 4in (101·6mm) blocks, 2in (50·8mm) cavity, outside rendered, inside plastered:		
aerated concrete blocks	0·21	1·19
clinker concrete blocks	0·19	1·08
expanded clay concrete blocks	0·17	0·95
foamed slag concrete blocks	0·16	0·91
9in (228·6mm) single skin hollow concrete block wall outside rendered, inside plastered:		
aerated concrete blocks	0·30	1·70
clinker concrete blocks	0·28	1·59
expanded clay concrete blocks	0·29	1·65
foamed slag concrete blocks	0·26	1·47
8in (203·2mm) in-situ lightweight concrete wall outside rendered, inside plastered:		
clinker concrete	0·23	1·30
foamed slag concrete	0·20	1·13
Corrugated asbestos cement sheets on steel framing	1·15	6·53
Do with lining on inside of steel frame:		
½in (12·7mm) asbestos insulating board	0·49	2·78
½in (12·7mm) fibreboard	0·36	2·04
⅜in (9·5mm) insulating plasterboard	0·37	2·10
2in (50·8mm) strawboard or wood wool slabs	0·21	1·19
3in (76·2mm) aerated concrete blocks	0·37	2·10
Roofs, pitched		
Corrugated asbestos cement sheets	1·40	7·95
with ½in (12·7mm) boarding	0·38	2·16
with 2in (50·8mm) strawboard or wood wool slabs	0·22	1·25
with 1in (25·4mm) quilt on ½in (12·7mm) boarding	0·15	0·85
Corrugated iron sheets	1·50	8·52
Tiles on battens	1·50	8·52
with felt underlay	0·70	3·98
with boarding and felt underlay	0·35	1·98
with plaster ceiling	0·56	3·18
Tiles or slates on boarding and felt, plaster ceiling	0·30	1·70
Aluminium deck, ½in (12·7mm) fibreboard, two layers bitumen felt	0·38	2·16
Do but 2in (50·8mm) wood wool slabs or strawboard, no fibreboard	0·22	1·25
Roofs, flat		
4in (101·6mm) reinforced concrete slab, screed 2½in (63·5mm) to ½in (12·7mm), three layers bitumen felt	0·59	3·35
Do with insulation on the screed:		
1in (25·4mm) cork	0·19	1·08
2in (50·8mm) strawboard or wood wool slabs	0·20	1·13
two ½in (12·7mm) fibreboards	0·22	1·25
Do but lightweight instead of normal screed:		
5in (127mm) to 3in (76·2mm) aerated concrete	0·24	1·36
5in (127mm) to 3in (76·2mm) foamed slag concrete	0·26	1·47
1in (25·4mm) boards on 7in × 2in (177·8 × 50·8mm) joists, three layers bitumen felt, ⅜in (9·5mm) plaster ceiling	0·32	1·82
Do with insulating slabs on boarding:		
1in (25·4mm) cork	0·15	0·85
½in (12·7mm) fibreboard	0·22	1·25
2in (50·8mm) strawboard or wood wool slabs	0·16	0·91
Do with insulation draped over joists:		
aluminium foil (½in (12·7mm) corrugated and plain combined)	0·16	0·91
1in (25·4mm) eel grass, glass-, slag- or rockwool quilt	0·13	0·74
Floors		
Concrete on ground or hardcore fill	0·20	1·13
with grano, terrazzo or tile finish	0·20	1·13
with wood block finish	0·15	0·85
Timber boards on joists, space ventilated one side	0·30	1·70
with parquet, lino or rubber cover	0·25	1·42
Timber boards on joists, space ventilated on more sides	0·40	2·27
with parquet, lino or rubber cover	0·35	1·98
with 1in (25·4mm) fibreboard under boarding	0·19	1·08
with 1in (25·4mm) corkboard under boarding	0·17	0·95
with 1in (25·4mm) corkboard under joists or forming cavity	0·14	0·79
with 2in (50·8mm) strawboard forming cavity (between joists)	0·15	0·85
with double sided aluminium foil draped over joists	0·25	1·42

Table II continued

			Btu/ft²h degF	W/m² degC
Windows				
exposure S sheltered	— single glazing		0·70	**3·97**
	— double, ¼in (6·4mm) space		0·47	**2·67**
	— ¾in (19·1mm) or more space		0·41	**2·32**
S normal, W, SW	— single glazing		0·79	**4·48**
SE sheltered	— double, ¼in (6·4mm) space		0·51	**2·90**
	— ¾in (19·1mm) or more space		0·44	**2·50**
S severe, W, SW,	— single glazing		0·88	**5·00**
SE normal or NW, N,	— double, ¼in (6·4mm) space		0·54	**3·06**
NE, E sheltered	— ¾in (19·1mm) or more space		0·47	**2·67**
W, SW, SE severe,	— single glazing		1·00	**5·67**
NW, N, NE, E normal	— double, ¼in (6·4mm) space		0·58	**3·29**
	— ¾in (19·1mm) or more space		0·50	**2·84**
exposure NW severe	— single glazing		1·14	**6·47**
	— double, ¼in (6·4mm) space		0·63	**3·58**
	— ¾in (19·1mm) or more space		0·53	**3·00**
exposure N severe	— single glazing		1·30	**7·38**
	— double, ¼in (6·4mm) space		0·67	**3·80**
	— ¾in (19·1mm) or more space		0·56	**3·18**

Note The conversion factor from British to SI units is 5·67 (the figure in British units is multiplied by this factor to get the SI value)
The above values are based on Nash: Comrie: Broughton *Thermal insulation of buildings* and on the IHVE Guide, 1955

Table III   Conductivities and resistivities of building materials

Material	Conductivity: k		Resistivity: 1/k	
	Btu in/ ft²h degF	W/m degC	ft²h degF/ Btu in	m degC/W
Mineral wool: felt	0·26	**0·037**	3·85	**27·00**
rigid slab	0·34	**0·049**	2·95	**20·40**
Glass wool: quilt	0·24	**0·034**	4·15	**29·40**
blanket	0·29	**0·042**	3·45	**23·80**
Asbestos, sprayed	0·32	**0·046**	3·13	**21·75**
Eel grass blanket	0·30	**0·043**	3·33	**23·25**
Strawboard	0·65	**0·093**	1·54	**10·75**
Wood wool slab	0·65	**0·093**	1·54	**10·75**
Cork slab	0·34	**0·049**	2·94	**20·40**
Polystyrene foam slab	0·23	**0·033**	4·35	**30·30**
Onozote (expanded ebonite)	0·20	**0·029**	5·00	**34·50**
Asbestos cement sheet:				
light	1·50	**0·216**	0·67	**4·63**
average	2·50	**0·360**	0·40	**2·78**
dense	4·00	**0·576**	0·25	**1·74**
Asphalt	4·00	**0·576**	0·25	**1·74**
Brickwork: commons				
light	5·60	**0·806**	0·18	**1·24**
average	8·40	**1·210**	0·12	**0·83**
dense	10·20	**1·470**	0·10	**0·68**
in lightweight bricks	2·60	**0·374**	0·39	**2·68**
in engineering bricks	8·00	**1·150**	0·12	**0·87**
Concrete, dense	10·00	**1·440**	0·10	**0·69**
clinker aggregate	2·80	**0·403**	0·36	**2·48**
expanded clay aggregate	2·40	**0·345**	0·42	**2·90**
foamed slag aggregate	1·70	**0·245**	0·59	**4·08**
Plasterboard, gypsum	1·10	**0·158**	0·91	**6·33**
Plastering: gypsum	3·20	**0·461**	0·31	**2·17**
vermiculite	1·40	**0·201**	0·71	**4·98**
Rendering, sand-cement	3·70	**0·532**	0·27	**1·88**
Stone: granite	20·30	**2·920**	0·05	**0·34**
limestone	10·60	**1·530**	0·09	**0·65**
sandstone	9·00	**1·295**	0·11	**0·77**
Timber: softwood	0·96	**0·138**	1·04	**7·25**
hardwood	1·11	**0·160**	0·90	**6·25**
Plywood	0·96	**0·138**	1·04	**7·25**
Wood chipboard	0·75	**0·108**	1·33	**9·26**
Wood fibre softboard	0·45	**0·065**	2·22	**15·38**

Average values, based on IHVE *Guide*, 1965

Table IV   Conductances and resistances for unventilated cavities

Cavity type	Conductance: C		Resistance: 1/C	
	Btu/ft²h degF	W/m² degC	ft²h degF/ Btu	m² degC/ W
Vertical				
⅛in (3·2mm) * wide	2·56	**14·50**	0·39	**0·069**
¼in (6·4mm) wide	1·54	**8·74**	0·65	**0·114**
½in (12·7mm) wide	1·24	**7·04**	0·81	**0·142**
¾in (19·1mm) wide	1·17	**6·63**	0·85	**0·151**
1in (25·4mm) wide	1·15	**6·52**	0·87	**0·153**
1½in (38·1mm) wide	1·15	**6·52**	0·87	**0·153**
Horizontal, 3in (76·2mm)				
Heat flow up	1·32	**7·48**	0·76	**0·133**
Heat flow down	0·94	**5·32**	1·06	**0·188**
Values generally used in UK				
For normal 2in (50·8mm)				
cavity walls	1·00	**5·67**	1·00	**0·176**
lined with aluminium foil	0·50	**2·84**	2·00	**0·352**

* Current sizes of materials and constructions are indicated with exact metric equivalents in brackets

Table V   Surface conductances and resistances

Surface	Conductance: f		Resistance: 1/f	
	Btu/ft²h degF	W/m² degC	ft²h degF/ Btu	m² degC/ W
Internal surfaces: (f_i)				
Walls	1·43	**8·12**	0·70	**0·123**
Floor, ceiling, heat flow up	1·67	**9·48**	0·60	**0·105**
Floor, ceiling, heat flow down	1·18	**6·70**	0·85	**0·149**
Underside of roof	1·67	**9·48**	0·60	**0·105**
External surfaces: (f_e)				
Walls S:				
sheltered	1·37	**7·78**	0·73	**0·128**
normal	1·76	**10·00**	0·57	**0·100**
severe exposure	2·32	**13·18**	0·43	**0·076**
Walls W, SW, SE:				
sheltered	1·76	**10·00**	0·57	**0·100**
normal	2·32	**13·18**	0·43	**0·076**
severe exposure	3·33	**18·90**	0·30	**0·053**
Walls NW:				
sheltered	2·32	**13·18**	0·43	**0·076**
normal	3·33	**18·90**	0·30	**0·053**
severe exposure	5·55	**31·50**	0·18	**0·032**
Walls N, NE, E:				
sheltered	2·32	**13·18**	0·43	**0·076**
normal	3·33	**18·90**	0·30	**0·053**
severe exposure	14·30	**81·20**	0·07	**0·012**
Roofs:				
sheltered	2·50	**14·20**	0·40	**0·070**
normal	4·00	**22·70**	0·25	**0·044**
severe exposure	10·00	**56·70**	0·10	**0·018**

Figures based on Nash: Comrie: Broughton *Thermal insulation of buildings*

Table VI   Solar radiation incident on vertical surfaces of eight different orientations, in UK, in W/m²

May 21 and July 23

Hours	6	7	8	9	10	11	12	13	14	15	16	17	18
N	72	38	41	44	47	47	47	47	47	47	47	47	72
NE	126	227	264	246	186	135	107	85	72	66	57	50	38
E	154	284	378	412	384	284	214	154	126	104	85	69	53
SE	69	91	202	302	356	381	356	296	217	160	113	91	69
S	22	35	44	107	176	236	274	300	296	274	224	154	104
SW	22	35	41	44	47	60	170	262	337	384	384	337	226
W	22	35	41	44	47	47	47	79	164	277	372	428	425
NW	22	35	41	44	47	47	47	47	53	110	195	264	296

April 20 and August 24

Hours	6	7	8	9	10	11	12	13	14	15	16	17	18
N	28	28	35	41	44	47	47	47	44	41	38	38	28
NE	101	183	211	198	151	107	85	69	60	53	47	41	31
E	148	274	365	400	372	290	208	148	120	101	82	66	50
SE	79	98	218	322	390	420	390	324	240	173	126	98	79
S	16	31	60	138	230	249	359	391	388	359	293	198	135
SW	16	28	35	41	44	66	186	290	368	422	422	368	252
W	16	28	35	41	44	47	47	79	160	268	366	416	410
NW	16	28	35	41	44	47	47	47	44	88	157	211	240

Based on IHVE *Guide*, 1965

Table VII   Absorptivity of surfaces for solar radiation

Surface	Absorptivity
Black, non-metallic	0·85 to 0·98
Red brick, tile, concrete, stone, dark paint	0·65 to 0·80
Yellow and buff brick and stone	0·50 to 0·70
White or cream brick, tile and paint, plaster	0·30 to 0·50
Window glass	transparent
Bright aluminium, gilt, bronze paint	0·30 to 0·50
Dull brass, copper, aluminium, galvanised steel	0·40 to 0·65
Polished brass, copper	0·30 to 0·50
High polished aluminium, tin, nickel or chromium	0·10 to 0·40

Note that the solar radiation absorbed by a building surface is a function of its colour and surface finish which can be expressed in terms of absorptivity. Unit absorptivity is obtained from a matt black surface.

Table IX   Permeability and vapour resistivity

	Ranges of values	
	Permeability (μgm/Ns)	Vapour resistivity MNs/g m
Air, still	0·182	5·5
Mineral woo	0·168	6
Concrete	0·005 to 0·035	200 to 28·3
Brickwork	0·006 to 0·042	167 to 23·8
Cement render	0·010	100
Plaster	0·017	59
Plasterboard	0·017 to 0·023	59 to 43·5
Urea formaldehyde foam	0·031 to 0·053	32·3 to 18·9
Fibreboard	0·02 to 0·07	50 to 14·3
Wood wool slab	0·024 to 0·07	41·6 to 14·3
Strawboard	0·014 to 0·022	71 to 45·5
Hardboard	0·001 to 0·002	1000 to 500
Timber: air dry	0·014 to 0·022	71 to 45·5
wet	0·001 to 0·008	1000 to 125
Plywood	0·002 to 0·007	500 to 143
Corkboard	0·003 to 0·004	333 to 250
Expanded polystyrene	0·002 to 0·007	500 to 143
Polyurethane foam: closed cell	0·001	1000
open cell	0·035	28·6
Expanded ebonite (Onozote)	0·000 02 to 0·000 09	50 000 to 11 100

Table X   Surface coefficients

	Surface coefficient of vapour transfer (μg/Ns)	Surface vapour resistance (MNs/g)
Still air (if thermal surface conductance (f) = 4·5 W/m² degC)	25·5	0·039
Moving air (if thermal surface conductance (f) = 11·4 W/m² degC)	62·3	0·016
Moving air (if thermal surface conductance (f) = 17 W/m² degC)	96·3	0·010

Table VIII   Permeance and vapour resistance

	Permeance (μg/Ns)		Vapour resistance (MNs/g)	
Concrete blockwork, 200mm, hollow	0·14		7·15	
Brickwork, 100mm	0·04	to 0·06	25	to 16·70
Sand-cement render or screed, 25mm: 4:1	0·67		1·49	
1:1	0·40		2·50	
Plasterboard, 10mm	2·00	to 2·86	0·50 to 0·35	
Plaster on lath: 25mm	0·63		1·59	
20mm	0·83		1·20	
12mm	0·93		1·08	
Wood wool slab, 25mm	3·08	to 4·14	0·32 to 0·24	
Fibreboard: 12mm	1·20	to 3·34	0·83 to 0·30	
25mm	0·93	to 2·68	1·08 to 0·37	
Strawboard, 50mm	0·13	to 0·26	7·70 to 3·85	
Softwood (pine): 25mm	0·08		12·50	
12mm	0·10	to 0·17	10	to 5·90
Plywood, 6mm: external quality	0·026	to 0·041	38·50 to 24·40	
internal quality	0·106	to 0·370	9·43 to 2·70	
Corkboard, 25mm	0·40	to 0·54	2·50 to 1·85	
Bitumen impregnated paper	0·09		11·10	
Roofing felts	0·01	to 0·23	100	to 4·35
Bitumen laminated fibreboard, 12mm	0·007		143	
Polythene film, 0·06mm	0·004		250	
Aluminium foil	0·0001	to 0·0057	10 000	to 175
Kraft paper: single	4·54		0·22	
double	2·81		0·36	
three-ply	2·00		0·50	
four-ply	1·74		0·57	
five-ply	1·60		0·62	
Two coats emulsion paint (on fibreboard)	1·71	to 4·85	0·58 to 0·21	
Two coats flat oil paint (on plaster)	0·09	to 0·17	11·10 to 5·90	
Two coats gloss oil paint (on plaster)	0·03	to 0·13	33·40 to 7·70	
Three coats lead or zinc oil paint (on wood)	0·02	to 0·06	50	to 16·70

Note In the US a material is considered to be a vapour barrier if its permeance is not more than 0·067 μg/Ns or if its vapour-resistance is at least 15 MNs/g.

28 **Lighting**

Contents

1 Introduction
This section on lighting is simplified by the fact that the recently published 1968 edition of the IES Code is in SI units, and IES Technical Report 10 deals with the problem of glare (see para 4 below). Also most of the design data currently in use can be re-used in metric terms provided that room dimensions are expressed in metres, as explained below.

The basic reference (in metric units) for artificial lighting is the AJ Handbook **Building environment** Section 9: Electric lighting. This handbook was published in weekly instalments in the AJ during the latter half of 1968 and the first half of 1969. See p 186 reference 53.

2 Definitions and terms
2.1 Luminous intensity is the light-giving power of a source. It is expressed in candelas (cd) which are one of the six basic SI units. It is already in general use in the UK: there is no change.

2.2 Luminous flux is a measure of the flow of light and is measured in lumens (lm). The unit is already in general use in the UK: there is no change.

2.3 Illumination is the light falling on unit area of a surface, formerly measured in lumens/ft^2 and now to be measured in lux which is 1 lm/m^2.

$$1 \text{ lux} = \frac{1}{10 \cdot 76} \text{ lumens/ft}^2 \text{ or } \frac{1}{11} \text{ for mental calculations.}$$

2.4 Luminance is the light emitted by unit area of a surface. Two types of unit are in common use:

(*a*) The brightness of a source of light, eg lighting fitting, is conveniently measured in candelas/unit area: formerly candelas/ft^2 or /in^2; now candelas/m^2. The candela/m^2 is

an SI unit. $1 \text{ cd/m}^2 = \dfrac{1}{10 \cdot 76} \text{cd/ft}^2$.

(*b*) When discussing a non-luminous surface such as a wall or table top it is often convenient to measure its luminance in a unit related to its illumination, eg a matt surface receiving an illumination E with a reflectance ρ has a luminance E ρ measured in suitable units. (This simple expression is not applicable to polished or glossy surfaces.) The imperial version of this type of luminance unit was the foot lambert and referred to a source which was reflecting or emitting 1 lumen for each sq ft of surface. The metric unit is

the apostilb and refers to a surface which emits or reflects 1 lumen/m^2.

The relationship, therefore, is $1 \text{ apostilb} = \dfrac{1}{10 \cdot 76}$ lamberts

(or $\dfrac{1}{11}$ for mental calculations).

Note that the candela/m^2 is an SI unit whereas the apostilb is not. However, despite the disadvantage of having two units it seems better to permit its use on occasions.

1 candela/m^2 = π apostilbs (π = 3·142)

3 Design calculations
In calculation of illumination and so on the present (imperial) calculation is completely unchanged and all utilisation coefficients, room indices and other current data can be retained.

Where dimensions of rooms are in metres, the answer will be given in lux or apostilbs.

This applies to the lumen method of design, luminance design and point by point calculations.

4 Glare
The IES glare index system is substantially unchanged except that the terms in the basic formula representing luminance of the source and background will now be in apostilbs (equal to 0·0929 ft lamberts) or candelas/m^2. This necessitates the use of a multiplying factor to obtain the same glare constant as with imperial units and, in practice, when using the IES glare index tables, different values are required for the correction terms for the luminous area of the source and for the height above eye-level.

Tables giving these correction terms for metric dimensions are included in the appendix to IES Technical Report 10. A circular calculator for glare index is available from Equipment News Ltd, 35 Red Lion Square, London, WC1. Both metric and imperial versions are available price 58s each. The values given by this calculator are not exactly the same as by the IES method but many consider it to be good enough.

5 References and sources
ILLUMINATING ENGINEERING SOCIETY Recommendations for lighting building interiors. London, 1968, The Society *Popularly known as the IES Code.* CI/SfB (N)

AJ Handbook of Building environment, section 9 *Electric lighting*. AJ 1969; 4, 11, 18 June; 9, 16, 23, 30 July. CI/SfB (N8)

ILLUMINATING ENGINEERING SOCIETY Technical Report 10 The IES glare index system and the evaluation of discomfort glare in artificial lighting. London, 1967, The Society CI/SfB (N)

29 Sound

Contents

1 Introduction

The metrication of acoustics is very simple, since none of the units is changed. The 'hertz' is just an alternative name for 'cycles per second':

1 hertz (Hz) = 1 cycle per second (c/s)

Though decibels have not changed, their reference quantities have been converted to SI units.

For sound power, the reference quantity is 10^{-12} watts

$$\text{Sound power (dB)} = 10 \log_{10} \frac{\text{actual power (w)}}{10^{-12}}$$

The reference quantity for sound intensity is 10^{-12} W/m²

$$\text{Sound intensity (dB)} = 10 \log_{10} \frac{\text{actual intensity W/m}^2}{10^{-12}}$$

For sound pressure, the reference quantity is 2×10^{-5} N/m²

$$\text{Sound pressure (dB)} = 10 \log_{10} \frac{\text{actual pressure (N/m}^2)}{2 \times 10^{-5}}$$

Other units such as dBA, phons, sones, dBN, are all in standard use on the Continent.

Absorption coefficients, again, are ratios and therefore not affected, except for those listed in Table I.

Table I List of absorption coefficients affected by the metric change showing new values

| | Frequency (Hz) | | | | | | |
	63	125	250	500	1000	2000	4000
Air (per m³)	nil	nil	nil	nil	0·003	0·007	0·02
Audience seated in fully upholstered seats (per person)	0·15	0·18	0·4	0·47	0·45	0·51	0·47
Audience seated in wood or padded seat (per person)		0·16		0·4		0·44	0·4
Seats (unoccupied), fully upholstered (per seat)		0·12		0·28		0·32	0·37
Seats (unoccupied), wood or padded (per seat)		0·08		0·15		0·18	0·2
Orchestral player with instrument (average)	0·18	0·37	0·8	1·1	1·3	1·2	1·1

2 Reverberation time

There are two formulas in common use for calculating the reverberation times of enclosed spaces: Sabine's formula and the Norris-Eyring formula.

The Sabine equation is satisfactory for small and average-sized halls (volume up to about 1000m³). The metric version of the equasion is

$$T = \frac{0·16V}{A}$$

where T is reverberation time in seconds,

 V is room volume in m³,

 A is total room absorption in m² absorption units.*

For halls larger than approximately 1000m³, the Norris-Eyring formula should preferably be used. The metric version of this equation is

$$T = \frac{0·161V}{S(-2·30\log_{10}(1-\overline{\propto}))}$$

where T is reverberation time in seconds,

 V is volume of room in m³,

 S is surface area of room in m²,

 $\overline{\propto}$ is average absorption coefficient of room.

Before this formula can be used, it will be necessary to calculate $\overline{\propto}$, and this is done by using the equation

$$\overline{\propto} = \frac{S_1\propto_1 + S_2\propto_2 + \ldots\ldots + S_n\propto_n}{S_1 + S_2 + \ldots\ldots + S_n}$$

where $\overline{\propto}$ is average of absorption coefficients in the room,

 S_1, S_2 etc are the areas of the specific sound absorbing surfaces in the room,

 $\propto_1$, $\propto_2$ etc are the absorption coefficients of the materials constituting the absorbing surfaces.

---

* The imperial absorption unit (based on one square foot of surface area) is termed a Sabine, and some writers have carried this term over into metric usage, referring to a metric Sabine (m² Sabine) which is based on one square metre of surface area. This seems to invite confusion, and the term *absorption unit* is preferable when metric measurements are being used.

See also AJ Handbook **Building environment** Section 5. This handbook was published in weekly instalments in the AJ during the latter half of 1968 and the first half of 1969. See p186 reference 53.

3 Standards of performance

Simple definitions are unchanged. Thus a 9in brick wall, plastered both sides and weighing about 100 lb/ft² will still give an average sound insulation of 50dB even though it may be described as a 230mm brick wall, plastered both sides, and weighing about 488 kg/m². On the other hand, there are more detailed specifications as, for example, insulation 'grade' curves, which are national in origin. Thus, the grade 1 insulation for party walls in the UK is not precisely the same as a German specification for the same element and, in fact, the specified requirements (where they exist) vary from country to country throughout Europe. An effort is being made by the International Standardisation Organization (ISO) to reach agreement, but nothing has yet been agreed.

Similarly, noise climates and permissible levels are often quoted as NC (noise criterion) numbers. These were invented in the US, but are in current use here, and to some extent on the Continent. Again the ISO is trying to supplant these numbers by a slightly different (although very similar) specification, called noise rating (NR) number.

30 **Structural design**

Contents

Tables

1 Types of unit

The units with which these notes are primarily concerned are those adopted by a resolution of the tenth 'Conférence générale des poids et mesures', and formalised by the eleventh conference under the name 'Système International d'Unités', abbreviated SI.

The system of units at present in use in this country is usually referred to as the 'foot-pound-second' system, abbreviated fps.

The changeover to metric will make use of Continental references and textbooks easier in this country. The units used in these vary both from SI and also from one country to another. While all countries are supposed to adopt the SI units in due course, it seemed desirable to give examples of the common metric units now current, and these are referred to as 'Metric Technical' or MT.

1.1 LENGTH

System	Units
fps	yard (yd); foot (ft); inch (in)
MT	metre (m); centimetre (cm); millimetre (mm)
SI	metre (m); millimetre (mm)

Conversions

1yd = 0·9144m	1m = 1.094yd = 3.281ft = 39.37in
1ft = 0·0348m = 30·48cm	1cm = 0.3937in
1in = 2·54cm = 25·4mm	1mm = 0.0394in

1.2 AREA

System	Units
fps	sq yard (sq yd); sq foot (sq ft); sq inch (sq in)
MT	sq metre (m²); sq centimetre (cm²) sq millimetre (mm²)
SI	sq metre (m²); sq millimetre (mm²)

Conversions

1 sq yd = 0·836 127 m²	1m² = 1·196 sq yd = 10·764 sq ft
1 sq ft = 0·092 903m²	1cm² = 0·155 sq in
1 sq in = 6·4516cm² = 645·16mm²	1mm² = 0·001 55 sq in

Note 1m² = 10 000cm² = 1 000 000mm²
$\quad$ 1cm² = 100mm²

1.3 VOLUME

System	Units
fps	cubic yard (cu yd); cubic foot (cu ft) cubic inch (cu in)
MT	cubic metre (m³); litre or cubic decimetre (l or dm³) cubic centimetre (cm³)
SI	cubic metre (m³); cubic millimetre (mm³)

Conversions

1 cu yd = 0·764 555m³	1m³ = 1.308 cu yd = 35.315 cu ft
1 cu ft = 0·028 317m³	1cm³ = 0.061 024 cu in
1 cu in = 16·3871cm³ = 16 387·1mm³	1mm³ = 0.000 061 cu in

Note 1m³ = 1000 litre = 1 × 10⁶ cm³ = 1 × 10⁹ mm³
$\quad$ 1 litre = 1000cm³
$\quad$ 1cm³ = 1000mm³

1.4 MASS

System	Units
fps	ton (ton); kip (kip); pound (lb)
MT	tonne (t); kilogramme (kg)
SI	kilogramme (kg)

Conversions

1 ton = 1·016 047t = 1016·0469kg
1 kip = 0·453 59t = 453·59kg = 1000lb
1lb = 0·453 59kg
1t = 0.984 2065 tons = 2.204 62 kip = 2204.62lb
1kg = 2.204 62lb
1 lb/ft run = 1·488 16 kg/m   1 kg/m = 0.671 969 lb/ft
1 lb/sq ft = 4·882 43 kg/m²   1 kg/m² = 0.204 816 lb/sq ft
Note 1t = 1000kg and 1kip = 1000lb

1.5 FORCE

System	Units
fps	ton-force (tonf); kip-force (kipf): pound-force (lbf)
MT	tonne-force (tf); kilogramme-force (kgf)
SI	meganewton (MN); kilonewton (kN); newton (N)

Conversions fps—MT

as in 'Mass' above with the addition of the -force (f) suffix
in each case

Conversions fps—SI

1 tonf = 9964 02N = 9·964 kN	1MN = 100 361 tonf = 224.809 kipf
1 kipf = 4448·22N = 4·448kN	1kN = 224.809 lbf
1lbf = 4·448N	1N = 0.225 lbf

Conversions MT—SI

1 tf = 9·806 65kN	1MN = 101·972 tf
1 kgf = 9·806 65N	1kN = 101·972 kgf
	1N = 0·101 972 kgf
Also 1 lb/ft run = 14·593 N/m	1 N/m 0.068 525 lb/ft
1 lb/sq ft = 47·880 N/m²	1 N/m² = 0.020 885 lb/sq ft

1.6 STRESS AND PRESSURE (force divided by area)

System	Units
fps	tonf/sq in; tonf/sq ft; lbf/sq in
MT	tf/cm²; kgf/cm²; hectobar (hbar)
SI	N/m²; kN/m²; N/mm² (strictly this should be written MN/m² but the alternative has been adopted as standard)

Conversions fps—MT

1 tonf/sq in = 0·157 488 tf/cm² = 1·544 43 hbar
1 tonf/sq ft = 1·094   kgf/cm² = 1·073 mbar
1 lbf/sq in = 0·070 31 kgf/cm² = 68·948 mbar
1 tf/cm² = 6.3497 tonf/sq in
1 kgf/cm² = 14.2233 lbf/sq in
1 mbar = 0.014 504 lbf/sq in
1 bar = 14.504 lbf/sq in
1 hbar = 1.450 38 kipf/sq in = 0.647 49 tonf/sq in = 93.239 tonf/sq ft

Conversions fps—SI

1 tonf/sq in = 15·444 3 N/mm²
1 tonf/sq ft = 107·3 ' kN/m²
1 lbf/sq in = 6894·8 N/m²
1 N/mm² = 0.064 749 tonf/sq in
       = 9.323 850 tonf/sq ft
       = 145.038 lbf/sq in

Conversions MT—SI

1 tf/cm² = 98·0665N/mm²
1 kgf/cm² = 98 066·5 N/m²
1 mbar = 100 N/m²
1 bar = 0·1 N/mm²
1 hbar = 10 N/mm²
1 N/mm² = 10·1972 kgf/cm²
       = 10 bar
Note 1 tonf/sq ft = 15·556 lbf/sq in
       1 bar = 1·019 72 kgf/cm²

1.7 MOMENT (force multiplied by length)

System	Units
fps	tonf-ft; tonf-in; kipf-in; lbf-ft; lbf-in
MT	kgfm; tfm
SI	Nm; kNm; MNm

Conversions fps—MT

1 tonf-ft = 0·309 691 tfm	1 tfm = 3.229 01 tonf-ft
1 tonf-in = 25·808 kgfm		= 38.748 12 tonf-in
1 kipf-in = 11·521 kgfm		= 86.7966 kipf-in
1 lbf-ft = 0·138 255 kgfm	1 kgfm = 7.233 lbf-ft
1 lbf-in = 0 011 521 kgfm		= 86.796 lbf-in

Conversions fps—SI

1 tonf-ft = 3037·03Nm	1MNm = 329.269 tonf-ft
1 tonf ft = 253·09Nm		= 3951.228 tonf-in
1 kipf-in = 112·985Nm		= 850.75 kipf-in
1 lbf-ft = 1·355 82Nm	1Nm = 0.7376 lbf-ft
1 lbf-in = 0·112 985Nm		= 8.8508 lbf-in

Notes on para 1

1 Yards, square yards and cubic yards are used in structural engineering only for calculating costs (see para 1.1, 1.2 and 1.3).

2 The kilogramme-force unit is also referred to as the kilopond (kp), and the tonne-force unit as the megapond (Mp) (see 1.5).

3 For an explanation of the use of the newton see para 2 Force.

4 The newton per square metre (N/m²) is also referred to as the pascal (Pa) (see 1.6).

5 The abbreviation for litre is l, but as this is liable to be confused with the figure 1, the alternative lt may be used, or it may be written out in full.

6 Full conversion tables for lbf, kgf and newtons are contained in Supplement 1 (1967) to British Standard 350: Part 2: 1962 *Additional tables for SI conversion*. PD 6203, 20s. See also p201, 202 of this Handbook.

2 Force

The unit of force in the SI system of units is the newton (N). This is defined as the force required to give a mass of one kilogramme an acceleration of one metre per second per second (1 m/s²). As weight is a force this is also measured in newtons, and the weight of a kilogramme mass is generally taken as 9·806 65N.

2.1 Mass and weight It is important to distinguish clearly between mass and weight, as in SI they are measured in different units and are not even numerically equal.

Mass is that property of a body that never changes (except in nuclear physics). It can be measured by comparing the weight of a body with the weight of a known mass on a pair of scales, and it is measured in kilogrammes (kg).

Weight, on the other hand, is a force that is proportional to the mass of the body and to the intensity of the gravitational field. It is reduced by moving away from the centre of the earth. Transference from the earth to, for example, the moon will reduce the weight of a given mass to a fraction of its terrestrial value. Weight can be directly measured with a spring balance; and the SI unit is, as already mentioned, the newton. Continental engineers however have been in the habit of using the kilogramme-force unit (kgf) which is called the kilopond (kp). This is similar to the pound-force unit in that the weight in kgf units is numerically equal to the mass of the body in kg under standard gravity (ie normal terrestrial conditions). In the calculations that follow, the Metric Technical column will use the kgf unit. It is often difficult to decide whether the property of the body in which one is primarily interested is its weight or its mass.

In normal commerce one is not normally concerned with the forces that a body imposes on a structure, but in the bulk of that body. For example, a shipper of tea is primarily interested in the amount of tea in the consignment, and his customer in the number of cups of tea that can be made from it. Therefore it is the mass of the tea that both of them are concerned about, and this will be measured in kilogrammes.

When an engineer is designing a warehouse, however, he is concerned with the weight of the goods to be stored. If he is told that the client wishes to store say. 765 kg/m² he must convert this into a force of 7·5 kN/m² to include in his calculations. The 'loads' on a structure should in general be considered as masses, the weights of which are the 'forces' on that structure. Unfortunately, there is at present no consistent ruling on this point, and in some cases the term 'load' is taken to be synonymous with 'weight'. This is true of the new BS CP 3: Chapter V: Part 1: 1967 where all the loadings are given in force units*. On the other hand, tables giving the densities of common building and other materials in kg/m³ are often headed 'Weights of materials'.

2.2 Stresses Once in the realm of stresses in the materials of construction, there is no doubt that we are talking about forces. All stresses must be in force units. The standard SI unit is the newton per square metre, N/m², but this unit is equivalent to only 0.000 145 lb/sq in. It is therefore too small for structural calculations. Even the kN/m² is only 0.145 lb/sq in; so the accepted unit is the MN/m², but this is to be written in the form N/mm² although it is not really correct in SI units to have in the denominator anything other than a basic unit. The N/m² can be called a pascal (Pa), so that the N/mm² would be the megapascal (MPa). However this is not in common use and for the moment is not recommended.

2.3 The bar unit It should be noted that in certain instances, for example in the steel industry, there is already a considerable use of a unit called the bar. The value of this unit is within 2 per cent of the value of the kgf/cm² and has therefore won considerable support on the Continent. It may well be adopted in this country in certain fields in preference to the N/mm². The millibar (mbar or mb) is already accepted as the international unit of atmospheric pressure. The other multiple that may on occasion be employed is the hectobar (hbar) which is equivalent to 1·019 72 kgf/mm² or 0.647 49 tonf/sq in.

3 Loading

Table I shows uses and loads for a variety of buildings and structures. It has been extracted from BS CP 3: Chapter V: Part 1: 1967.

Part 1 is concerned with dead and imposed loads. Part 2, to be published later, will deal with wind loads.

Where no figures are given for concentrated load, it can be assumed that the tabulated distributed load is adequate for design purposes.

Table II shows mass densities in kg/m³ for a wide range of materials.

Table III gives safe bearing capacities of soils.

Table I Uses and loads

Use to which building or structure is to be put	Intensity of distributed load			Concentrated load to be applied, unless otherwise stated, over any square with a 300mm (1ft) side		
	kN/m²	kgf/m²	lbf/ft²	kN	kgf	lbf
Art gallery (see 'Museum floors') **Assembly buildings** such as public halls and theatres, but excluding drill halls, places of worship, public lounges, schools and toilet rooms:						
with fixed seating†	4·0	408	83.5	—	—	—
without fixed seating	5·0	510	104	3·6	367	809
Balconies	Same as the rooms to which they give access			1·5 per metre run	153 per metre run	103 per foot run concentrated at the edge

*Now generally agreed to express loads in force units (newtons) rather than mass units (kilogrammes)

†Fixed seating implies that removal of the seating and use of the space for other purposes is improbable

Table 1 Uses and loads *continued*

Use to which building or structure is to be put	Intensity of distributed load			Concentrated load to be applied, unless otherwise stated, over any square with a 300mm (1ft) side		
	kN/m²	kgf/m²	lbf/ft²	kN	kgf	lbf
Banking halls	3·0	306	62.7	—	—	—
Bedrooms:						
Domestic buildings	1·5	153	31.3	1·4	143	315
Hotels and motels	2·0	204	41.8	1·8	184	405
Institutional buildings	1·5	153	31.3	1·8	184	405
Billiard rooms	2·0	204	41.8	2·7	275	603
Boiler rooms	7·5	765	157	To be determined		
Book stores	2·4 for each metre of storage height	245 for each metre of storage height	15.3 for each foot of storage height	To be determined		
Broadcasting studios:						
Corridors (see 'Corridors')						
Dressing-rooms	2·0	204	41.8	1·8	184	405
Fly galleries	4·5kN per metre run uniformly distributed over the width	459kgf per metre run	308lbf per foot run	—	—	—
Grids	2·5	255	52.2	—	—	—
Stages	7·5	765	157	4·5	459	1012
Studios	4·0	408	83.5	—	—	—
Toilet rooms	2·0	204	41.8	1·4	143	315
Bungalows	1·5	153	31.3	1·0 at 1·0m centres	102 at 1·0m centres	225 at 3ft centres
Catwalks	Concentrated loads only			0·9 on any joist	91·8	202
Ceilings	Concentrated loads only					
Chapels and churches	3·0	306	62.7	2·7	275	603
Cinemas (see 'Assembly buildings and 'Broadcasting studios')						
Classrooms	3·0	306	62.7	2·7	275	603
Clubs:						
Assembly areas with fixed seating*	4·0	408	83.5	—	—	—
Assembly areas without fixed seating	5·0	510	104	3·6	367	809
Bedrooms	1·5	153	31.3	1·8	184	405
Billiard rooms	2·0	204	41.8	2·7	275	603
Corridors (see 'Corridors')						
Dining-rooms	2·0	204	41.8	2·7	275	603
Kitchens	To be determined but not less than 3·0	306	62.7	4·5	459	1012
Lounges	2·0	204	41.8	2·7	275	603
Laundries	3·0	306	62.7	4·5	459	1012
Toilet rooms	2·0	204	41.8	—	—	—
Cold storage	5·0 for each metre of storage height, with a minimum of 15·0	510 for each metre of storage height, with a minimum of 1530	31.8 for each foot of storage height, with a minimum of 313	To be determined		
Colleges:						
Assembly areas with fixed seating*	4·0	408	83.5	—	—	—
Assembly areas without fixed seating	5·0	510	104	3·6	367	809
Bedrooms	1·5	153	31.3	1·8	184	405
Classrooms	3·0	306	62.7	2·7	275	603
Corridors (see Corridors)						
Dining-rooms	2·0	204	41.8	2·7	275	603
Dormitories	1·5	153	31.3	1·8	184	405
Gymnasia	5·0	510	104	3·6	367	809
Kitchens	To be determined but not less than 3·0	306	62.7	4·5	459	1012
Laboratories, including equipment	To be determined but not less than 3·0	306	62.7	To be determined but not less than 4·5	459	1012
Stages	5·0	510	104	3·6	367	809
Toilet rooms	2·0	204	41.8	—	—	—
Corridors, hallways, passageways, aisles, public spaces and footbridges between buildings:						
Buildings subject to crowd loading, except grandstands	4·0	408	83.5	4·5	459	1012
Buildings subject to loads greater than from crowds, including wheeled vehicles, trolleys, and the like	To be determined but not less than 5·0	510	104	To be determined but not less than 4·5	459	1012
All other buildings	Same as the rooms to which they give access					
Dance halls	5·0	510	104	3·6	367	809
Department stores:						
Shop floors for the display and sale of merchandise	4·0	408	83.5	3·6	367	809
Dormitories	1·5	153	31.3	1·8	184	405
Drill rooms and drill halls	5·0	510	104	To be determined but not less than 9·0	918	2023
Driveways and vehicle ramps other than in garages for the parking only of passenger vehicles and light vans not exceeding 2500kg (2½tons) gross weight	To be determined but not less than 5·0	510	104	To be determined but not less than 9·0	918	2023
Dwellings	1·5	153	31.3	1·4	143	315
Factories and similar buildings	5·0; 7·5 or 10·0 as appropriate	510; 765 or 1020	104; 157 or 209	To be determined		
File rooms in offices	5·0	510	104	To be determined		
Flats	1·5	153	31.3	1·4	143	315
Footpaths, terraces and plazas leading from ground level:						
No obstruction to vehicular traffic	5·0	510	104	9·0	918	2023
Used only for pedestrian traffic	4·0	408	83.5	4·5	459	1012
Foundries	To be determined but not less than 20	2040	418	—	—	—
Garages:						
Car parking only, for passenger vehicles and light vans not exceeding 2500kg (2½tons) gross weight including driveways and ramps	2·5	255	52.2	9·0	918	2023
All repair workshops for all types of vehicles and parking for vehicles exceeding 2500kg (2½tons) gross weight, including driveways and ramps	To be determined but not less than 5·0	510	104	Worst possible combination of wheel loads		
Grandstands:						
Assembly areas with fixed seating*	4·0	408	83.5	—	—	—
Assembly areas without fixed seating	5·0	510	104	3·6	367	809
Corridors and passageways	5·0	510	104	4·5	459	1012
Toilet rooms	2·0	204	41.8	—	—	—

* Fixed seating implies that removal of the seating and use of the space for other purposes is improbable

Table I Uses and loads *continued*

Use to which building or structure is to be put	Intensity of distributed load			Concentrated load to be applied, unless otherwise stated, over any square with a 300mm (1ft) side		
	kN/m²	kgf/m²	lbf/ft²	kN	kgf	lbf
Gymnasia	5·0	510	104	3·6	367	809
Halls:						
Corridors, hallways and passageways (see 'Corridors')						
Dressing-rooms	2·0	204	41.8	1·8	184	405
Fly galleries	4·5kN per metre run uniformly distributed over the width	459kgf per metre run	308lbf per foot run	—	—	—
Grids	2·5	255	52.2	—	—	—
Projection rooms	5·0	510	104	—	—	—
Stages	5·0	510	104	3·6	367	809
Toilet rooms	2·0	204	41.8	—	—	—
Hospitals:						
Bedrooms and wards	2·0	204	41.8	1·8	184	405
Corridors, hallways and passageways (see 'Corridors')						
Dining-rooms	2·0	204	4.18	2·7	275	603
Kitchens	To be determined but not less than 3·0	306	62.7	4·5	459	1012
Laundries	3·0	306	62.7	4·5	459	1012
Toilet rooms	2·0	204	41.8	—	—	—
Utility rooms	2·0	204	41.8	4·5	459	1012
X-ray rooms and operating theatres	2·0	204	41.8	4·5	459	1012
Hotels and motels:						
Bars and vestibules	5·0	510	104	—	—	—
Bedrooms	2·0	204	41.8	1·8	184	405
Corridors, hallways and passageways (see 'Corridors')						
Dining-rooms	2·0	204	41.8	2·7	275	603
Kitchens	To be determined but not less than 3·0	306	62.7	4·5	459	1012
Laundries	3·0	306	62.7	4·5	459	1012
Lounges	2·0	204	41.8	2·7	275	603
Toilet rooms	2·0	204	41.8	—	—	—
Houses	1·5	153	31.3	1·4	143	315
Indoor sporting facilities:						
Areas for equipment	To be determined but not less than 2·0	204	41.8	To be determined		
Assembly areas with fixed seating*	4·0	408	83.5	—	—	—
Assembly areas without fixed seating	5·0	510	104	3·6	367	809
Corridors (see 'Corridors')						
Dressing-rooms	2·0	204	41.8	1·8	184	405
Gymnasia	5·0	510	104	3·6	367	809
Toilet rooms	2·0	204	41.8	—	—	—
Institutional buildings:						
Bedrooms	1·5	153	31.3	1·8	184	405
Communal kitchens	To be determined but not less than 3·0	306	62.7	4·5	459	1012
Corridors, hallways and passageways (see 'Corridors')						
Dining-rooms	2·0	204	41.8	2·7	275	603
Dormitories	1·5	153	31.3	1·8	184	405
Laundries	3·0	306	62.7	4·5	459	1012
Lounges	2·0	204	41.8	2·7	275	603
Toilet rooms	2·0	204	41.8	—	—	—
Kitchens other than in domestic buildings, including normal equipment	To be determined but not less than 3·0	306	62.7	4·5	459	1012
Laboratories, including equipment	To be determined but not less than 3·0	306	62.7	4·5	459	1012

*Fixed seating implies that removal of the seating and use of the space for other purposes is improbable

Table I Uses and loads *continued*

Use to which building or structure is to be put	Intensity of distributed load			Concentrated load to be applied, unless otherwise stated, over any square with a 300mm (1ft) side		
	kN/m²	kgf/m²	lbf/ft²	kN	kgf	lbf
Landings	Same as the floors to which they give access					
Laundries other than in domestic buildings, excluding equipment	To be determined but not less than 3·0	306	62.7	4·5	459	1012
Libraries:						
Reading-rooms without book storage	2·5	255	52.2	4·5	459	1012
Rooms with book storage (e.g. public lending libraries)	4·0	408	83.5	4·5	459	1012
Stack rooms	2·4 for each metre of room height with a minimum of 6·5	245 for each metre of room height with a minimum of 663	15.3 for each foot of room height with a minimum of 136	To be determined		
Dense mobile stacking on mobile trucks	To manufacturer's recommendations					
Corridors	4·0	408	83.5	4·5	459	1012
Toilet rooms	2·0	204	41.8	—	—	—
Machinery halls circulation spaces therein	4·0	408	83.5	To be determined		
Maisonettes	1·5	153	31.3	1·4	143	315
Motor rooms, fan rooms and the like, including weight of machinery	To be determined but not less than 7·5	765	157	To be determined		
Museum floors and art galleries for exhibition purposes	To be determined but not less than 4·0	408	83.5	To be determined		
Offices:						
Corridors and public spaces (see 'Corridors')						
Filing and storage spaces	5·0	510	104	To be determined		
Offices for general use	2·5	255	52.2	2·7	275	603
Offices with computing, data processing and similar equipment	3·5	357	73.1	To be determined		
Toilet rooms	2·0	204	41.8	—	—	—
Pavement lights	To be determined but not less than 5·0	510	104	1½ times the wheel load but not less than 9·0	918	2023
Places of worship	3·0	306	62.7	2·7	275	603
Printing plants:						
Paper storage	To be determined but not less than 4·0 for each metre of storage height	408 for each metre of storage height	25.5 for each foot of storage height	To be determined		
Type storage and other areas	To be determined but not less than 12·5	1275	261	To be determined		
Public halls (see 'Halls')						
Public lounges	2·0	204	41.8	2·7	275	603
Residential buildings such as apartment houses, boarding houses, guest houses, hostels, lodging houses and residential clubs, but excluding hotels and motels:						
Bedrooms	1·5	153	31.3	1·8	184	405
Communal kitchens	To be determined but not less than 3·0	306	62.7	4·5	459	1012

Table I Uses and loads *continued*

Use to which building or structure is to be put	Intensity of distributed load			Concentrated load to be applied, unless otherwise stated, over any square with a 300mm (1ft) side		
	kN/m²	kgf/m²	lbf/ft²	kN	kgf	lbf
Corridors, hallways and passageways (see 'Corridors')						
Dining-rooms and public rooms	2·0	204	41.8	2·7	275	603
Dormitories	1·5	153	31.3	1·8	184	405
Laundries	3·0	306	62.7	4·5	459	1012
Toilet rooms	2·0	204	41.8	—	—	—
Schools (see 'Colleges')						
Shop floors for the display and sale of merchandise	4·0	408	83.5	3·6	367	809
Stairs:						
Dwellings not over three-storey	1·5	153	31.3	1·8	184	405
All other buildings	Same as the floors to which they give access, but not less than			Same as the floors to which they give access		
	3·0	306	62.7			
	and not more than					
	5·0	510	104			
Stationery stores	4·0	408	25.5	To be determined		
	for each for each for each metre of metre of foot of storage storage storage height height height					
Storage other than types listed separately	To be determined but not less than			To be determined		
	2·4	245	15.3			
	for each for each for each metre of metre of foot of storage storage storage height height height					
Television studios (see 'Broadcasting studios')						
Theatres (see 'Assembly buildings' and 'Broadcasting studios')						
Universities (see 'Colleges' and 'Libraries')						
Warehouses (see 'Storage')						
Workrooms, light, without storage	2·5	255	52.2	1·8	184	405
Workshops (see 'Factories')						

Table II Mass densities of Materials

Material	kg/m³
Adamantine clinkers, stacked	2082
Aerated concrete	801–961
Aggregates:	
coarse	1522
fine	850
Alabaster	2691
Alcohol:	
absolute	785
commercial proof spiri	913
wood—barrels	449
Alluvium, undisturbed	1602
Aluminium:	
cast	2771
rolled	2675
bronze	7545
DTD alloys	2675–2787
paint	1201
paste	1474
powder	721–801
sheet, per mm thickness	kg/m²
	2·8
	kg/m³
sulphate, bags	721
Ancaster stone	2499
Animal food, cases	400
Anthracite, broken	805
Antimony:	
pure	6680
ore, bags	1442
Apples, barrels	400
Argentine	7208
Asbestos:	
crude	897
fibre, cases	673
natural	3044
pressed	961
cement	1922–2082
sand	961
felt	150
Ash:	
English	689
Canadian	737
Ashes, dry	641
Asphalt:	
natural	1009
paving	2082
Automatic machines, cases	160
Automobiles, cases	128
Aviation spirit	753
Axles and wheels	513
Baggage	128
Ballast:	
loose, graded	1602
undisturbed	1922
Balsa wood	112
Bamboo	352
Barbed wire	384
Barium oxide, solid	4645–5446
Barley:	
grain	705
bags	593
ground	529
Barrels, empty	128
Bars, steel, bundles	2723
Basic slag, crushed	1794
Bath stone	2082
Baths, iron, cases	208
Bauxite	2563
crushed	1281
ore, bags	1201
Beech	769
Beer	1025
bottled, cases	449
barrels	529
Beeswax	961
Bell metal	8490
Benzene	881
Benzol	881
Bicycles, crates	128
Birch:	
American	641
logs	449
squares	625
yellow	705

Material	kg/m³
Bitumen:	
natural	1089
prepared	1362
emulsion	1121
Blood	1057
dried, casks	561
Bolts and nuts, bags	1201
Bone	1762–2002
manure, bags	513
meal, bags	801
Books:	
on shelves	641
bulk	961
Boots and shoes, cases	384
Bottled goods, cases	897
Bottles, empty. crates	416
Boxwood	929
Brass:	
cast	8330
rolled	8570
casks	721
tubes, bundles	897
Brewer's grains:	
wet	497
desiccated	256
Bricks:	
(common burnt clay)	
stacked	1602–1920
sand cement	1840
sand lime	2080
ballast	1200
brickwork	1920
British Columbia pine	529
Bronze:	
cast	8330
drawn, sheet	8794
Cadmium	8618
Calcium carbide, solid	2211
Canvas, bales	769
Carpets, rolls	256
Casein	1346
Casks, empty	128
Cedar, western red	384
Celluloid	1346–1602
Goods, cases	160
Cement:	
bags	1281
bulk	1281–1442
casks	961
slurry	1442
Chalk	1602–2723
broken, barrels	961
Cheese, cases	513
Cherry wood	721
Chestnut:	
horse	513
sweet	561
Chromium	7096
Cigarettes, cases	240
Cinders	641
Clay: Fill:	
dry, lumps	1041
dry, compact	1442
damp, compact	1762
wet, compact	2082
undisturbed	1922
undisturbed, gravelly	2082
china, compact	2243
Clinker, furnace	1025
Coal:	
loose lumps	897
slurry	993
Cobalt	8586
Coke	481–561
Columbian pine	529
Concrete: cement, plain:	
aerated	961
brick aggregate	1840
clinker	1440
stone ballast	2240
Concrete: cement, reinforced:	
1 per cent steel	2370
2 per cent steel	2420
5 per cent steel	2580

Table II Mass densities of materials *continued*

Material	kg/m³
Copper:	
cast	8762
drawn or sheet	8938
ingots	3588
Cork	128–240
bales	80
Corn, bulk	721
Cotton:	
raw, compressed	400–577
bales, American	272
pressed bales, Egyptian	529
Cupro-nickel (60 per cent to 80 per cent Cu)	8938
Cypress wood	593
Deal, yellow	432
Delta metal	8602
Diatomaceous brick	481
Diesel oil	881
Doors, crates	320
Douglas fir	529
Dry goods, average	481
Duralumin	2787
Dutch clinkers, stacked	1602
Dynamite	1233
Earth:	
dry, loose	1280
dry, compact	1550
moist, loose	1440–1600
moist, compact	1760–1840
Earthenware, packed	320
Ebonite	1201–1281
Ebony	1185–1330
Elm:	
American	673
Canadian	673
Dutch	577
English	577
wych	689
Felt:	
hair	272
roofing, rolls	593
Fibreboard	160–400
Files, etc, cases	897
Fir:	
Douglas	529
silver	481
Firebrick, Stourbridge	2002
Fish, boxes	721
Flint	2563
Flour	705
sacks	641
barrels	545
Foam slag	700
Forest of Dean stone	2435
Freestone	2243–2483
masonry, dressed	2403
rubble	2243
Fuller's earth, natural	1762–2403
Galvanised sheets, bundles	897
Glass:	
bottle	2723
common green	2515
crown, extra white	2451
silicate	2195
flint, best	3076
heavy	4966–5927
optical	3524
plate	2787
crates	801
Pyrex	2243
bottles, crates	416
refuse (broken)	1522
silk	160–208
Gold	19318
Grain:	
barley	625
oats	416
rye	721
Granite	2643
chippings	1442
dressed, cases	2243

Material	kg/m³
Granolithic	2243
Gravel:	
loose	1602
undisturbed	1922–2162
Gunmetal:	
cast	8458
rolled	8794
Gunpowder	897
Gypklith	449
Gypsum:	
crushed	1041–1602
bags	833
solid	2563
plaster	737
Hardcore	1922
Hemlock, western	497
Hiduminium	2803
Hoggin	1762
Hosiery, cased	224
Ice	913
Implements, agricultural bundles	256
Indiarubber	1121
Iroko	657
Iron:	
cast	7208
malleable cast	7368–7497
wrought	7689
corrugated, bundles	897
pig,	
random	2723
stacked	4485
pyrites,	
ground	2883
solid (60 per cent Fe)	4806–5126
sulphate, powdered	1121
wire, coils	897
Ironstone:	
Cleveland, lumps	2162
Spanish, lumps	2403
Swedish, lumps	3684
Ironmongery, packages	897
Ironwood	1137
Ivory	1842
Jointing compo, for tanks	801
Jute:	
bales	481
bales, compressed	641
Kentish rag	2675
crushed	1602
Kupfernickel	7208–7609
Larch wood	593
Lead:	
cast or rolled	11325
pigs	3588
bronze (Cu 70 Pb 30)	9771
red, powder	2082
white, powder	1378
paste in drums	2787
Leather	961
hides, compressed	368
rolls	160
Lime:	
acetate of, bags	1281
Blue Lias,	
ground	849
lump	993
carbonate of, barrels	1281
chloride of, lead lined cases	449
grey chalk, lump	705
grey stone, lump	881
hydrate, bags	513
hydraulic	721
quick, ground	1025
slaked,	
ground, dry	561
ground, wet	1522
Lime mortar:	
dry	1650
wet	1746

Material	kg/m³
Lime wood	561
American	416
Linoleum, rolls	481
Loam (sandy clay):	
dry, loose	1201
dry, compact	1602
wet, compac'	1922
Logwood	913
Macadam	2082
Magnesia, solid	2403
Magnesite	3044
Magnesium	1730
alloys, about	1842
Magnetic oxide of iron	4966
Magnetite	4966
Mahogany:	
African	561
Honduras	545
Spanish	689
Manganese	7368
bronze	8602
Manganite	4325
Maple:	
Canadian	737
English	689
Marble	2595–2835
Mastic	1121
Mercury	13536
Mica	2723–3044
Millstone grit	2323
Molybdenum	9980
Mortar:	
cement, set	1922–2082
lime, set	1602–1762
Mud	1762–1922
Muntz metal:	
cast	8394
sheet	8922
Nails, wire, bags	1201
Neoprene	1201
Nickel	8810
silver	8730
Oak:	
African	961
American red	721
white	769
Austrian	721
English	801–881
Ore. See individual kinds	
Oregon pine	529
Padauk	785
Paint:	
aluminium	1201
bituminous emulsion	1121
red lead	3123
red lead dispersed	1522
white lead	2803
zinc	2403
Paper:	
blotting, bales	400
printing, reels	897
wall, rolls	384
writing	961
Paraffin:	
oil	801
wax	897
Peat:	
dry, stacked	561
sandy, compact	801
wet, compact	1362
Perspex	1346
Peruvian bark, bales	240
Petrol	689–769
cans or drums	721–801
Pine:	
American red	529
British Columbian	529
Christiania	689
Columbian	529
Dantzig	577
Memel	545

Material	kg/m³
Pine—cont:	
Kauri, Queensland	481
New Zealand	609
Oregon	529
pitch	657
Riga	545–753
Pipes:	
brass, bundles	897
cast iron, stacked	961–1281
earthenware, loose	320
salt-glazed, stacked	400
wrought iron,	
stacked ⅜in (9·5mm)	3204
3in (76·2mm)	1442
6in (152·4mm)	801
Pitch	1089
Plaster of Paris:	
loose	929
set	1281
Platinum	21465
Plywood	481–641
plastic-bonded	721–1442
Polystyrene	1057
Polvinyl chlor. acetate	1201–1346
Poplar	449
Porcelain	2323
Porphyry	2803
Portland cement:	
loose	1201–1362
bags	1121–1281
drums	1201
Portland stone	2243
Potatoes	641
Pulp, wood:	
dry	561
wet	721
Pumice stone	481–913
Purbeck stone	2707
Pyrites:	
iron,	
ground	2883
solid (60 per cent Fe)	4806–5126
copper, solid	4085–4325
Quartz	2643
loose	1442–1682
Quartzite	2723
Quicklime, ground, dry	1025
Quilt, eel grass	176
Ragstone	2403
Rails, railway	2403
Redwood:	
American	529
Baltic	497
non-graded	432
Rhodesian	913
Resin:	
lumps	1073
barrels	769
Resin bonded plywood	721–1362
Resin oil	993
Rubber:	
crepe, cases	400
processed sheet	1121
raw	929
sponge	48–160
vulcanised	1201
Salt, bulk	961
Salt-glazed ware	2243
Sand:	
saturated	1922
undisturbed dry	1682
saturated	2002
Satinwood	961
Sawdust	208
Screws, iron, packages	1602
Sea water	1009–1041
Shale	2563
granulated	1121
oil, Scottish	945
Silica, fused transparent	2211
translucent	2050
Silicon, pure	291
Silk, bales	352
Silver:	
cast	10444
pure	10492
glance	7208

Table II Mass densities of materials *continued*

Material	Kg/m³	Material	Kg/m³	Material	Kg/m³	Material	Kg/m³
Sirapite, powder	1025	Stone:		Tar	1137–1233	White lead:	
Slag:		Ancaster	2499	barrels	801	powder	1378
coarse	1442	Bath	2082	Tarmacadam	2082	paste in drums	2787
granulated	961	Caen	2002	Tarpaulins, bundles	721	paint	2803
Slag wool	224–288	Darley Dale	2371	Teak, Burma African	657	White metal	7368
Slate:		Forest of Dean	2435	Terracotta	1794	Whitewood	465
Welsh	2803	freestone	2243–2483	Tetraethyl lead	1602	Willow:	
Westmorland	2995	granite	2643	Timbers, See individual kinds		American	577
Sludge cake, pressed, 50 per cent		Ham Hill	2162	Tinned goods, cases	481–641	English	449
water	929	Hopton Wood	2531	Tinplate, boxes	3204–4485	Wine:	
Snow:		Kentish rag	2675	Tinstone	6407–7048	bulk	977
fresh	96	Mansfield	2259	Tinware, cases	192	bottles in cases	593
wet compact	320	marble	2723	Titanium	4485	Wire	
Soap, boxed	913	millstone grit	2323	oxide, solid	3684	iron, coils	1185
Soapstone	2723	Portland	2243	Tools, hand, cases	897	nails, bags	1201
Soda, bags	657	Purbeck	2707	Treetex	208	rod, coils	801
Solder, pigs	2723	slate, Welsh	2803	Tubes, see 'Pipes'		rope, coils	1442
Soot	352	Westmorland	2995	Tungsten	19222	Wolfram (Wolframite)	7368
Spar:		York	2243	Tyres, rubber	176–256	Wood block paving	897
calcareous	2723	Stoneware	2243			Wool:	
feld	2691	Straw:		Vanadium	5991	compressed bales	769
fluor	3204	pressed	96	Varnish:		uncompressed	208
Spirits of wine	785	compressed bales	304	barrels	593		
Sponge rubber	48–160	Strawboards, bundles	593	tins in cases	721	Yew	673–801
Spruce:		Strontium white:				York stone	2243
Canadian	465	solid	3844	Walnut	657		
Norway	465	ground	1762	Waste paper	352	Zinc:	
Sitka	449	Sulphate of:		pressed packed	449–513	cast	6804
Stationery cases	513	aluminium, bags	721	Water:		rolled	7192
Steel:	7833	ammonia, bags	641	fresh	1001	sheets packed	897
balls, barrels	1201	copper, cryst	1346	salt	1009–1201	Zincblende	4085
punchings	4806	iron, powder	1121	Wax:			
		Sulphur, pure solid	1922–2082	bees'	961		
		Sulphuric acid, 100 per cent	1970	Brazil	993		
		commercial	1682–1794	cases of barrels	593		
		Sycamore	609	paraffin	897		

Table III Safe bearing capacities of soils
from British Standard Code of Practice CP 101 : 1963 table I

	FPS	MT(1)	MT(2)	SI
	tonf/sq ft	tf/m²	kgf/cm²	kN/m²
Rocks				
1 Igneous and gneissic rocks in sound condition	100	1090	109	10700
2 Massively-bedded limestones and hard sandstones	40	440	44	4300
3 Schists and slates	30	330	33	3200
4 Hard shales, mudstones and soft sandstones	20	220	22	2200
5 Clay shales	10	110	11	1100
6 Hard solid chalk	6	66	6·6	650
7 Thinly-bedded limestones and sandstones }	To be assessed after inspection			
8 Heavily shattered rocks and the softer chalks }				
Non-cohesive soils				
9 Compact well graded sands and gravel-sand mixtures:				
dry	4 to 6	44 to 66	4·4 to 6·6	430 to 650
submerged	2 to 3	22 to 33	2·2 to 3·3	220 to 320
10 Loose well graded sands and gravel-sand mixtures:				
dry	2 to 4	22 to 44	2·2 to 4·4	220 to 430
submerged	1 to 2	11 to 22	1·1 to 2·2	110 to 220
11 Compact uniform sands:				
dry	2 to 4	22 to 44	2·2 to 4·4	220 to 430
submerged	1 to 2	11 to 22	1·1 to 2·2	110 to 220
12 Loose uniform sands:				
dry	1 to 2	11 to 22	1·1 to 2·2	110 to 220
submerged	½ to 1	5·5 to 11	0·55 to 1·1	55 to 110
Cohesive soils				
13 Very stiff boulder clays and hard clays with a shaly structure	4 to 6	44 to 66	4·4 to 6·6	430 to 650
14 Stiff clays and sandy clays	2 to 4	22 to 44	2·2 to 4·4	220 to 430
15 Firm clays and sandy clays	1 to 2	11 to 22	1·1 to 2·2	110 to 220
16 Soft clays and silts	½ to 1	5·5 to 11	0·55 to 1·1	55 to 110
17 Very soft clays and silts	½ to nil	5·5 to 0	0·55 to 0	55 to 0

4 General structural principles

The principles of structural engineering are not at all dependent on the units in which quantities are measured, and so long as the system of units is consistent, it does not matter which are substituted.

The following worked examples illustrate the effect of this.

In each case the relevant formula is quoted. The various quantities to be substituted in the formula are then given in their normal units. They are then converted into the units chosen as consistent for that formula, and the answer is obtained in the same units. Calculations are to slide-rule degree of accuracy.

4.1 Stress Case of uniform stress

FPS
Question What is the compressive stress in a short column 1ft 6in square with a central load of 100 tons?

Answer

$$p = \frac{W}{A}$$

$$W = 100 \text{ tonf}$$

$$A = 18\text{in} \times 18\text{in} = 324 \text{ sq in}$$

$$\therefore \quad p = \frac{100}{324}$$

$$= 0.309 \text{ tonf/sq in}$$

MT
Question What is the compressive stress in a short column 450mm square with a central load of 100 000kg?

Answer

$$p = \frac{W}{A}$$

$$W = 100\ 000 \text{ kgf}$$

$$A = 45\text{cm} \times 45\text{cm} = 2025\text{cm}^2$$

$$\therefore \quad p = \frac{100\ 000}{2025}$$

$$= 49\cdot4 \text{ kgf/cm}^2$$

SI
Question What is the compressive stress in a short column 450mm square with a central load of 100 000kg?

Answer

$$p = \frac{W}{A}$$

$$W = 100\ 000 \times 9\cdot81\text{N}$$

$$= 981\ 000\text{N}$$

$$A = 450\text{mm} \times 450\text{mm}$$

$$= 202\ 500\text{mm}^2$$

$$\therefore \quad p = \frac{981\ 000}{202\ 500}$$

$$= 4\cdot85 \text{ N/mm}^2$$

4.2 Strain is a completely non-dimensional quantity

FPS
Question What is the strain when a 12in long steel bar in tension stretches $\frac{1}{8}$in?

Answer

$$\text{strain} = \frac{e}{L}$$

$$e = \text{extension}$$

$$= 0.125\text{in}$$

$$L = \text{original length}$$

$$= 12\text{in}$$

$$\therefore \quad \text{strain} = \frac{0.125}{12}$$

$$= 0.0104$$

MT and SI
Question What is the strain when a 300mm long steel bar in tension stretches 3mm?

Answer

$$\text{strain} = \frac{e}{L}$$

$$e = \text{extension}$$

$$= 3\text{mm}$$

$$L = \text{original length}$$

$$= 300\text{mm}$$

$$\therefore \quad \text{strain} = \frac{3}{300}$$

$$= 0\cdot01$$

4.3 Bending moment

FPS
Question What is the bending moment at the midpoint of a simply supported beam of 20ft span carrying a load of 5 kip per ft run?

Answer

$$M = \frac{wL^2}{8}$$

$$w = \text{wt per ft run}$$

$$= 5 \text{ kipf/ft}$$

$$L = \text{span}$$

$$= 20\text{ft}$$

$$\therefore \quad M = \frac{5 \times 20^2}{8}$$

$$= 250 \text{ kipf-ft}$$

Bending moments are generally expressed in kipf-in, so this answer must be multiplied by 12

$$M = 250 \times 12$$

$$= 3000 \text{ kipf-in}$$

MT
Question What is the bending moment at the midpoint of a simply supported beam of 6m span carrying a load of 7500kg per metre run?

Answer

$$M = \frac{wL^2}{8}$$

$$w = \text{wt per m run}$$

$$= 7\cdot5 \text{ tf/m}$$

$$L = \text{span}$$

$$= 6\text{m}$$

$$\therefore \quad M = \frac{7\cdot5 \times 6^2}{8}$$

$$= 33\cdot7 \text{ tfm}$$

SI
Question What is the bending moment at the midpoint of a simply supported beam of 6m span carrying a load of 7500kg per metre run?

Answer

$$M = \frac{wL^2}{8}$$

$$w = \text{wt per m run}$$

$$= 7500 \times 9\cdot81 \text{ N/m}$$

$$= 73\cdot7 \text{ kN/m}$$

$$L = \text{span}$$

$$= 6\text{m}$$

$$\therefore \quad M = \frac{73\cdot7 \times 6^2}{8}$$

$$= 331 \text{ kNm}$$

4.4 Deflection

FPS

Question What is the deflection at midspan in the previous example if the beam is a mild steel universal beam 27in $\times$ 10in $\times$ 84lb per ft run?

Answer

$$\text{deflection} = \frac{5}{384} \times \frac{wL^4}{EI}$$

w = wt per ft run
  = 5 kipf/ft
L = span
  = 20ft
E = Young's modulus (see **4.6**)
  = 30 000 kipf/sq in
  = 30 000 $\times$ 12^2 kipf/sq ft
I = second moment of area of section about the neutral axis (obtained from Steel Tables)
  = 2828in^4
  = 2828 $\times$ 12^{-4} ft^4

$\therefore$ deflection

$$= \frac{5}{384} \times \frac{5 \times 20^4 \times 12^3}{30\ 000 \times 2828}$$

$$= 0.0174\text{ft}$$
$$= 0.209\text{i}.$$

MT

Question What is the deflection at midspan in the previous example, if the beam is a mild steel universal beam 678mm $\times$ 253mm $\times$ 125 kg/m?

Answer

$$\text{deflection} = \frac{5}{384} \times \frac{wL^4}{EI}$$

w = wt per m run
  = 7·5 tf/m
L = span
  = 6m
E = Young's modulus (see **4.6**)
  = 2110 tf/cm^2
  = 2110 $\times$ 10^4 tf/m^2
I = second moment of area of section about the neutral axis (obtained from Steel Tables, yellow pages)
  = 117 700cm^4
  = 117 700 $\times$ 10^{-8}m^4

$\therefore$ deflection

$$= \frac{5}{384} \times \frac{7·5 \times 6^4 \times 10^4}{2110 \times 117\ 700}$$

$$= 0·005\ 15\text{m}$$
$$= 0·515\text{cm}$$

SI

Answer

$$\text{deflection} = \frac{5}{384} \times \frac{wL^4}{EI}$$

w = wt per m run
  = 7500 $\times$ 9·81 N/m
  = 73 700 N/m
L = span
  = 6m
E = Young's modulus (see **4.6**)
  = 207 000 N/mm^2
  = 207 000 $\times$ 10^6 N/m^2
I = second moment of area of section about the neutral axis (obtained from Steel Tables, yellow pages)
  = 117 700 $\times$ 10^{-8}m^4

$\therefore$ deflection

$$= \frac{5}{384} \times \frac{73\ 700 \times 6^4 \times 10^2}{207\ 000 \times 117\ 700}$$

$$= 0·005\ 15\text{m}$$
$$= 5·15\text{mm}$$

4.5 Properties of sections

4.5.1 Second moment of area sometimes called the moment of inertia of the section

FPS

Question What is the second moment of area of a timber beam section 9in $\times$ 2in (assume the stated size is the actual size)?

Answer

$$I = \frac{bd^3}{12}$$

b = section breadth
  = 2in
d = section depth
  = 9in

$\therefore$ $I = \dfrac{2 \times 9^3}{12}$

$$= 122\text{in}^4$$

MT

Question What is the second moment of area of a timber beam section 230mm $\times$ 51mm?

Answer

$$I = \frac{bd^3}{12}$$

b = section breadth
  = 5·1cm
d = section depth
  = 23cm

$\therefore$ $I = \dfrac{5·1 \times 23^3}{12}$

$$= 5190\text{cm}^4$$

SI

Answer

$$I = \frac{bd^3}{12}$$

b = section breadth
  = 51mm
d = section depth
  = 230mm

$\therefore$ $I = \dfrac{51 \times 230^3}{12}$

$$= 51·9 \times 10^6\text{mm}^4$$

4.5.2 Section modulus

FPS

Question What is the section modulus of the above section (see **4.5.1**)?

Answer

$$Z = \frac{I}{y_1}$$

$I = 122\text{in}^4$ from above

y_1 = distance from the neutral axis to the extreme fibre

$$= \frac{9\text{in}}{2} = 4.5\text{in}$$

$$\therefore \quad Z = \frac{122}{4.5}$$

$$= 27\text{in}^3$$

MT

Question What is the section modulus of the above section (see **4.5.1**)?

Answer

$$Z = \frac{I}{y_1}$$

$I = 5190\text{cm}^4$ from above

y_1 = distance from the neutral axis to the extreme fibre

$$= \frac{23}{2}\text{ cm} = 11.5\text{cm}$$

$$\therefore \quad Z = \frac{5190}{11.5}$$

$$= 450\text{cm}^3$$

SI

Answer

$$Z = \frac{I}{y_1}$$

$I = 51.9 \times 10^6\text{mm}^4$ from above

y_1 = distance from the neutral axis to the extreme fibre

$$= \frac{230}{2}\text{ mm} = 115\text{mm}$$

$$\therefore \quad Z = \frac{51.9 \times 10^6}{115}$$

$$= 450 \times 10^3\text{mm}^3$$

4.5.3 Radius of gyration

FPS

Question What is the radius of gyration of the above section?

Answer

$$k^2 = \frac{I}{A}$$

$I = 122\text{in}^4$ from above

A = section area
$= 9\text{in} \times 2\text{in}$
$= 18 \text{ sq in}$

$$\therefore \quad k^2 = \frac{122}{18}$$

$$= 6.8\text{in}^2$$

$$\therefore \quad k = 2.6\text{in}$$

MT

Question What is the radius of gyration of the above section?

Answer

$$k^2 = \frac{I}{A}$$

$I = 5190\text{cm}^4$ from above

A = section area
$= 23\text{cm} \times 5.1\text{cm}$
$= 117\text{cm}^2$

$$\therefore \quad k^2 = \frac{5190}{117}$$

$$= 44.2\text{cm}^2$$

$$\therefore \quad k = 6.6\text{cm}$$

SI

Answer

$$k^2 = \frac{I}{A}$$

$I = 51.9 \times 10^6\text{mm}^4$ from above

A = section area
$= 230\text{mm} \times 51\text{mm}$
$= 11\ 700\text{mm}^2$

$$\therefore \quad k^2 = \frac{51.9 \times 10^6}{11\ 700}$$

$$= 4420\text{mm}^2$$

$$\therefore \quad k = 66\text{mm}$$

4.6 Moduli of elasticity

	FPS	MT	SI
	Kipf/sq in	tf/cm²	N/mm²
Concrete	2 000	141	13 900
Steel	30 000	2 110	207 000

Note 1 kipf/sq in = 1000 lbf/sq in

5 Structural steel

Structural steel design at present is controlled by BS 449: 1959 as reset and reprinted in 1965. All the steel stresses in that publication are quoted in tonf/in², and all the tables of stress reductions for slenderness and so on are also directly in tonf/in². Many of the formulae are not non-dimensional and metric quantities cannot be substituted in them. For these reasons it is not recommended to use this Standard with metric design. In any case it is an extremely complicated document, and is intended for use only by professional engineers. The architect who requires to size members in a structure is recommended to refer instead to AJ information sheet 1463 Design of simple mild steel structural members (AJ 15.2.67) [Ab3], part of a larger series on structural design. The following is based on that sheet.

5.1 Bending stress The basic allowable bending stress for joists in mild steel for tension and compression is

FPS	10.5	tonf/sq in
MT	1·65	tf/cm²
SI	162	N/mm²

5.2 Shear stress The allowable maximum intensity of shear stress for mild steel is

FPS	7	tonf/sq in
MT	1·1	tf/cm²
SI	108	N/mm²

5.3 Web crushing To prevent local crushing of the web at the root the bearing stress on B1 (see drawing below) must not exceed

FPS	12	tonf/sq in
MT	1·89	tf/cm²
SI	186	N/mm²

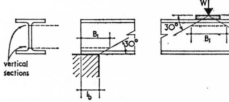

Critical area of web over which local crushing may occur

5.4 Design example

FPS

Question What is the stress in the extreme fibres of a universal beam 10in × 4in × 15lb/ft under a bending moment of 145 tonf-in?

Answer

$$\frac{M}{I} = \frac{f}{y}$$

M = bending moment

= 145tonf-in

I = second moment (from Steel Tables)

= 68.8in⁴

y = distance of extreme fibre from neutral axis

$= \frac{10}{2} = 5$in

$\therefore$ $f = \frac{145}{68.8} \times 5$

= 10.6 tonf/sq in

MT

Question What is the stress in the extreme fibres of a universal beam 254mm × 101·6mm × 22kg/m under a bending moment of 3720 kgfm?

Answer

$$\frac{M}{I} = \frac{f}{y}$$

M = bending moment

= 3720 kgfm

= 372 000 kgf-cm

I = second moment (from yellow pages of Steel Tables)

= 2863cm⁴

y = distance of extreme fibre from neutral axis

$= \frac{25·4}{2} = 12·7$cm

$\therefore$ $f = \frac{372\,000}{2863} \times 12·7$

= 1650 kgf/cm²

SI

Question What is the stress in the extreme fibre of a universal beam 254mm × 101·6mm × 22kg/m under a bending moment of 36·8 kNm?

Answer

$$\frac{M}{I} = \frac{f}{y}$$

M = bending moment

= 36 800 Nm

= 36 800 × 10³ Nmm

I = second moment (from yellow pages of Steel Tables)

= 2863 × 10⁴mm⁴

y = distance of extreme fibre from neutral axis

$= \frac{254}{2} = 127$mm

$\therefore$ $f = \frac{36\,800}{2863} \times 10^{-1} \times 127$

= 163 N/mm²

6 Reinforced concrete

6.1 Concrete strength Modern classification of the strengths of concrete is based on its crushing strength. This is obtained in this country by testing a 6in cube to failure. Continental practice is not uniform, cubes of 100mm and 200mm are used, and so are cylinders 150mm diam × 300mm long (among other sizes). The *Recommendations for an international code of practice* suggest the latter cylinder test as the standard; the cylinder strength can be obtained by multiplying the cube strength by 0·8.
The British Standard is to be a 150mm cube.

Table IV Cube and cylinder strengths of concrete in works tests twenty-eight days after mixing (compare with table I in BS CP 114: 1957)

Mix	Cube strength (150mm cube)			Cylinder strength (150mm diam × 300mm)		
	FPS lbf/sq in	MT kgf/cm²	SI N/mm²	FPS lbf/sq in	MT kgf/cm²	SI N/mm²
1 :2 :4	3000	210	21·0	2400	170	16·5
1:1½:3	3750	265	25·5	3000	210	20·5
	4000	280	27·5	3200	225	22
1 :1 :2	4500	315	30·0	3600	255	25
	5000	350	34·5	4000	280	27·5
	6000	420	41·5	4800	340	33

Table V Permissible concrete stresses (compare with table VI in BS CP 114: 1957)

Concrete mix (nominal)	1:1:2			1:1½:3			1:2:4		
Units	**FPS lbf/sq in**	**MT kgf/cm²**	**SI N/mm²**	**FPS lbf/sq in**	**MT kgf/cm²**	**SI N/mm²**	**FPS lbf/sq in**	**MT kgf/cm²**	**SI N/mm²**
Compressive, direct	1140	80	7·6	950	67	6·5	760	53	5·3
Compressive, due to bending	1500	105	10·0	1250	88	8·5	1000	70	7·0
Shear	130	9·1	0·9	115	8·1	0·8	100	7·0	0·7
Bond, average	150	10·5	1·0	135	9·5	0·93	120	8·4	0·83
Bond, local	220	15·2	1·5	200	14·1	1·4	180	12·6	1·25

6.2 Reinforcement

The following programme has been agreed for manufacture and supply of metric sized steel reinforcement.

Until March 1969 Only imperial sizes will be rolled as standard.

During March, April and May 1969 Both imperial and metric sizes will be rolled as standard.

From June 1969 Only metric sizes will be *rolled* as standard.

From January 1970 Only metric will be *supplied* as standard.

Available sizes are shown in table VI. Length will be expressed in millimetres to the nearest 5mm and the standard length will be 12 000mm (12m). The density of steel will be taken as exactly 0·00785 kg/mm²/m.

Table VI Metric sizes of reinforcing bars

Size (mm)	Area (mm²)	Weight (kg/m)	Approx equivalent inch size
6	28	0·222	¼
8	50	0·395	5/16
10	79	0·617	⅜
12	113	0·888	½
16	201	1·58	⅝
20	314	2·47	¾
25	491	3·86	1
32	804	6·31	1¼
40	1260	9·87	1½
50	1964	15·41	2

See also Table XII p180

6.3 Stiffness of members (cf Clause 309 of BS CP 114: 1957)

Reinforced concrete should possess adequate stiffness to prevent such deflection or deformation as might impair the strength or efficiency of the structure, or produce cracks in finishes or partitions.

For all normal cases it may be assumed that the stiffness will be satisfactory if the ratio of span to overall depth does not exceed the values given in the table below. Where lightweight aggregate concrete is used, these provisions may need modification.

Table VIII Span to depth ratios

	A	B	C
Beams			
Simply supported beams	20	18	17
Continuous beams	25	23	21
Cantilever beams	10	9	8
Slabs			
Slabs spanning in one direction, simply supported	30	27	25
Slabs spanning in one direction, continuous	35	31	30
Slabs spanning in two directions, simply supported	35	31	30
Slabs spanning in two directions, continuous	40	36	34
Cantilever slabs	12	11	10

Notes

Column A Members with steel stresses notmore than 20 kipf/sq in, 1400 kgf/cm², 140 N/mm² and concrete stresses not more than 1500 lbf/sq in, 105 kgf/cm², 10·0 N/mm²

Column B Members with either steel stresses greater than in column A or concrete stresses greater, but not both

Column C Members with both steel and concrete stresses greater than in column A

Table VII Permissible stresses in steel reinforcement (compare with table XI in BS CP 114: 1957)

	Tensile stress in other than shear reinforcement			Tensile stress in shear reinforcement			Compressive stress		
Units	**FPS Kipf/sq in**	**MT kgf/cm²**	**SI N/mm²**	**FPS kipf/sq in**	**MT kgf/cm²**	**SI N/mm²**	**FPS kipf/sq in**	**MT kgf/cm²**	**SI N/mm²**
Mild steel, diameter not exceeding ½in/40mm	20	1400	140	20	1400	140	18	1260	125
Mild steel, diameter exceeding 1½in/40mm	18	1260	125	18	1260	125	16	1130	110
High yield bars f_y not less than 60kipf/sq in, 4200 kgf/cm², 418 N/mm², and diameter not exceeding ⅞in/20mm	33	2320	230	25	1760	175	25	1760	175
High yield bars f_y not less than 60 kipf/sq in, 4200 kgf/cm², 418 N/mm² and diameter exceeding ⅞in/20mm	30	2110	210	25	1760	175	25	1760	175

Table IX Areas of round bar reinforcement (mm²)

Diam (mm)	Weight (kg/m)	Areas in mm² for numbers of bars											
		1	2	3	4	5	6	7	8	9	10	11	12
6	0·222	28	57	85	113	142	170	198	226	255	283	311	340
8	0·395	50	101	151	201	252	302	352	402	453	502	552	604
10	0·617	79	157	236	314	393	471	550	628	707	785	864	942
12	0·888	113	226	339	452	565	678	791	904	1 017	1 130	1 243	1 356
16	1·58	201	402	603	804	1 005	1 206	1 407	1 608	1 809	2 010	2 211	2 412
20	2·47	314	628	942	1 256	1 570	1 884	2 198	2 512	2 826	3 140	3 454	3 768
25	3·86	491	983	1 474	1 966	2 457	2 948	3 439	3 932	4 423	4 915	5 406	5 896
32	6·31	804	1 608	2 412	3 216	4 020	4 824	5 628	6 432	7 236	8 040	8 844	9 648
40	9·87	1 260	2 520	3 780	5 040	6 300	7 560	8 820	10 080	11 340	12 600	13 860	15 120

Diam (mm)	Areas in mm²/m for spacings in mm								
	50	75	100	125	150	175	200	250	300
6	566	376	283	226	188	162	141	113	94
8	1 006	670	503	402	335	287	251	201	168
10	1 570	1 048	785	628	524	449	393	314	262
12	2 262	1 508	1 131	904	754	646	565	452	377
16	4 020	2 680	2 010	1 608	1 340	1 149	1 005	804	670
20	6 284	4 190	3 142	2 514	2 095	1 795	1 571	1 257	1 047
25	9 830	6 552	4 915	3 932	3 276	2 809	2 457	1 966	1 638
32		10 720	8 040	6 432	5 360	4 594	4 020	3 216	2 680
40			12 600	10 080	8 400	7 200	6 300	5 040	4 200

Table X Reinforcement: Cross-sectional areas in sq in: Plain round bars

These bar sizes are no longer available but are included as a reference for the imperial columns of the worked examples

Diam (in)	$\frac{3}{16}$	$\frac{1}{4}$	$\frac{5}{16}$	$\frac{3}{8}$	$\frac{7}{16}$	$\frac{1}{2}$	$\frac{5}{8}$	$\frac{3}{4}$	$\frac{7}{8}$	1	$1\frac{1}{8}$	$1\frac{1}{4}$	$1\frac{3}{8}$	$1\frac{1}{2}$
Cross-sectional areas for specified number of bars (sq in)														
1	0·028	0·049	0·077	0·110	0·150	0·196	0·307	0·442	0·601	0·785	0·994	1·227	1·484	1·767
2	0·055	0·098	0·153	0·221	0·301	0·393	0·614	0·884	1·203	1·571	1·988	2·45	2·97	3·53
3	0·082	0·147	0·230	0·331	0·451	0·589	0·920	1·325	1·804	2·36	2·98	3·68	4·45	5·30
4	0·110	0·196	0·307	0·442	0·601	0·785	1·227	1·767	2·41	3·14	3·98	4·91	5·94	7·07
5	0·138	0·245	0·384	0·552	0·752	0·982	1·534	2·21	3·01	3·93	4·97	6·14	7·42	8·84
6	0·165	0·295	0·460	0·663	0·902	1·178	1·841	2·65	3·61	4·71	5·96	7·36	8·91	10·60
7	0·193	0·344	0·537	0·773	1·052	1·374	2·15	3·09	4·21	5·50	6·96	8·59	10·39	12·37
8	0·221	0·393	0·614	0·884	1·202	1·571	2·45	3·53	4·81	6·28	7·95	9·82	11·88	14·14
9	0·248	0·442	0·690	0·994	1·353	1·767	2·76	3·98	5·41	7·07	8·95	11·04	13·36	15·90
10	0·276	0·491	0·767	1·104	1·503	1·963	3·07	4·42	6·01	7·85	9·94	12·27	14·85	17·67
11	0·304	0·540	0·844	1·215	1·654	2·16	3·37	4·86	6·61	8·64	10·93	13·50	16·33	19·44
12	0·331	0·589	0·920	1·325	1·804	2·36	3·68	5·30	7·22	9·42	11·93	14·73	17·82	21·21
13	0·359	0·638	0·997	1·436	1·954	2·55	3·99	5·74	7·82	10·21	12·92	15·95	19·30	22·97
14	0·387	0·687	1·074	1·547	2·10	2·75	4·30	6·19	8·42	11·00	13·92	17·18	20·79	24·74
15	0·414	0·736	1·151	1·657	2·25	2·95	4·60	6·63	9·02	11·78	14·91	18·41	22·27	26·51
16	0·442	0·785	1·227	1·768	2·41	3·14	4·91	7·07	9·62	12·57	15·90	19·64	23·76	28·27
17	0·469	0·835	1·304	1·878	2·56	3·34	5·22	7·51	10·22	13·35	16·90	20·86	25·24	30·04
18	0·497	0·884	1·381	1·989	2·71	3·53	5·52	7·95	10·82	14·14	17·89	22·09	26·73	31·81
19	0·525	0·933	1·457	2·10	2·86	3·73	5·83	8·39	11·43	14·92	18·89	23·32	28·21	33·57
20	0·552	0·982	1·534	2·21	3·01	3·93	6·14	8·84	12·03	15·71	19·88	24·54	29·70	35·34

	$\frac{3}{16}$	$\frac{1}{4}$	$\frac{5}{16}$	$\frac{3}{8}$	$\frac{7}{16}$	$\frac{1}{2}$	$\frac{5}{8}$	$\frac{3}{4}$	$\frac{7}{8}$	1	
Cross-sectional areas of bars at specified spacings (sq in per ft)											
3in	0·110	0·196	0·307	0·442	0·601	0·785	1·227	1·767	2·405	3·142	A_{st} = cross-sectional area (sq in)
3½in	0·095	0·168	0·263	0·379	0·515	0·673	1·052	1·515	2·060	2·690	
4in	0·083	0·147	0·230	0·331	0·451	0·589	0·920	1·325	1·804	2·356	D = diameter of bar (in)
4½in	0·074	0·131	0·205	0·295	0·401	0·524	0·818	1·178	1·604	2·090	
5in	0·066	0·118	0·184	0·265	0·361	0·471	0·736	1·060	1·443	1·885	**Specified number**
5½in	0·060	0·107	0·167	0·241	0·328	0·428	0·669	0·964	1·312	1·714	N = number of bars
6in	0·055	0·098	0·153	0·221	0·301	0·393	0·614	0·884	1·203	1·571	
6½in	0·051	0·091	0·142	0·204	0·278	0·363	0·566	0·816	1·110	1·450	$A_{st} = \frac{1}{4} \pi D^2 N$
7in	0·047	0·084	0·131	0·189	0·258	0·337	0·526	0·757	1·031	1·346	$= 0\cdot7854\ D^2 N$ (sq in)
7½in	0·044	0·079	0·123	0·177	0·241	0·314	0·491	0·707	0·962	1·257	
8in	0·041	0·074	0·115	0·166	0·225	0·295	0·461	0·663	0·902	1·178	**Specified spacing**
8½in	0·039	0·069	0·108	0·156	0·212	0·277	0·433	0·624	0·849	1·109	S = spacing (or pitch) of bars (in)
9in	0·037	0·065	0·102	0·147	0·200	0·262	0·409	0·589	0·802	1·047	
9½in	0·035	0·062	0·097	0·140	0·190	0·248	0·388	0·558	0·760	0·992	$A_{st} = \frac{12}{S}(0\cdot7854\ D^2)$
10in	0·033	0·059	0·092	0·133	0·180	0·236	0·368	0·530	0·722	0·942	
10½in	0·032	0·056	0·088	0·126	0·172	0·224	0·351	0·505	0·687	0·898	
11in	0·030	0·054	0·084	0·120	0·164	0·214	0·335	0·482	0·656	0·857	$= \frac{9\cdot425}{S} D^2$
12in	0·028	0·049	0·077	0·110	0·150	0·196	0·307	0·442	0·601	0·785	
15in	0·022	0·039	0·061	0·088	0·120	0·157	0·245	0·353	0·481	0·628	(sq in per ft)
18in	0·018	0·033	0·051	0·074	0·100	0·131	0·205	0·295	0·401	0·524	
24in	0·014	0·025	0·038	0·055	0·075	0·098	0·153	0·221	0·301	0·393	

6.4 Cover (cf clause 307 of BS CP 114: 1957)

Reinforcement should have concrete cover and the thickness of such cover (exclusive of plaster or other decorative finish) should be: (*metric figures not necessarily exact equivalents*).

1 for each end of a reinforcing bar, not less than 1in/25mm nor less than twice the diameter of such bar;

2 for a longitudinal reinforcing bar in a column, not less than 1½in/40mm nor less than the diameter of such bar. In the case of columns with a minimum dimension of 7½in/200mm or under, whose bars do not exceed ½in/15mm diameter, 1in/25mm cover may be used;

3 for a longitudinal reinforcing bar in a beam, not less than 1in/25mm nor less than the diameter of such bar;

4 for tensile, compressive, shear or other reinforcements in a slab, not less than ½in/15mm nor less than the diameter of such reinforcement;

5 for any other reinforcement not less than ½in/15mm nor less than the diameter of such reinforcement.

For all external work, for work against earth faces and also for internal work where there are particularly corrosive conditions, the cover of the concrete should not be less than 1½in/40mm for all steel, including stirrups, links and so on, except where the face of the concrete is adequately protected by a suitable cladding or by a protective coating, which may indeed be advisable where the corrosive conditions are unduly severe.

Additional cover may be necessary where lightweight or porous aggregates are used, or to comply with fire resistance requirements.

6.5 Distance between bars (cf Clause 308 of BS CP 114: 1957)

The horizontal distance between two parallel steel reinforcements in reinforced concrete should usually, except at splices, be not less than the greatest of the three following distances:

1 the diameter of either bar if their diameters be equal;

2 the diameter of the larger bar if the diameters be unequal;

3 ¼in/5mm more than the nominal maximum size of the coarse aggregate used in the concrete.

A greater distance should be provided where convenient. Where immersion vibrators are intended to be used, however, the horizontal distance between bars of a group may be reduced to two-thirds of the nominal maximum size of the coarse aggregate provided that enough space is left between groups of bars to enable the vibrator to be inserted; this would normally be a space of 3in/75mm.

The vertical distance between two horizontal main steel reinforcements, or the corresponding distance at right-angles to two inclined main steel reinforcements, should be not less than ½in/15mm or the nominal maximum size of aggregate, whichever is the greater, except at splices or where one of such reinforcements is transverse to the other.

The pitch of the main bars in a reinforced concrete solid slab should be not more than three times the effective depth of such slab.

The pitch of distributing bars in a reinforced concrete slab should not be more than five times the effective depth of such slab.

6.6 Load-factor design method
Most reinforced concrete design of slabs and beams is now done by the load-factor method. A simplified version of this method is given below followed by examples showing the method in use.

6.6.1 First find the concrete moment of resistance

$$M_r = \frac{p_{cb}}{4} \, bd_1{}^2$$

where

M_r = concrete moment of resistance
p_{cb} = permissible compressive stress in concrete in bending
b = breadth of beam
d_1 = effective depth of beam, ie the depth from the top of the beam to the centre of the reinforcement

6.6.2 Next ensure that the applied moment M is less than the resistance If it is not, increase the size of the beam until it is. Alternatively, compression reinforcement can be used: but it is desirable that this is done only under the supervision of a structural engineer.

6.6.3 Now calculate the amount of tensile reinforcement required

$$A_t = \frac{M}{p_{st} \, l_a}$$

where

A_t = area of tension reinforcement
M = applied moment
p_{st} = permissible tensile stress in reinforcement
l_a = lever arm
The lever arm is calculated from the ratio of the applied and resistance moments:

$\dfrac{M}{M_r}$	1	0·9	0·8	0·7	0·6	0·5 and below
$\dfrac{l_a}{d_1}$	0·75	0·78	0·81	0·84	0·87	0·9

6.6.4 The shear stress is now calculated

$$q = \frac{Q}{bl_a}$$

where

q = shear stress
Q = shear force
If this stress is greater than the permitted shear stress, shear reinforcement must be used. This can be in the form of bent-up bars or in the form of stirrups. The shear values for single bars bent-up at an angle of 45deg are given in table XI. When using stirrups the following formula can be used:

$$A_s = \frac{h\,Q}{p_{ss}l_aN}$$

where

A_s = area of shear steel in length 'h'

p_{ss} = Permissible tensile stress in shear reinforcement

N = number of arms in links.

Table XI Bent-up bars as shear reinforcement. Singles bars bent-up at 45 deg. in single system: shear value

Size	Mild steel			High yield steel		
mm	FPS lbf	MT kgf	SI N	FPS lbf	MT kgf	SI N
12	2 480	1 120	11 200	3 110	1 410	14 000
16	4 440	2 000	19 900	5 540	2 500	24 900
20	6 900	3 100	31 100	8 650	3 900	38 900
25	10 900	4 900	48 600	13 600	6 100	60 800
32	17 700	8 000	79 600	22 200	10 000	99 500
	25 000	11 250	125 000	34 700	15 700	156 000

6.7 Design examples All concrete in the examples below is assumed to be 1:1½:3.

6.7.1 Simply supported slab

FPS

Problem Design a simply supported slab spanning 16ft to carry a superimposed load (including finishes) of 300lb/sq ft.

Answer If high yield reinforcement is to be used, the span: depth ratio must not exceed 27 (table VIII)

$$\therefore \text{ depth} = \frac{\text{span}}{27}$$

$$\text{span} = 16\text{ft} = 16 \times 12\text{in}$$

$$\therefore \text{ depth} = \frac{16 \times 12}{27}$$

$$= 7\cdot1\text{in}$$

assume slab is 8in deep
density of concrete is always taken as 150lb/cuft

$$\therefore \text{ self-weight of slab} = \frac{8}{12} \times 150$$

$$= 100 \text{ lbf/sq ft}$$
$$\text{superload} = 300 \text{ lbf/sq ft}$$
$$\therefore \text{ total load} = 400 \text{ lbf/sq ft}$$

$$M = \frac{wL^2}{8} \text{ for a simply supported beam}$$

Consider 1ft width of slab as a beam

MT

Problem Design a simply supported slab spanning 5m to carry a superimposed load (including finishes) of 1500 kg/m².

Answer If high yield reinforcement is to be used the span: depth ratio must not exceed 27 (table VIII)

$$\therefore \text{ depth} = \frac{\text{span}}{27}$$

$$\text{span} = 5\text{m} = 500\text{cm}$$

$$\therefore \text{ depth} = \frac{500}{27}$$

$$= 18\cdot5\text{cm}$$

assume slab is 20cm deep
density of concrete is 2500kg/m³

$$\therefore \text{ self-weight of slab} = 0\cdot2 \times 2500$$
$$= 500 \text{ kgf/m}^2$$
$$\text{superload} = 1500 \text{ kgf/m}^2$$
$$\therefore \text{ total load} = 2000 \text{ kgf/m}^2$$

$$M = \frac{wL^2}{8} \text{ for a simply supported beam}$$

Consider 1m width of slab as a beam

SI

Problem Design a simply supported slab spanning 5m to carry a superimposed load (including finishes) of 1500 kg/m².

Answer If high yield reinforcement is to be used the span: depth ratio must not exceed 27 (table VIII)

$$\therefore \text{ depth} = \frac{\text{span}}{27}$$

$$\text{span} = 5\text{m} = 5000\text{mm}$$

$$\therefore \text{ depth} = \frac{5000}{27}$$

$$= 185\text{mm}$$

assume slab is 200mm deep
mass density of concrete is 2500kg/m³

$$\therefore \text{ self-weight of slab} = 0\cdot2 \times 2500$$
$$\times 9\cdot81$$
$$= 4900 \text{ N/m}^2$$
$$\text{superload} = 1500 \times 9\cdot81$$
$$= 14\,700 \text{ N/m}^2$$
$$\therefore \text{ total load} = 19\cdot6 \text{ kN/m}^2$$

$$M = \frac{wL^2}{8} \text{ for a simply supported beam}$$

Consider 1m width of slab as a beam

FPS

$$w = 400 \text{ lbf/ft}$$
$$L = 16\text{ft}$$
$$\therefore \ M = \frac{400 \times 16^2}{8}$$
$$= 12\,800 \text{ lbf-ft}$$
$$= 154\,000 \text{ lbf-in/ft width}$$

assume the reinforcement will be in the form of $\frac{3}{4}$in diam bars
cover required $= \frac{3}{4}$in (para **6.4**)

d_1 = effective depth of slab
$\quad$ = total depth − cover − $\frac{1}{2}$ bar diameter
$\quad = 8 - 0.75 - 0.37$
$\quad = 6.88$in

now the concrete moment of resistance

$$M_r = \frac{p_{cb}}{4}\,bd_1{}^2 \text{ (para \textbf{6.6.1})}$$

$$p_{cb} = 1250 \text{ lb/sq in (table \textbf{v})}$$

$$\therefore \ \frac{p_{cb}}{4} = 312 \text{ lb/sq in}$$

$\quad$ this value may be retained and substituted directly in future calculations.
$\quad b = 12$in
$\quad d_1 = 6.88$in
$$\therefore \ M_r = 312 \times 12 \times 6.88^2$$
$$= 177\,000 \text{ lbf-in/ft width}$$

The concrete moment of resistance is greater than the applied moment (para **6.6.2**). The depth of the slab is therefore satisfactory.

$$\frac{M}{M_r} = \frac{154\,000}{177\,000}$$
$$= 0.88$$

from this the lever arm is found (para **6.6.3**).

$$\frac{l_a}{d_1} = 0.78$$
$$\therefore \ l_a = 0.78 \times 6.88$$
$$= 5.4\text{in}$$

the area of steel is now calculated (para **6.6.3**).

$$A_t = \frac{M}{p_{st}l_a}$$

$$p_{st} = 33 \text{ kipf/sq in (table \textbf{vii})}$$

$$\therefore \ A_t = \frac{154\,000}{33\,000 \times 5.4}$$
$$= 0.87 \text{ sq in/ft}$$

$\frac{3}{4}$in diam high yield at 6in centres **=** 0.884 sq in/ft
(See table **x**)

MT

$$w = 2000 \text{ kgf/m}$$
$$L = 5\text{m}$$
$$\therefore \ M = \frac{2000 \times 5^2}{8}$$
$$= 6250 \text{ kgfm/m width}$$

assume the reinforcement to be in the form of 20mm diam bars
cover required = 20mm (para **6.4**)

d_1 = effective depth of slab
$\quad$ = total depth − cover − $\frac{1}{2}$ bar diam
$\quad = 20 - 2 - 1$
$\quad = 17$cm

now the concrete moment of resistance

$$M_r = \frac{p_{cb}}{4}\,bd_1{}^2 \text{ (para \textbf{6.6.1})}$$

$$p_{cb} = 88 \text{ kgf/cm}^2 \text{ (table \textbf{v})}$$

$$\therefore \ \frac{p_{cb}}{4} = 22 \text{ kgf/cm}^2$$

$\quad$ this value may be retained and substituted directly in future calculations.
$\quad b = 100$cm
$\quad d_1 = 17$cm
$$\therefore \ M_r = 22 \times 100 \times 17^2$$
$$= 638\,000 \text{ kgfcm}$$
$$= 6380 \text{ kgfm/m width}$$

The concrete moment of resistance is greater than the applied moment (para **6.6.2**). The depth of the slab is therefore satisfactory.

$$\frac{M}{M_r} = \frac{6250}{6380}$$
$$= 0.98$$

from this the lever arm is found (para **6.6.3**)

$$\frac{l_a}{d_1} = 0.75$$
$$\therefore \ l_a = 0.75 \times 17$$
$$= 12.7\text{cm}$$

the area of steel is now calculated (para **6.6.3**)

$$A_t = \frac{M}{p_{st}l_a}$$

$$p_{st} = 2320 \text{ kgf/cm}^2 \text{ (table \textbf{vii})}$$
$$\therefore \ M = 6250 \text{ kgfm/m width}$$
$$= 625\,000 \text{ kgfcm/m}$$

$$\therefore \ A_t = \frac{625\,000}{2320 \times 12.7}$$
$$= 21.2\text{cm}^2\text{/m width}$$

20mm diam high yield at 12.5cm centres = 25.4cm²/m
(See table **ix**)

SI

$$w = 19.6 \text{ kN/m}$$
$$L = 5\text{m}$$
$$\therefore \ M = \frac{19.6 \times 5^2}{8}$$
$$= 61.5 \text{ kNm/m width}$$

assume the reinforcement to be in the form of 20mm diam bars
cover required = 20mm (para **6.4**)

d_1 = effective depth of slab
$\quad$ = total depth − cover − $\frac{1}{2}$ bar diam
$\quad = 200 - 20 - 10$
$\quad = 170$mm

now the concrete moment of resistance

$$M_r = \frac{p_{cb}}{4}\,bd_1{}^2 \text{ (para \textbf{6.6.1})}$$

$$p_{cb} = 8.5 \text{ N/mm}^2 \text{ (table \textbf{v})}$$

$$\therefore \ \frac{p_{cb}}{4} = 2.12 \text{ N/mm}^2$$

$\quad$ this value may be retained and substituted directly in future calculations.
$\quad b = 1000$mm
$\quad d_1 = 170$mm
$$\therefore \ M_r = 2.12 \times 1000 \times 170^2$$
$$= 61\,300\,000 \text{Nmm}$$
$$= 61.3 \text{ kNm/m width}$$

The concrete moment of resistance is greater than the applied moment (para **6.6.2**). The depth of the slab is therefore satisfactory.

$$\frac{M}{M_r} = \frac{61.5}{61.3}$$
$$= 1.00$$

from this the lever arm is found (para **6.6.3**)

$$\frac{l_a}{d_1} = 0.75$$
$$\therefore \ l_a = 0.75 \times 170$$
$$= 127\text{mm}$$

the area of steel is now calculated (para **6.6.3**)

$$A_t = \frac{M}{p_{st}l_a}$$

$$p_{st} = 230 \text{ N/mm}^2 \text{ (table \textbf{vii})}$$
$$M = 61.5 \text{ kNm/m width}$$
$$= 61.5 \times 10^6 \text{ Nmm/m}$$

$$\therefore \ A_t = \frac{61.5 \times 10^6}{230 \times 127}$$
$$= 2105\text{mm}^2\text{/m width}$$

20mm diam high yield at 125mm centres = 2514 mm²/m
(See table **ix**)

FPS

6.7.2 Rectangular reinforced concrete beam

Problem Design a rectangular reinforced concrete beam 30in × 9in simply supported with a span of 25ft carrying a central point load of 10 tons. (Self-weight may be neglected.)

Answer First find the applied moment

$$M = \frac{WL}{4}$$ for a central point load on a simply supported beam

$$W = 10 \text{ tonf}$$
$$= 10 \times 2.24$$
$$= 22.4 \text{ kipf}$$
$$L = 25\text{ft}$$
$$= 25 \times 12\text{in}$$
$$\therefore M = \frac{22.4 \times 25 \times 12}{4}$$
$$= 1680 \text{ kipf-in}$$

If the reinforcement is to be 1in diam high yield bars

cover = 1in (para **6.4**)

effective depth = total depth − cover − $\frac{1}{2}$ bar diameter

$$\therefore d_1 = 30 - 1 - 0.5$$
$$= 28.5\text{in}$$

beam breadth b = 9in

now the concrete moment of resistance

$$M_r = \tfrac{1}{2}p_{cb}\, bd_1{}^2 \text{ (para } \mathbf{6.6.1})$$
$$= 312 \times 9 \times 28.5^2$$
$$= 2\ 280\ 000 \text{ lbf-in}$$
$$= 2280 \text{ kipf-in}$$
$$\therefore \frac{M}{M_r} = \frac{1680}{2280} = 0.74$$

∴ from para **6.6.3**

$$l_a = 0.83 \times d_1$$
$$= 0.83 \times 28.5$$
$$= 23.5\text{in}$$

now

$$A_t = \frac{M}{p_{st}l_a} \text{ (para } \mathbf{6.6.3})$$
$$p_{st} = 30 \text{ kipf/sq in (table VII)}$$
$$\therefore A_t = \frac{1680}{30 \times 23.5}$$
$$= 2.38 \text{ sq in}$$

Three 1in diam high yield bars = 2.36 sq in (table x)

Shear force $Q = \dfrac{W}{2}$

$$= 11.2 \text{ kipf} = 11\ 200\text{lbf}$$

(from para **6.6.4**)

shear stress $q = \dfrac{Q}{bl_a}$

$$= \frac{11\ 200}{9 \times 23.5}$$
$$= 53 \text{ lbf/sq in}$$

permissible stress from table v
$$= 115 \text{ lbf/sq in}$$

As shear stress is less than permitted amount, nominal shear reinforcement only is required.

This is normally taken as 0.15 per cent

MT

Problem Design a rectangular reinforced concrete beam 75cm × 22·5cm simply supported with a span of 7·5m carrying a central point load of 10t. (Self-weight may be neglected.)

Answer First find the applied moment

$$M = \frac{WL}{4}$$ for a central point load on a simply supported beam

$$W = 10\text{tf}$$
$$L = 7\cdot5\text{m}$$
$$\therefore M = \frac{10 \times 7\cdot5}{4}$$
$$= 18\cdot7 \text{ tfm}$$

If the reinforcement is to be 25mm diam bars

cover = 25mm (para **6.4**)

effective depth = total depth − cover − $\frac{1}{2}$ bar diameter

$$\therefore d_1 = 75 - 2\cdot5 - 1\cdot3$$
$$= 71\text{cm}$$

beam breadth b = 22·5cm

now the concrete moment of resistance

$$M_r = \tfrac{1}{2}p_{cb}\, bd_1{}^2 \text{ (para } \mathbf{6.6.1})$$
$$= 22 \times 22\cdot5 \times 71^2$$
$$= 2\ 500\ 000 \text{ kgfcm}$$
$$= 25\cdot0 \text{ tfm}$$
$$\therefore \frac{M}{M_r} = \frac{18\cdot7}{25\cdot0} = 0\cdot75$$

∴ from para **6.6.3**

$$l_a = 0\cdot83 \times d_1$$
$$= 0\cdot83 \times 71$$
$$= 59\text{cm}$$

now $A_t = \dfrac{M}{p_{st}l_a}$ (para **6.6.3**)

$$p_{st} = 2110 \text{ kgf/cm}^2 \text{ (table VII)}$$
$$= 2\cdot11 \text{ tf/cm}^2$$
$$M = 18\cdot7 \text{ tfm} = 1870 \text{ tfcm}$$
$$\therefore A_t = \frac{1870}{2\cdot11 \times 59}$$
$$= 15\cdot0\text{cm}^2$$

Four 25mm diam high yield bars = 19·66cm² (table IX)

Shear force $Q = \dfrac{W}{2}$

$$= 5 \text{ tf} = 5000 \text{ kgf}$$

(from para **6.6.4**)

shear stress $q = \dfrac{Q}{bl_a}$

$$= \frac{5000}{22\cdot5 \times 59}$$
$$= 3\cdot8 \text{ kgf/cm}^2$$

permissible stress from table v
$$= 8\cdot1 \text{ kgf/cm}^2$$

As shear stress is less than permitted amount, nominal shear reinforcement only is required.

This is normally taken as 0·15 per cent

SI

Problem Design a rectangular reinforced concrete beam 750mm × 225mm simply supported with a span of 7·5m carrying a central point load weighing 0·1MN. (Self-weight may be neglected).

Answer First find the applied moment

$$M = \frac{WL}{4}$$ for a central point load on a simply supported beam

$$W = 100\text{kN}$$
$$L = 7\cdot5\text{m}$$
$$\therefore M = \frac{100 \times 7\cdot5}{4}$$
$$= 187\text{kNm}$$

If the reinforcement is to be 25mm diam bars

cover = 25mm (para **6.4**)

effective depth = total depth − cover − $\frac{1}{2}$ bar diameter

$$\therefore d_1 = 750 - 25 - 13$$
$$= 712\text{mm}$$

beam breadth b = 225mm

now the concrete moment of resistance

$$M_r = \tfrac{1}{2}p_{cb}\, bd_1{}^2 \text{ (para } \mathbf{6.6.1})$$
$$= 2\cdot12 \times 225 \times 712^2$$
$$= 245\ 000\ 000 \text{ Nmm}$$
$$= 242 \text{ kNm}$$
$$\therefore \frac{M}{M_r} = \frac{187}{242} = 0\cdot77$$

∴ from para **6 .6.3**

$$l_a = 0\cdot81 \times d_1$$
$$= 0\cdot81 \times 712$$
$$= 580\text{mm}$$

now $A_t = \dfrac{M}{p_{st}l_a}$ (para **6.6.3**)

$$p_{st} = 210 \text{ N/mm}^2 \text{ (table VII)}$$
$$= 0\cdot210 \text{ kN/mm}^2$$
$$M = 187 \text{ kNm} = 187\ 000 \text{ kNmm}$$
$$\therefore A_t = \frac{187\ 000}{0\cdot210 \times 580}$$
$$= 1535\text{mm}^2$$

Four 25mm diam high yield bars = 1966mm² (table IX)

Shear force $Q = \dfrac{W}{2}$

$$= 50\text{kN} = 50\ 000\text{N}$$

(from para **6.6.4**)

shear stress $q = \dfrac{Q}{bl_a}$

$$= \frac{50\ 000}{225 \times 580}$$
$$= 0\cdot383 \text{ N/mm}^2$$

permissible stress from table v
$$= 0\cdot8 \text{ N/mm}^2$$

As shear stress is less than permitted amount, nominal shear reinforcement only is required.

This is normally taken as 0·15 per cent

FPS

of the horizontal section of beam that would be required to take shear force on the concrete alone.
If b_e is the required breadth of this beam and q_p is the maximum permitted shear stress on the concrete alone, then

$$b_e = \frac{Q}{q_p l_a}$$

now $Q = 11200$ lbf (above)
$q_p = 115$ lbf/sq in (table v)
$l_a = 23.5$in (above)

$$\therefore b_e = \frac{11200}{115 \times 23.5}$$

$$= 4.2\text{in}$$

the horizontal area of 1ft run of the assumed beam is then

$$4.2 \times 12 \text{ sq in} = 50 \text{ sq in}$$

the required area of reinforcement is

$$\frac{0.15}{100} \times 50 = 0.075 \text{ sq in/ft}$$

each link has two arms

$$\frac{0.075}{2} = 0.038 \text{ sq in/ft run}$$

$\frac{1}{4}$in diam bars at 15in centres = 0.039 sq in/ft (table x)
$\frac{1}{4}$in diam links at 15in centres = 0.078 sq in/ft
The maximum permitted spacing is $l_a = 23.5$in

MT

of the horizontal section of beam that would be required to take shear force on the concrete alone.
If b_e is the required breadth of this beam and q_p is the maximum permitted shear stress on the concrete alone, then

$$b_e = \frac{Q}{q_p l_a}$$

now $Q = 5000$ kgf (above)
$q_p = 8.1$ kgf/cm² (table v)
$l_a = 59$cm (above)

$$\therefore b_e = \frac{5000}{8.1 \times 59}$$

$$= 10\text{cm}$$

the horizontal area of 1m run of the assumed beam is then

$$10 \times 100 = 1000\text{cm}^2$$

the required area of reinforcement is

$$\frac{0.15}{100} \times 1000 = 1.5 \text{ cm}^2/\text{m}$$

each link has two arms

$$\frac{1.5}{2} = 0.75 \text{ cm}^2/\text{m}$$

6mm diam bars at 35cm = 0.81 cm²/m

6mm diam links at 35cm = 1.62 cm²/m

The maximum permitted spacing is $l_a = 59$cm

SI

of the horizontal section of beam that would be required to take shear force on the concrete alone.
If b_e is the required breadth of this beam and q_p is the maximum permitted shear stress on the concrete alone, then

$$b_e = \frac{Q}{q_p l_a}$$

now $Q = 50\,000$N (above)
$q_p = 0.8$ N/mm² (table v)
$l_a = 580$mm (above)

$$\therefore b_e = \frac{50\,000}{0.8 \times 580}$$

$$= 108\text{mm}$$

the horizontal area of 1m run of the assumed beam is then

$$108 \times 1000 = 108\,000\text{mm}^2$$

the required area of reinforcement is

$$\frac{0.15}{100} \times 108\,000 = 162\text{mm}^2/\text{m}$$

each link has two arms

$$\frac{162}{2} = 81\text{mm}^2/\text{m}$$

6mm diam bars at 350mm = 81mm²/m

6mm diam links at 350mm = 162mm²/m

The maximum permitted spacing is $l_a = 580$mm

6.7.3 Reinforced concrete T-beam

Problem Design a simply supported T-beam spanning 28ft; consisting of a 6in slab with a downstand of 18in × 9in width. The beam is to carry a total distributed load of 4 kip/ft run including self-weight.

Answer
applied bending moment

$$M = \frac{wL^2}{8} \text{ for a simply supported beam under a distributed load}$$

load $w = 4$ kipf/ft
span $L = 28$ft

$$\therefore M = \frac{4 \times 28^2}{8} = 391 \text{ kipf-ft}$$

$$= 4700 \text{ kipf-in}$$

from BS CP 114:1957 clause 311e the maximum breadth of a T-beam flange is the least of the following three figures:

1 span $\div$ 3 $= \frac{28}{3} = 9$ft 4in

2 distance between centres of beams (not applicable in this case)
3 breadth of rib + 12 × depth of flange $= 9 + (12 \times 6) = 81$in
this last figure 81in is the maximum permitted flange breadth

Problem Design a simply supported T-beam spanning 8.5m; consisting of a 15cm slab with downstand 45cm × 22.5cm width. The beam is to carry a total distributed load of 6t/m including self-weight.

Answer
applied bending moment

$$M = \frac{wL^2}{8} \text{ for a simply supported beam under a distributed load}$$

load $w = 6$ tf/m
span $L = 8.5$m

$$\therefore M = \frac{6 \times 8.5^2}{8} = 54.2 \text{ tfm}$$

from BS CP 114:1957 clause 311e the maximum breadth of a T-beam flange is the least of the following three figures:

1 span $\div$ 3 $= \frac{8.5}{3} = 2.8$m

2 distance between centres of beams (not applicable in this case)
3 breadth of rib + 12 × depth of flange $= 22.5 + (12 \times 15) = 202.5$cm
this last figure 202.5cm is the maximum permitted flange breadth

Problem Design a simply supported T-beam spanning 8.5m; consisting of a 150mm slab with downstand 450mm × 225mm width. The beam is to carry a total distributed load of 6000kg/m including self-weight.

Answer
applied bending moment

$$M = \frac{wL^2}{8} \text{ for a simply supported beam under a distributed load}$$

load $w = 6000 \times 9.81$
$= 59\,000$ N/m $= 59$ kN/m
span $L = 8.5$m

$$\therefore M = \frac{59 \times 8.5^2}{8} = 532 \text{ kNm}$$

from BS CP 114:1957 clause 311e the maximum breadth of a T-beam flange is the least of the following three figures:

1 span $\div$ 3 $= \frac{8.5}{3} = 2.8$m

2 distance between centres of beams (not applicable in this case)
3 breadth of rib + 12 × depth of flange $= 225 + (12 \times 150) = 2025$mm
this last figure 2025 mm is the maximum permitted flange breadth

FPS

assume reinforcement to be 1in diam high yield bars in three layers

then effective depth

d_1 = total depth − cover − bar diam − vertical spacing − $\frac{1}{2}$ bar diam

(from **6.4**) cover = 1in

(from **6.5**) vertical spacing = $\frac{1}{2}$in or nominal max size of aggregate if larger

nominal maximum aggregate size = $\frac{3}{4}$in

∴ vertical spacing = $\frac{3}{4}$in

∴ d_1 = (6 + 18) − 1 − 1 − $\frac{3}{4}$ − $\frac{1}{2}$
= 20.75in

now consider a rectangular beam

breadth = 81in, d_1 = 20.75

moment of resistance

$M_r = \frac{1}{4}p_{cb}\ bd_1^2$
= 312 × 81 × 20.75²
= 10 800 000 lbf/in
= 10 800 kipf-in

∴ $\frac{M}{M_r} = \frac{4700}{10\ 800} = 0.44$

now if n is the depth of the neutral axis below the top of the beam

$\frac{n}{d_1} = 1 - \sqrt{1 - \frac{3\ M}{4 M_r}}$
= $1 - \sqrt{0.67} = 0.18$

∴ n = 0.18 × 20.75 = 3.7in

∴ compression does not extend below the soffit of the flange and assumption above was justified

now referring back to para **6.6.3**

$\frac{M}{M_r} = 0.44$

∴ $l_a = 0.9 × d_1$
= 0.9 × 20.75 = 18.7in

now $A_t = \frac{M}{p_{st}l_a}$

p_{st} = 30 kipf/sq in (table VII)

M = 4700 kipf-in (above)

∴ $A_t = \frac{4700}{30 × 18.7} = 8.4$ sq in

Eleven 1in diam high yield bars
= 8.64 sq in

(See table X)

Shear force Q = $\frac{W}{2}$

W = total load on span
= 4 × 28 = 112 kipf

∴ Q = 56 kipf = 56 000 lbf

then shear stress (para **6.6.4**)

$q = \frac{Q}{bl_a}$

now in this case

b = breadth of rib
= 9in

l_a = 18.7in (above)

∴ $q = \frac{56\ 000}{9 × 18.7} = 335$ lbf/sq in

permitted shear stress on concrete alone (table V) = 115 lbf/sq in

∴ shear reinforcement is required

MT

assume reinforcement to be 25mm ϕ high yield bars in three layers

then effective depth

d_1 = total depth − cover − bar diam − vertical spacing − $\frac{1}{2}$ bar diam

(from **6.4**) cover = 2.5cm

(from **6.5**) vertical spacing = 12mm or nominal max size of aggregate if larger

nominal max aggregate size = 19mm

∴ vertical spacing = 1.9cm

∴ d_1 = (15 + 45) − 2.5 − 2.5 − 1.9 − 1.3
= 51.8cm

now consider a rectangular beam

breadth = 202.5cm, d_1 = 51.8cm

moment of resistance

$M_r = \frac{1}{4}p_{cb}\ bd_1^2$
= 22 × 202.5 × 51.8²
= 11 900 000 kgf-cm
= 119 tfm

∴ $\frac{M}{M_r} = \frac{54.2}{119} = 0.46$

now if n is the depth of the neutral axis below the top of the beam

$\frac{n}{d_1} = 1 - \sqrt{1 - \frac{3\ M}{4\ M_r}}$
= $1 - \sqrt{0.66} = 0.19$

∴ n = 0.19 × 51.8 = 9.8cm

∴ compression does not extend below the soffit of the flange and assumption above was justified

now referring back to para **6.6.3**

$\frac{M}{M_r} = 0.46$

∴ $l_a = 0.9 × d_1$
= 0.9 × 51.8 = 46.5cm

now $A_t = \frac{M}{p_{st}l_a}$

p_{st} = 2110 kgf/cm² (table VII)
= 2.11 tf/cm²

M = 54.2 tfm (above)

∴ $A_t = \frac{54.2}{2.11 × 46.5} = 55$cm²

Twelve 25mm ϕ high yield bars
= 58.96cm²

(See table IX)

Shear force Q = $\frac{W}{2}$

W = total load on span
= 6 × 8.5 = 51 tf

∴ Q = 25.5 tf = 25 500 kgf

then shear stress (para **6.6.4**)

$q = \frac{Q}{bl_a}$

now in this case b = breadth of rib
= 22.5cm

l_a = 46.5 cm (above)

∴ $q = \frac{25\ 500}{22.5 × 46.5} = 24$ kgf/cm²

permitted shear stress on concrete alone (table V) = 8.1 kgf/cm²

∴ shear reinforcement is required

SI

assume reinforcement to be 25mm ϕ high yield bars in three layers

then effective depth

d_1 = total depth − cover − bar diam − vertical spacing − $\frac{1}{2}$ bar diam

(from **6.4**) cover = 25mm

(from **6.5**) vertical spacing = 12mm or nominal max size of aggregate if larger

nominal max size of aggregate = 19mm

∴ vertical spacing = 19mm

∴ d_1 = (150 + 450) − 25 − 25 − 19 − 13
= 518mm

now consider a rectangular beam

breadth = 2025mm, d_1 = 518mm

moment of resistance

$M_r = \frac{1}{4}p_{cb}\ bd_1^2$
= 2.12 × 2025 × 518²
= 1 160 000 000 Nmm
= 1160 kNm

∴ $\frac{M}{M_r} = \frac{532}{1160} = 0.46$

now if n is the depth of the neutral axis below the top of the beam

$\frac{n}{d_1} = 1 - \sqrt{1 - \frac{3\ M}{4\ M_r}}$
= $1 - \sqrt{0.66} = 0.19$

∴ n = 0.19 × 518 = 98mm

∴ compression does not extend below the soffit of the flange and assumption above was justified

now referring back to para **6.6.3**

$\frac{M}{M_r} = 0.46$

∴ $l_a = 0.9 × d_1$
= 0.9 × 518 = 465mm

now $A_t = \frac{M}{p_{st}l_a}$

p_{st} = 210 N/mm² (table VII)
= 0.210 kN/mm²

M = 532 kNm (above)

∴ $A_t = \frac{532}{0.210 × 465} = 5450$mm²

Twelve 25mm ϕ high yield bars
= 5896mm²

(See table IX)

Shear force Q = $\frac{W}{2}$

W = total load on span
= 59 × 8.5 = 500kN

∴ Q = 250kN = 250 000N

then shear stress (para **6.6.4**)

$q = \frac{Q}{bl_a}$

now in this case b = breadth or rib
= 25mm

l_a = 465mm

∴ $q = \frac{250\ 000}{225 × 465} = 2.4$ N/mm²

permitted shear stress on concrete alone (table V) = 0.8 N/mm²

∴ shear reinforcements is required

FPS

First some bent-up bars will be used. Bend-up two 1in diam bars in single system.

shear force carried
$$= 27 \text{ kipf}$$

shear force left to be carried by links
$$= Q - 27$$
$$= 56 - 27 = 29 \text{ kipf}$$

from para **6.6.4**
$$A_s = \frac{hQ}{p_{ss}l_aN}$$

take
$$h = 1\text{ft} = 12\text{in}$$
$$p_{ss} = 20 \text{ kipf/sq in (table VII)}$$
$$l_a = 18.7\text{in (above)}$$
$$N = 2 \text{ (each link has two arms)}$$
$$\therefore A_s = \frac{12 \times 29}{20 \times 18.7 \times 2}$$
$$= 0.465 \text{ sq in/ft}$$

½in diam mild steel links at 5in centres
$$= 0.471 \text{ sq in/ft (table x)}$$

MT

First some bent-up bars will be used. Bend-up two 25mm ϕ bars in single system.

shear force carried (table XI)
$$= 12\text{tf}$$

shear force left to be carried by links
$$= Q - 12$$
$$= 25.5 - 12 = 13.5\text{tf}$$

from para **6.6.4**
$$A_s = \frac{hQ}{p_{ss}l_aN}$$

take
$$h = 1\text{m} = 100\text{cm}$$
$$p_{ss} = 1400 \text{ kgf/cm}^2 = 1.4 \text{ tf/cm}^2$$
$$l_a = 46.5\text{cm (above)}$$
$$N = 2 \text{ (each link has two arms)}$$
$$\therefore A_s = \frac{100 \times 13.5}{1.4 \times 46.5 \times 2}$$
$$= 10.4\text{cm}^2/\text{m}$$

10mm diam mild steel links at 7.5cm
$$= 10.5\text{cm}^2/\text{m (table IX)}$$

SI

First some bent-up bars will be used. Bend-up two 25mm ϕ bars in single system.

shear force carried (table XI)
$$= 122\text{kN}$$

shear force left to be carried by links
$$= Q - 122$$
$$= 250 - 122 = 128\text{kN}$$
$$= 128\,000\text{N}$$

from para **6.6.4**
$$A_s = \frac{hQ}{p_{ss}l_aN}$$

take
$$h = 1\text{m} = 1000\text{mm}$$
$$p_{ss} = 140 \text{ N/mm}^2 \text{ (table VII)}$$
$$l_a = 465\text{mm (above)}$$
$$N = 2 \text{ (each link has two arms)}$$
$$\therefore A_s = \frac{1000 \times 128\,000}{140 \times 465 \times 2}$$
$$= 983\text{mm}^2/\text{m}$$

10mm diam mild steel links at 75mm
$$= 1048\text{mm}^2/\text{m (table IX)}$$

Table XII Sizes of reinforcing meshes

BS reference	Mesh sizes Nominal pitch of wires (mm)		Size of wires (mm)		Cross sectional area (mm²) per metre width		Nominal mass kg per m²
Square mesh fabric							
	Main	Cross	Main	Cross	Main	Cross	
A 393	200	200	10	10	393	393	6·16
A 252	200	200	8	8	252	252	3·95
A 193	200	200	7	7	193	193	3·02
A 142	200	200	6	6	142	142	2·22
A 98	200	200	5	5	98	98	1·54
Structural fabric							
	Main	Cross	Main	Cross	Main	Cross	
B 1131	100	200	12	8	1131	252	10·90
B 785	100	200	10	8	785	252	8·14
B 503	100	200	8	8	503	252	5·93
B 385	100	200	7	7	385	193	4·53
B 283	100	200	6	7	283	193	3·73
B 196	100	200	5	7	196	193	3·05
Long mesh fabric*							
	Main	Cross	Main	Cross	Main	Cross	
C 785	100	400	10	6	785	70·8	6·72
C 503	100	400	8	5	503	49·0	4·34
C 385	100	400	7	5	385	49·0	3·41
C 283	100	400	6	5	283	49·0	2·61
Wrapping fabric							
	Main	Cross	Main	Cross	Main	Cross	
D 49	100	100	2·5	2·5	49·0	49·0	0·760
D 31	100	100	2·0	2·0	31·0	31·0	0·480
Recommended fabric for carriageways							
	Main	Cross	Main	Cross	Main	Cross	
C 636	80–130	400	8–10	6	636	70·8	5·55

*Cross wires for all types of long mesh may be of plain hard drawn steel wire

7 Loadbearing walls

Design of loadbearing walls is covered by BS CP 111: 1964. Most material in that code is non-dimensional and can be used directly in any system of units. This applies to the reduction factors on the basic stresses. However, the table III *Basic stresses* is given only in FPS units. Table XIII below gives the same table in MT and SI units. Using this in conjunction with BS CP 111 the design of loadbearing walls can be carried out in metric units, as the following examples will show.

Table XIII Basic compressive stresses for brickwork or blockwork members (at and after the stated times)

For this table in FPS see BS CP 111 : 1964 table III from which the MT and SI values have been derived

Metric technical (MT)

Description of mortar	Mix (parts by vol)			Hardening time after completion of work† (days)	Basic stress in kgf/cm² corresponding to: units with crushing strengths in lbf/sq in‡								
	Cement	Lime	Sand		400	1000	1500	3000	4000	5000	7500	10 000	14 000 or greater
					units with crushing strengths in kgf/cm²‡								
					28	70	105	210	280	350	525	703	983 or greater
Cement	1	0–¼*	3	7	2·8	7·0	10·5	16·8	21·0	25·3	35·8	46·4	59·8
	1	¼	4½	14	2·8	7·0	9·8	14·8	17·6	21·0	28·8	36·5	45·7
Cement-lime	1	1	6	} 14	2·8	7·0	9·8	13·4	16·2	19·0	25·4	31·6	38·6
Cement with plasticiser§	1	–	6										
Masonry cement‖	–	–	–										
Cement-lime	1	2	9	} 14	2·8	5·6	8·4	11·9	14·8	16·8	21·0	25·4	31·6
Cement with plasticiser§	1	–	8										
Masonry cement‖	–	–	–										
Cement-lime	1	3	12	14	2·1	4·9	7·0	9·8	11·9	14·0	17·6	21·0	24·5
Hydraulic lime	–	1	2	14	2·1	4·9	7·0	9·8	11·9	14·0	17·6	21·0	24·5
Non-hydraulic	–	1	3	28 ¶	2·1	4·2	5·6	7·0	7·7	8·4	10·5	11·9	14·0

Système international (SI)

Description of mortar	Mix (parts by vol)			Hardening time after completion of work† (days)	Basic stress in N/mm² corresponding to: units with crushing strengths in lbf/sq in‡								
	Cement	Lime	Sand		400	1000	1500	3000	4000	5000	7500	10 000	14 000 or greater
					units with crushing strengths in N/mm²‡								
					2·75	6·90	10·3	20·6	27·5	34·4	51·5	69·0	96·0 or greater
Cement	1	0–¼*	3	7	0·27	0·69	1·03	1·65	2·06	2·48	3·50	4·54	5·84
	1	¼	4½	14	0·27	0·69	0·96	1·44	1·72	2·06	2·82	3·58	4·46
Cement-lime	1	1	6	} 14	0·27	0·69	0·96	1·30	1·58	1·86	2·48	3·10	3·78
Cement with plasticiser§	1	–	6										
Masonry cement‖	–	–	–										
Cement-lime	1	2	9	} 14	0·27	0·55	0·82	1·17	1·44	1·65	2·06	2·48	3·10
Cement with plasticiser§	1	–	8										
Masonry cement‖	–	–	–										
Cement-lime	1	3	12	14	0·21	0·48	0·69	0·96	1·17	1·38	1·72	2·06	2·40
Hydraulic lime	–	1	2	14	0·21	0·48	0·69	0·96	1·17	1·38	1·72	2·06	2·40
Non-hydraulic	–	1	3	28	0·21	0·41	0·55	0·69	0·75	0·82	1·03	1·17	1·38

For notes see BS CP 111 : 1964 table III

FPS

MT

SI

7.1.1 Loadbearing brick wall

Question A long brick wall at ground floor level in a multi-storey loadbearing brick building carries a load of 45 kipf/ft run. The storey height floor to floor is 10ft and the floors are of reinforced concrete. If the wall is 9in thick and built of bricks of crushing strength 7500 lbf/sq in set in 1-¼-3 cement mortar, will this be satisfactory?

Question A long brick wall at ground floor level in a multi-storey loadbearing brick building carries a load of 65 tf/m. The storey height floor to floor is 3m and the floors are of reinforced concrete. If the wall is 22·5cm thick and built of bricks of crushing strength 525 kgf/cm² set in 1-¼-3 cement mortar, will this be satisfactory?

Question A long brick wall at ground floor level in a multi-storey loadbearing brick building carries a load of 650 kN/m. The storey height floor to floor is 3m and the floors are of reinforced concrete. If the wall is 225mm thick and built of bricks of crushing strength 51·5 N/mm² set in 1-¼-3 cement mortar, will this be satisfactory?

Answer (All clause references are to BS CP 111:1964.)

Compressive stress in brickwork

$$= \frac{W}{A}$$

where

W = load
A = area over which the load acts

consider 1ft run of wall

W = 45 kipf = 45000 lbf
A = 9 × 12 = 108 sq in

∴ compressive stress $= \dfrac{45000}{108}$

$$= 415 \text{ lbf/sq in}$$

Units of crushing strength 7500 lb/sq in in 1-¼-3 cement mortar can be used with a basic stress of 510 lbf/sq in (BS CP 111, table III)

effective height of wall (cl 305a (i))

$$= 0.75 \times 10\text{ft}$$
$$= 7.5\text{ft}$$

effective thickness of wall (cl 307a)

$$= 9\text{in} = 0.75\text{ft}$$

∴ slenderness ratio $= \dfrac{7.5}{0.75} = 10$

∴ from cl 315c
reduction factor = 0.84

∴ maximum permitted compressive

stress = 0.84 × 510
$$= 430 \text{ lbf/sq in}$$

as this is greater than the actual stress the design is satisfactory.

Answer (All clause references are to BS CP 111:1964.)

Compressive stress in brickwork

$$= \frac{W}{A}$$

where

W = load
A = area over which the load acts

consider 1m run of wall

W = 65 tf = 65 000 kgf
A = 22·5 × 100 = 2250cm²

∴ compressive stress $= \dfrac{65\,000}{2250}$

$$= 29 \text{ kgf/cm}^2$$

Units of crushing strength 525 kgf/cm² in 1-¼-3 cement mortar can be used with a basic stress of 35·8 kgf/cm² (table XIII).

effective height of wall (cl 305a(i))

$$= 0·75 \times 3\text{m}$$
$$= 225\text{cm}$$

effective thickness of wall (cl 307a)

$$= 22·5\text{cm}$$

∴ slenderness ratio $= \dfrac{225}{22·5} = 10$

∴ from cl 315c
reduction factor = 0·84

∴ maximum permitted compressive

stress = 0·84 × 35·8
$$= 30 \text{ kgf/cm}^2$$

as this is greater than the actual stress the design is satisfactory.

Answer (All clause references are to BS CP 111:1964.)

Compressive stress in brickwork

$$= \frac{W}{A}$$

where

W = load
A = area over which the load acts

consider 1m run of wall

W = 650 kN = 650 000N
A = 225 × 1000 = 225 000mm²

∴ compressive stress $= \dfrac{650\,000}{225\,000}$

$$= 2·9 \text{ N/mm}^2$$

Units of crushing strength 51·5 N/mm² in 1-¼-3 cement mortar can be used with a basic stress of 3·50 N/mm² (table XIII)

effective height of wall (cl 305a(i))

$$= 0·75 \times 3\text{m}$$
$$= 2250\text{mm}$$

effective thickness of wall (cl 307a)

$$= 225\text{mm}$$

∴ slenderness ratio $= \dfrac{2250}{225} = 10$

∴ from cl 315c
reduction factor = 0·84

∴ maximum permitted compressive

stress = 0·84 × 3·50
$$= 2·94 \text{ N/mm}^2$$

as this is greater than the actual stress the design is satisfactory.

7.1.2 Loadbearing brick pier

Question: An isolated brick pier 13¼ in square carries a steel stanchion base 5¼ in square on its top which is 9 ft above its base. The stanchion is 2¼ in eccentric from the centre of the brick pier, and its load is 9 tons. What bricks set in 1-¼-3 cement mortar will do for the pier shaft, and what bricks in the same mortar will be required in lieu of a padstone at the top of the pier? (Ignore pier self-weight).

Question : An isolated brick pier 34·25cm square carries a steel stanchion base 13·5cm square on its top which is 2·75m above its base. The stanchion is 5·7cm eccentric from the centre of the brick pier, and its load is 9 tf. What bricks set in 1-¼-3 cement mortar will do for the pier shaft, and what bricks in the same mortar will be required in lieu of a padstone at the top of the pier? (Ignore pier self-weight)

Question An isolated brick pier 342·5 mm square carries a steel stanchion base 135mm square on its top which is 2·75m above its base. The stanchion is 57mm eccentric from the centre of the brick pier, and its load is 90 kN. What bricks set in 1-¼-3 cement mortar will do for the pier shaft, and what bricks in the same mortar will be required in lieu of a padstone at the top of the pier? (Ignore pier self-weight)

FPS

Answer all clause references are to CP 111:1964.

the eccentricity of loading is

$$\frac{2.25}{13.5} = \frac{1}{6}$$

the distribution of stress is therefore triangular, and the maximum stress is twice the average value.

$$\therefore \text{ maximum stress} = 2 \times \frac{W}{A}$$

$$\text{load } W = 9 \text{ tonf}$$
$$= 9 \times 2240 = 20\,100 \text{ lbf}$$
$$\text{area } A = 13.5 \times 13.5 = 182 \text{ sq in}$$
$$\therefore \text{ maximum stress} = 2 \times \frac{20\,100}{182}$$

$$= 221 \text{ lbf/sq in}$$

Now the cross-sectional plan area of the pier is 182 sq in, which is less than 500 sq in (see cl 315b)

$$\therefore \text{ reduction factor} = 0.75 + \frac{182}{2\,000}$$

$$= 0.84$$

Effective height of pier (cl 305a (iii))
$$= 1.5 \times 9 = 13.5 \text{ ft}$$
Effective thickness of pier (cl 307a)
$$= 13.5 \text{ in} = 1.125 \text{ ft}$$

$$\therefore \text{ slenderness ratio} = \frac{13.5}{1.125} = 12$$

eccentricity of loading (above) $= \dfrac{1}{6}$

$\therefore$ from cl 315c reduction factor
$$= 0.72$$

From cl 315d as the load is eccentric the maximum stress can be subject to a basic stress increased by 1.25

$\therefore$ Total reduction factor
$$= 0.84 \times 0.72 \times 1.25 = 0.75$$
$\therefore$ basic stress that is required

$$= \frac{\text{actual stress}}{0.75} = \frac{221}{0.75}$$

$$= 295 \text{ lbf/sq in}$$

Bricks of crushing strength 4000 lbf/sq in set in cement mortar
basic stress = 300 lbf/sq in
The stress at the underside of the stanchion base is

$$\frac{W}{A}$$

where

$$W = 9 \text{ tonf}$$
$$= 9 \times 2240 = 20\,100 \text{ lbf}$$
$$A = 5.25 \times 5.25 = 27.5 \text{ sq in}$$

$$\therefore \text{ stress} = \frac{20\,100}{27.5}$$

$$= 730 \text{ lbf/sq in}$$

cl 315e provides that underneath a concentrated load the stress can be increased by a factor of 1.5

all the other factors as above still apply

MT

Answer All clause references are to CP 111:1964.

the eccentricity of loading is

$$\frac{5.7}{34.25} = \frac{1}{6}$$

the distribution of stress is therefore triangular, and the maximum stress is twice the average value.

$$\therefore \text{ maximum stress} = 2 \times \frac{W}{A}$$

$$\text{load } W = 9 \text{ tf}$$
$$= 9000 \text{ kgf}$$
$$\text{area } A = 34.25 \times 34.25$$
$$= 1170 \text{cm}^2$$

$$\therefore \text{ maximum stress} = 2 \times \frac{9000}{1170}$$

$$= 15.4 \text{ kgf/cm}^2$$

Now the cross-sectional plan area of the pier is 1170cm², which is less than 3225cm².
cl 315b in MT units would read:
Where the cross-sectional plan area of a wall or column does not exceed 3225cm², the basic stress should be multiplied by a reduction factor equal

to $0.75 + \dfrac{A}{12\,900}$ where A is the area

(in cm²) of the horizontal cross-section of the wall or column.

$$\therefore \text{ reduction factor} = 0.75 + \frac{1\,170}{12\,900}$$

$$= 0.84$$

Effective height of pier (cl 305a(iii))
$$= 1.5 \times 2.75 = 4.14 \text{m}$$
Effective thickness of pier (cl 307a)
$$= 34.25 \text{cm} = 0.3425 \text{m}$$

$$\therefore \text{ slenderness ratio} = \frac{4.14}{0.3425} = 12$$

eccentricity of loading (above) $= \dfrac{1}{6}$

from cl 315c reduction factor $= 0.72$
from cl 315d as the load is eccentric the maximum stress can be subject to a basic stress increased by 1.25
$\therefore$ Total reduction factor
$$= 0.84 \times 0.72 \times 1.25 = 0.75$$
$\therefore$ basic stress that is required

$$= \frac{\text{actual stress}}{0.75} = \frac{15.4}{0.75}$$

$$= 20.5 \text{ kgf/cm}^2$$

Bricks of crushing strength 280 kgf/cm² set in cement mortar
basic stress = 21.0 kgf/cm²
The stress at the underside of the stanchion base is

$$\frac{W}{A}$$

SI

Answer All clause references are to CP 111:1964

the eccentricity of loading is

$$\frac{57}{342.5} = \frac{1}{6}$$

the distribution of stress is therefore triangular, and the maximum stress is twice the average value.

$$\therefore \text{ maximum stress} = \frac{2 \times W}{A}$$

$$\text{load } W = 90 \text{ kN}$$
$$= 90\,000 \text{ N}$$
$$\text{area } A = 342.5 \times 342.5$$
$$= 117\,000 \text{mm}^2$$

$$\therefore \text{ maximum stress} = 2 \times \frac{90\,000}{117\,000}$$

$$= 1.54 \text{ N/mm}^2$$

Now the cross-sectional plan area of the pier is 117 000mm², which is less than 322 500mm²
cl 315b in SI units would read:
Where the cross-sectional plan area of a wall or column does not exceed 322 500mm², the basic stress should be multiplied by a reduction factor

equal to $0.75 + \dfrac{A}{1\,290\,000}$ where A is

the area (in mm²) of the horizontal cross-section of the wall or column.
$\therefore$ reduction factor

$$= 0.75 + \frac{117\,000}{1\,290\,000}$$
$$= 0.84$$

Effective height of pier (cl 305a(iii))
$$= 1.5 \times 2.75 = 4.14 \text{m}$$
Effective thickness of pier (cl 307a)
$$= 342.5 \text{mm}$$
$$= 0.3425 \text{m}$$

$$\therefore \text{ slenderness ratio} = \frac{4.14}{0.3425} = 12$$

eccentricity of loading (above) $= \dfrac{1}{6}$

from cl 315c reduction factor $= 0.72$
from cl 315d as the load is eccentric the maximum stress can be subject to a basic stress increased by 1.25
$\therefore$ Total reduction factor
$$= 0.84 \times 0.72 \times 1.25 = 0.75$$
$\therefore$ basic stress that is required

$$= \frac{\text{actual stress}}{0.75} = \frac{1.54}{0.75}$$

$$= 2.05 \text{ N/mm}^2$$

Bricks of crushing strength 27.5 N/mm² set in cement mortar
basic stress = 2.06 N/mm²

The stress at the underside of the stanchion base is

$$\frac{W}{A}$$

so total factor $= 0.75 \times 1.5 = 1.12$
required basic stress

$$= \frac{\text{actual stress}}{1.12}$$

$$= \frac{730}{1.12}$$

$$= 650 \text{ lbf/sq in}$$

use three courses of bricks crushing stress 10 000 lbf/sq in at pier top basic stress $= 660$ lbf/sq in

where

$$W = 9 \text{ tf} = 9000 \text{ kgf}$$
$$A = 13\cdot5 \times 13\cdot5 = 182 \text{ cm}^2$$

$$\therefore \text{ stress} = \frac{9\,000}{182}$$

$$= 49\cdot5 \text{ kgf/cm}^2$$

cl 315e provides that underneath a concentrated load the stress can be increased by a factor of 1·5

all the other factors as above still apply so total factor $= 0\cdot75 \times 1\cdot5 = 1\cdot12$ required basic stress

$$= \frac{\text{actual stress}}{1\cdot12}$$

$$= \frac{49\cdot5}{1\cdot12}$$

$$= 44\cdot0 \text{ kgf/cm}^2$$

use three courses of bricks crushing strength 703 kgf/cm² at pier top basic stress $= 46\cdot4$ kgf/cm²

where

$$W = 90 \text{ kN} = 90\,000 \text{ N}$$
$$A = 135 \times 135 = 18\,200 \text{ mm}^2$$

$$\therefore \text{ stress} = \frac{90\,000}{18\,200}$$

$$= 4\cdot95 \text{ N/mm}^2$$

cl 315e provides that underneath a concentrated load the stress can be increased by a factor of 1·5

all the other factors as above still apply so total factor $= 0\cdot75 \times 1\cdot5 = 1\cdot12$ required basic stress

$$= \frac{\text{actual stress}}{1\cdot12}$$

$$= \frac{4\cdot95}{1\cdot12}$$

$$= 4\cdot40 \text{ N/mm}^2$$

use three courses of bricks crushing strength 69·0 N/mm² at pier top basic stress $= 4\cdot54$ N/mm²

8 References and sources

BRITISH STANDARDS INSTITUTION:

BS 3763:1964 International system (SI) units CI/SfB (F7)
BS 350 Conversion factors and tables. Part 1: 1959 Basis of tables. Conversion factors. Part 2: 1962 Detailed conversion tables CI/SfB (F7).
BS 2856:1957 Precise conversion of inch and metric sizes on engineering drawings CI/SfB (F7)
BS 4 Specification for structural steel sections: part 2 Hot rolled hollow sections [Hh2]
BS 449:1969 Use of structural steel in building: part 2 metric units (includes BS CP 113) CI/SfB (2–) Gh2 (K)
BS 4483 Steel fabric for the reinforcement of concrete (metric units) [Jh2]
BS 4461 Cold worked steel bar for the reinforcement of concrete (metric units) [Hh]
PD 6203:1967 Additional tables for SI conversions: metric and inch units CI/SfB (F7)
BS CP3 Chapter V: Part 1: 1967: Loading CI/SfB (K4)
BS CP 101:1963 Foundations and substructures for non-industrial buildings of not more than four storeys CI/SfB (1–)
BS CP 111:1964 Structural recommendations for loadbearing walls CI/SfB (21·1) (K)

BS CP 112:1967 Structural use of timber in buildings CI/SfB (2–) Yi (K)
BS CP 114:1969 Structural use of reinforced concrete in buildings Part 2 (metric units) CI/SfB (2–) Eq 4 (K)

General
Structural steelwork 1964. London, 1964, Redpath Brown CI/SfB (2–) Gh2 (K)
NAFT, S. and R. DE SOLA International conversion tables. London, 1965, Cassell CI/SfB (F7)
RAO, V. V. L. Guide to the metric system. London, 1961, Asia Publishing House CI/SfB (F7)
AMERICAN CONCRETE INSTITUTE and CEMENT AND CONCRETE ASSOCIATION Recommendations for an international code of practice for reinforced concrete. London, 1963, The Institute and The Association CI/SfB (2–) Eq 4 (K)
REYNOLDS, C. E. (editor) Reinforced concrete designer's handbook. London 1964, Concrete Publications, 6th edition CI/SfB (2–) Eq 4 (K)
Beton-Kalender 1967. Wilhelm Ernst & Sohn Verlag CI/SfB (K)
AJ information sheet 1463 Design of simple mild steel instructural members. AJ, 1967, February 15 CI/SfB Gh2

31 Selective bibliography

Contents

To keep abreast of new publications see regular metric features in the Architects' Journal especially publications file, reviews and metric progress reports.

Notes

1 No AJ information sheets originally published with imperial measurements are included in this list although they have often been used as source material throughout the handbook. References to them are included at the end of each section.
2 British Standards Institution publications may be obtained from the BSI Sales Office, Newton House, 101-113 Pentonville Road, London N1.
3 Only a selection from the many possible references (especially to articles in periodicals) has been included.
4 Prices are given only if they are over £1.

1 Introductory and general articles

1 SLIWA, J. No mystique about metric. *Architectural Review*, 1966, April, p314–318 CI/sfB (F7)
2 SLIWA, J. Opportunities for wider change. *Architects' Journal*, 1966, November 30, p1333–1336 CI/sfB (F7)
3 SLIWA, J. Metric measures. *Designer (Journal of Society of Industrial Artists and Designers)*, 1967, April, p4–5 CI/sfB (F7)
4 SLIWA, J. Towards conscious architecture. RIBA *Journal*, 1967, January, p13–14 CI/sfB (F7)
5 CLARKE, M. D. Guide for the use of the metric system in the construction industry. BSI *News*, 1967, June, p8–9 CI/sfB (F7)
6 DUNSTONE, P. F. Case for the millimetre. *Architects' Journal*, 1966, November 23, p1265–1266, 1268–1269 CI/sfB (F7)
7 DUNSTONE, P. F. Developments in metrication. *Chartered Surveyor*, 1969, January, p101 CI/sfB (F7) (A4t) *Introduces reprinted pages 5–8 of the* RICS *Metric Guide now amended in view of the second edition of* PD 6031
8 COCKE, P. Against the 'metric foot'. RIBA *Journal*, 1967, January, p13 CI/sfB (F7)
9 Standardised metric units: Modular Society proposals to BSI. RIBA *Journal*, 1967, July, p294 CI/sfB (F7)
10 Change to metric. RIBA *Journal*, 1967, November, p497 CI/sfB (F7)
11 Preliminary basis for timing and other aspects of the metrication projects. RIBA *Journal*, 1968, March, p122–123 CI/sfB (F7)
12 ROYAL INSTITUTE OF BRITISH ARCHITECTS The architect and the change to metric. London, 1969, The Institute CI/sfB (F7) *Pamphlet: expanded edition*
13 CLAMP, H. Change to metric. RIBA *Journal*, 1969, January, p22–27 CI/sfB (F7)
14 MARTIN, B. Change to metric. RIBA *Journal*, 1969, February, p74–78 CI/sfB (F7)
15 CLAMP, H. Change to metric. RIBA *Journal*, 1969, March, p106–111 CI/sfB (F7)
16 MARTIN, B. Use of metric measure in building. *System building and design*, 1968, April, p31–33 CI/sfB (F7)
17 KELLWAY, F. W. (editor) Metrication. London, 1968, Penguin Books CI/sfB (F7)
18 POOLEY, F. B. How a country plans to go metric. *Municipal and Public Services Journal*, 1969, February 28, p521 CI/sfB (F7)
19 MOLLISON, A. R. Metrication. *Surveyor Local Government Technology*, 1969, April 11, p35–38. CI/sfB (F7)

2 Guides to the metric change

A—British Standards Institution

See also under part 6 Dimensional co-ordination.

20 PD 6030:1967 Programme for the change to the metric system in the construction industry CI/sfB (F7)

21 PD 6031:1968 The use of the metric system in the construction industry (2nd edition) CI/sfB (F7)

22 PD 5686:1969 The use of SI units CI/sfB (F7)

23 PD 3763:1964 International System (SI) units CI/sfB (F7)

24 PD 6421:1968 Change to metric in the construction industry. Are you on the critical path? A guide to the programming of your changeover. CI/sfB (F7) *4s; also available in* A1 *size as* PD *6422 at* 6s

25 PD 6427: 1969 Adoption of the metric system in the electrical industry: Basic programme CI/sfB (62) (F7)

26 PD 6430:1969 Adoption of the metric system in the marine industry. Report, basic programme and guide CI/sfB 2752 (F7)

27 BS 1192:1969 Recommendations for building drawing practice. Metric units CI/sfB (A3t) (F7) *24s.*

B—Other guides

See also items 13–24

28 ROYAL INSTITUTE OF BRITISH ARCHITECTS The architect and the change to metric. London, 1969, The Institute CI/sfB (F7)

29 ROYAL INSTITUTION OF CHARTERED SURVEYORS Metric guide. Special issue of *Chartered Surveyor*, 1968, March, CI/sfB (F7) *Reprint 2s*

30 NATIONAL FEDERATION OF BUILDING TRADES EMPLOYERS Change to metric: guide for building contractors. London, 1968, The Federation CI/sfB (F4j)

31 BRIXTON SCHOOL OF BUILDING (ILEA) Going metric: SI. London, 1968, The School CI/sfB (F7)

32 GAS COUNCIL Metrication. The international system of metric units. H. M. Glass. London, 1968, The Council CI/sfB (F7) *Free. Special reference to the gas industry*

33 ILLUMINATING ENGINEERING SOCIETY Recommendations for lighting building interiors. IES Code 1968 (metric edition). London, 1968, The Society CI/sfB (N) *Obtainable from York House, Westminster Bridge Road, London* SE1, *30s (20s to members)*

34 Focus on metrication. Supplement to *Electrical Times*, 1969, March 27 CI/sfB (F7)

35 INSTITUTION OF ELECTRICAL ENGINEERS 1969 supplement to the regulations for the electrical equipment of buildings: the use of the fourteenth edition in metric terms. London, 1969, The Institution CI/sfB (F7)

36 INSTITUTION OF HEATING AND VENTILATING ENGINEERS Change to metric: a reference manual, London, 1968, The Institution CI/sfB (F7) *Preprinted from* IHVE *Guide 1970 edition. Obtainable from 49 Cadogan Square, London* SW1

37 FORESTRY COMMISSION Metric guide to forestry. 1969, HMSO CI/sfB 161 (F7)

3 Conversion factors, tables and aids

Many conversion kits of one sort or another (some used in sales promotion) are appearing, but not all conform to BSI *recommendations. Only a few are included here.*

38 BRITISH STANDARDS INSTITUTION BS 350 Conversion factors and tables: Part 1: 1959 Basis of tables. Conversion factors. Part 2: 1962 Detailed conversion tables. Supplement No 2: 1967 (PD 6203) Additional tables for SI conversions CI/sfB (F7)

39 BRITISH STANDARDS INSTITUTION BS 2856: 1957 Precise conversion of inch and metric sizes on engineering drawings CI/sfB (F7)

40 MINISTRY OF TECHNOLOGY, NATIONAL PHYSICAL LABORATORY Changing to the metric system; conversion factors, symbols and definitions. P. Anderton and P. H. Bigg 1969, HMSO, 3rd edition CI/sfB (F7)

41 BIGGS, A. J. Direct reading: two-way metric conversion tables. London, 1969. Pitman CI/sfB (F7)

42 NAFT, S. and R. DE SOLA International conversion tables. Revised by P. H. Bigg. London, 1965, Cassell CI/sfB (F7)

43 BROOK, J. Metric system and British equivalents. London, 1967, Spon CI/sfB (F7)

44 ROYAL INSTITUTION OF CHARTERED SURVEYORS Metric conversion tables. London, 1968, The Institution CI/sfB (F7) *21s*

45 CONSTRUCTION INDUSTRY RESEARCH AND INFORMATION ASSOCIATION (CIRIA), Metric conversion factors. London, 1968, The Association CI/sfB (F4j)

46 Visual aid to metric conversion. RIBA *Journal*, 1969, April, p155 CI/sfB (F7)

47 Metric conversion slide (plastic). British Standards Institution CI/sfB (F7) *21s*

48 Imperial/metric (SI) conversion scales consisting of four A4 sheets covering light, heat and mechanics. CI/sfB (F7) *Single sheets 1s each (reduction for quantity), obtainable from the registrar, The Polytechnic, College of Architecture and Advanced Building Technology, 309 Regent Street, London* W1

49 Conversion scales. RIBA *Journal*, 1967, September, p393 CI/sfB (F7)

50 Design engineering guide, metrigrams. London, 1968, Design Engineering Publications CI/sfB (F4j)

4 Architects' Journal handbooks and other series

See also part 6 Dimensional co-ordination. There is a continuing series in the AJ *dealing with problems arising out of the metric change, as outlined in 'What the* AJ *is doing about metric' and 'Introduction to* AJ *metric proposals' (see below). General metric news items are not included in this list but their substance is included throughout the handbook.*

51 What the Architects' Journal is doing about metric. *Architects' Journal*, 1969, March 12, p670-671 CI/sfB (F7)

52 SLIWA, J. Introduction to AJ metric proposals. *Architects' Journal*, 1969, March 19, p775-776 CI/sfB (F7) *Technical study*

53 AJ Handbook of building environment (metric) CI/sfB (E6):

Section 1 Climate and topography (AJ 2.10.68 and 9.10.68)

Section 2 Sunlight: Direct and diffused (AJ 16.10.68, 23.10.68, 30.10.68, 20.11.68)

Section 3 Air movement and natural ventilation (AJ 27.11.68, 4.12.68, 11.12.68)

Section 4 Thermal properties (AJ 18 & 25.12.68, 8.1.69)

Section 5 Sound (AJ 22.1.69, 29.1.69, 5.2.69, 12.2.69)

Section 6 Hygiene (AJ 19.2.69, 26.2.69)

Section 7 User requirements (AJ 6.8.69)

Section 8 Heating installations, mechanical ventilation and air-conditioning (AJ 2.4.69, 9.4.69, 16.4.69, 23.4.69, 7.5.69, 14.5.69, 21.5.69, 28.5.69)

Section 9 Electric lighting (AJ 4.6.69, 11.6.69, 18.6.69, 9.7.69, 16.7.69, 23.7.69, 30.7.69)

Appendixes: Collected references and index (AJ 13.8.69)

54 ELDER, A. J. Metric guide to the Building Regulations. London, 1968, Architectural Press CI/sfB (A3j) (F7)

55 Metric summaries of AJ information sheets. *Architects' Journal*, 1968, July 31, p191-210 CI/sfB (F7)

56 GRUNBERG, R. Component co-ordination and the change to metric: what the official agencies are doing. *Architects' Journal*, 1968, May 15, p1081-1088, CI/sfB (F4j)

57 Progress with metrication. *Architects' Journal*, 1967, October 15, p990-993 CI/sfB (F7)

58 Metric change: Progress report 1. *Architects' Journal*, 1969, January 22, p235-238 CI/sfB (F7) *Technical study*

59 Metric change: Progress report 2. *Architects' Journal*, 1969, January 29, p301-304 CI/sfB (F7) *Technical study*

60 SZOKOLAY, S. V. Condensation and moisture movement: metric quantities. *Architects' Journal*, 1969, February 19, p523-525, CI/sfB (I) *Technical study*

See also AJ 4.3.70 p569 for additional articles up to March 1970

5 Design
Only a few of the references are included here as most of the available design data in metric terms are published by official bodies, ministries and so on: See under items 13 to 24.

61 Manuale dell'architetto. Rome, 1962, Consiglio Regionale delle Ricerche, 3rd edition CI/sfB (E1)

62 BERGLUND, E. Bord for måltider och arbete i hemmet (tables). Stockholm, 1957, Svenska Slöjdföreningen (Swedish Society of Industrial Design) CI/sfB (82·2)

63 BERGLUND, E. Skåp (storage). Stockholm, 1960, Svenska Slöjdföreningen (Swedish Society of Industrial Design) CI/sfB (76)

64 NEUFERT, E. Bauentwurfslehre. Handbuch für den baufachmann, bauherrn, lehrenden und lernenden. Berlin, 1966, Verlag Ullstein. CI/sfB (E1) *This should be read with some caution as a 12·5 cm module is used and most data are based on that module*

65 GOLDSMITH, S. Designing for the disabled. London, 1967, RIBA, 2nd edition CI/sfB (E3p) *70s*

6 Dimensional co-ordination
This section includes related topics such as joints and jointing, accuracy, quality control and performance specification. The Architects' Journal references are to articles within a continuing series and the list is therefore by no means complete. Regular subscribers will receive a wide range of advice on these and other topics during the next few years.

British Standards Institution *(See also p 190)*
66 BS 4011: 1966 Recommendations for the co-ordination of dimensions in building: Basic sizes for building components and assemblies CI/sfB (F4j)

67 BS 4330: 1968 (metric units) Recommendations for the co-ordination of dimensions in building: Controlling dimensions CI/sfB (F4j)

68 PD 6249: 1967 Dimensional co-ordination in building: Estimate of timing for BSI work CI/sfB (F4j)

69 PD 6426 Co-ordination of dimensions in buildings: steps to basic spaces for building components. CI/sfB 9 (F4j) *Issued free with BS 4330*

70 PD 6432 Dimensional co-ordination in building. Arrangement of building components and assemblies within functional groups Part 1:1969 Functional groups 1, 2, 3 and 4 CI/sfB (F4j) *30s*

71 BS 4176:1967 Specification of floor to floor heights (metric units). CI/sfB 98 (F4j) *Reference to this BS is made in BS 4330 where the main recommendations are incorporated*

72 BS 2900: Modular co-ordination in building Part 1:1957 Glossary CI/sfB (F4j)

73 BS 3626:1963 Recommendations for a system of tolerances and fits for building CI/sfB (F4j)

74 BS 4318:1968 Preferred metric basic sizes for engineering: metric units CI/sfB (F7)

75 ROYAL INSTITUTE OF BRITISH ARCHITECTS Co-ordination of dimensions for building. London, 1965, The Institute. CI/sfB (F4j) *35s. Recommended for understanding principles of dimensional co-ordination*

76 SLIWA, J. Introduction and external envelope: Cladding. Information sheet: Dimensional co-ordination 1. *Architects' Journal*, 1969, March 19, p777-782, CI/sfB (F4j) *Continuing series*

77 SLIWA, J. Introduction to the design of joints. Information sheet: Joints and jointing 1. *Architects' Journal* 1969, April 23, p1125-1128. CI/sfB Yt4 (F6) *Continuing series*

78 HARRISON, H. W. Performance specifications for building components. *Architects' Journal*, 1969, June 25, p1705-1714, CI/sfB (A3u) *Research study*

79 PAUL, W. Theory and practice. Technical study: Performance specification 1. *Architects' Journal*, 1969, April 23, p1121-1124, CI/sfB (A3u)

80 ASHDOWN, D. Dimensional performance. Technical study: Performance specification 2. *Architects' Journal*, 1969, July 23, p177-182, CI/sfB (A3u)

7 Scales and measuring instruments
All metric scales and metric conversion scales are obtainable from most drawing instrument suppliers and from the RIBA, 66 Portland Place, London W1N 4AD

81 BRITISH STANDARDS INSTITUTION BS 1347 Architects', engineers' and surveyors' scales: Part 3: Metric scales, *and* BS 4484 Construction work measuring instruments: Part 1: Graduation and figuring of instruments for linear measurements (metric units) CI/sfB (F7)

82 GUNN, D. M. Metrication-measuring instruments in the construction industry. *Chartered Surveyor*, 1969, May 11, p557-559 CI/sfB (F7)

83 Conversion scales. (a) RIBA *Journal*, 1967, September, p393. (b) Metric conversion slide (plastic). British Standards Institution, 21s CI/sfB (F7)

8 Materials
The main sources of information on materials, products and components are the trade associations and individual manufacturers and suppliers. See section 4 of this handbook for details. Details can also be obtained from building centres throughout the country. The Architects' Journal publishes metric details of materials and products as information becomes available, and articles on the problems of component design and product manufacture.

9 Decimal currency
DECIMAL CURRENCY BOARD Pamphlets:

84 Britain's new coins. 1968

85 Decimal currency: expression of amounts in printing, writing and speech. 1968

86 Facts and forecasts. 1968

87 Cash transactions during the changeover. 1969

88 Points for businessmen. 1969

All HMSO CI/sfB (F7)

89 Decimal Currency Act 1967. HMSO CI/sfB (Y) (Ajk)

90 TREASURY Decimal currency in the UK Cmnd 3164, 1967, HMSO. CI/sfB (Y)

91 WOOD, D. N. Decimal currency for Britain. London, 1968, Ward, Lock CI/sfB (Y)

92 HERITAGE, R. S. Learning decimal currency. Harmondsworth, 1968, Penguin Books CI/sfB (Y)

93 DUNBAR, D. S. Planning and pricing for decimalisation. London, 1969, Gower Press CI/sfB (Y) 35s

94 GRIFFITHS, A. A. New money. London, 1968, Oliver & Boyd CI/sfB (Y)

95 NATIONAL CASH REGISTER CO LTD Getting ready for decimal currency. London, 1968, The Company CI/sfB (Y)

96 DEPARTMENT OF EDUCATION AND SCIENCE Schools Council. Change for a pound. Teaching guide for the introduction of decimal currency and the adoption of metric measures. 1968, HMSO CI/sfB (Y)

97 TREASURY Decimal currency: the changeover. Cmnd 3889. 1969, HMSO CI/sfB (Y)

98 HUMPHREYS, T. A. Decimal currency arithmetic (with answers). London, 1969, Pergamon Press CI/sfB (Y)

10 Quantity surveying

See also RICS *Guide in part 2.*

99 ROYAL INSTITUTION OF CHARTERED SURVEYORS (+ NFBTE). Standard method of measurement of building works. London, 1968, The Institution, 5th edition (metric) CI/sfB (A4s) (F7) 26s

100 ROYAL INSTITUTION OF CHARTERED SURVEYORS (+ NFBTE) Code for the measurement of building works in small dwellings. London, 1968, The Institution, 2nd edition (metric) CI/sfB (A4s) (F7)

101 INSTITUTION OF CIVIL ENGINEERS Standard method of measurement of civil engineering quantities. London, 1968, The Institution CI/sfB (A4s) (F7)

102 SEELEY, I. H. Building quantities explained (SI metric edition). London, 1969 Macmillan, metric edition, CI/sfB (A4s) (F7) 45s

103 WILLIS, A. J., and C. J. WILLIS Elements of quantity surveying (metric). London, 1969, Crosby Lockwood, 6th edition (metric) CI/sfB (A4s) (F7) 35s

11 Training

See also item 13 in this bibliography.

CONSTRUCTION INDUSTRY TRAINING BOARD:

104 Publication BT/2 Scales on metric drawings CI/sfB (F7)

105 MT/1 Metric construction: metric units for everyday use CI/sfB (F7)

Both the above are essential reading for architects. They are obtainable from CITB, *Radnor House, London Road, Norbury, London* SW16. Other CITB pamphlets for the guidance of architects and of the building industry generally are:

106 MSC Metric summary cards

107 T/5 Taking off in metric

108 B/1 The use of metric units in construction

109 GRA Guide to all CITB metrication retraining aids

110 LTG Metrication learning texts administration guide

111 SS/1 Metric in construction £2 5s

112 SS/2 Metric in construction £2 5s

113 SS/VA Metric in construction

114 T/4 Metric design for electrical services

115 T/3 Metric design for heating and ventilating services

116 BUILDING CENTRE Metric wall charts CI/sfB (F7)

Mounted on stiff card. Obtainable from Building Centre, 26 Store Street, London WC1, *£1 1s including postage and packing*

117 TIMBER TRADE FEDERATION OF THE UK Metrication Bulletin 6. Training in metric. London, 1969, The Federation, CI/sfB (F7)

12 British Standards Institution

British Standards and Codes of Practice in metric are being published in increasing numbers, some of them of marginal interest to architects. Those who receive additions to BS *Handbook 3,* Summaries of British Standards for building materials and components, *will receive summaries of the main standards. Subscribers to* BSI *will receive up-to-date information in the* BSI *News and, annually, in the* BSI *Yearbook (note—these are available to subscribing members only). Some relevant standards and codes are listed below:*

British Standards *(See also p190)*

118 BS Handbook 18:1966 Metric standards for engineering. CI/sfB (F7) 60s

119 PD 6286:1967 Metric standards published and in progress CI/sfB (F7)

120 BS 476 Fire tests on building materials and structures. Part 5:1968 Ignitability test for materials. Metric units. 2nd revision CI/sfB Yy (R4) (Aq)

121 BS 476 Fire tests on building materials and structures. Part 6:1968 Fire propagation test for materials. Metric units. 2nd revision CI/sfB Yy (R4) (Aq)

122 BS 1972:1967 Polythene pipe (type 32) for cold water services. Metric units. CI/sfB (53·1) In6

123 BS 2028, 1364:1968 Specification for precast concrete blocks: metric units CI/sfB Ff

124 BS 2494 Rubber joint rings for gas mains, water mains and drainage purposes. Part 2:1967 Rubber joint rings for drainage purposes. Metric units CI/sfB (52) Xt4

125 BS 2525-27:1969 Specifications for undercoating and finishing paints for protective purposes (white lead based). Metric units. CI/sfB Vu *Requirements for composition and testing*

126 BS 2790 Specification for shell boilers of welded construction (other than water-tube boilers). Part 1:1969 Class 1 welded construction (metric units). 1st revision, 40s CI/sfB (56) Xh *Requirements for boilers for land use providing steam or high pressure hot water*

127 BS 2870:1968 Rolled copper alloys: Sheet, strip and foil: Metric units. 1st revision CI/sfB Yh56

128 BS 3284:1967 Polythene pipe (type 50) for cold water services. Metric units CI/sfB (53·1) In6

129 BS 3456 Testing and approval of domestic electrical appliances: Part C: 1967 Electrical refrigerators and food freezers. Metric units. CI/sfB (73·3) (Aq) 20s

130 BS 3505:1968 Specification for unplasticised pvc pipe for cold water services: Metric units. CI/sfB (53·1) In6

131 BS 3692:1967 ISO metric precision hexagon bolts, screws and nuts. Metric units. CI/sfB Xt6 25s

132 BS 3958 Thermal insulating materials. Part 3: 1967 Metal mesh faced mineral wool mats and mattresses. Metric units. CI/sfB Kml (J2)

133 BS 3999 Methods of measuring the performance of domestic electrical appliances: Part 3: 1967 Food preparation machines. Metric units. CI/sfB (73·8) (Aq)

134 BS 4155:1967 Dimensions of pallet trucks. Metric units. CI/sfB (66·6) Xy

135 BS 4167 Electrically-heated catering equipment. Part 1: 1967 Ovens. Metric units. CI/sfB (73·4) Xy

136 BS 4197:1967 A precision sound level meter. Metric units CI/sfB (M)

137 BS 4198:1967 Method for calculating loudness. Metric units CI/sfB (M5)

138 BS 4207:1967 Monolithic linings for steel chimneys and flues. Metric units CI/sfB (21·8) Yh2

139 BS 4226 Teaching machines and programmes (interchangeability of programmes). Part 1: 1967 Linear machines and programmes. Metric units CI/sfB 7 (64·1)

140 BS 4229: Part 2: 1969 Ferrous bars CI/SfB Hh
141 BS 4254 Two-part polysulphide-based sealing compounds for the building industry. Metric units CI/SfB Yt4
142 BS 4255 Preformed rubber gaskets for weather exclusion from buildings. Part 1: 1967 Non-cellular gaskets. Metric Units CI/SfB Ht4
143 BS 4315 Methods of test for resistance to air and water penetration. Part 1: 1968 Windows and gasket glazing systems. Metric units CI/SfB (31) Hy (Aq)
144 BS 4332:1967 Surface finish of blast-cleaned steel for painting. Metric units CI/SfB Yh2
145 BS 4357:1968 Specification for precast terrazzo units Metric units CI/SfB Gq5
146 BS 4363:1968 Specification for distribution units for electricity supplies for construction and building sites. Metric units. CI/SfB (B2c)
147 BS 4391:1969 Recommendations for metric base sizes for metal wire, sheet and strip (metric units) CI/SfB Xh *Tables of standard diameters giving first, second and third choice sizes*
148 BS 4408 Recommendations for non-destructive methods of test for concrete: Part 1: 1969 Electromagnetic cover measuring devices (metric units) CI/SfB Yq (Aq) *Describes electromagnetic cover measuring devices, gives limits of accuracy for particular conditions and indicates limits of accuracy under site conditions*
149 BS 4408 Recommendations for non-destructive methods of test for concrete: Part 2: 1969 Strain gauges for concrete investigations (metric units) CI/SfB Yq (Aq) *Covers methods for taking strain measurements in investigations on concrete by using mechanical, electrical resistance, vibrating wire gauges or inductive displacement transducers*
150 BS 4438:1969 Filing cabinets and suspended filing pockets CI/SfB (A1u)
151 BS 4449:1969 Hot rolled steel bars for the reinforcement of concrete CI/SfB Hh2
152 BS 4467:1969 Anthropometric and ergonomic recommendations for dimensions in designing for the elderly CI/SfB (E3c)
153 BS 4471:1969 Dimensions for softwood CI/SfB Yi2

Codes of Practice

154 BS CP 114: Structural use of reinforced concrete in buildings: Part 2: 1969 Metric units CI/SfB (2-) Eq4 (K) *30s*
155 BS CP 115:1969 Structural use of reinforced concrete in buildings CI/SfB (2-) Yq (K)
156 BS CP 116: Part 2: 1969 Structural use of precast concrete (metric units) CI/SfB (2-) Gf (K) *40s*
157 BS CP 1004: Street lighting: Part 3: 1969 Lighting for lightly trafficked roads and footways (group B) (metric units). CI/SfB 12 (90·63) *Recommendations for the functions, appearance and design of group B lighting and group B1 and B2 lighting standards*
158 BS CP 1004: Street lighting Part 9: 1969 Lighting for town and city centres and areas of civic importance (metric units) CI/SfB 12 (90·63) *Recommends two standards of lighting for town centres. Also deals with lighting for pedestrian precincts, public car parks, subways and stairways*

13 Ministry of Public Building and Works

159 Dimensional co-ordination for industrialised building:
DC4 Recommended vertical dimensions for education, health, housing, office and single-storey general purpose industrial buildings. 1967 CI/SfB (F4j)
DC5 Recommended horizontal dimensions for educational, health, housing, office and single-storey general purpose industrial buildings. 1967 CI/SfB (F4j)
DC6 Guidance on the application of recommended vertical and horizontal dimensions for educational, health, housing, office and single-storey general purpose industrial buildings. 1967 CI/SfB (F4j)
DC7 Recommended intermediate vertical controlling dimensions for educational, health, housing and office buildings and guidance on their application. 1967 CI/SfB (F4j)
DC8 Recommended dimensions of spaces allocated for selected components and assemblies used in educational, health, housing and office buildings. 1968 CI/SfB (F4j)
DC9 Performance specification writing for building components. 1969, CI/SfB [(F4j)]
DC10 Recommended dimensions of basic specifications for selected building components and assemblies used in educational, health, housing and office buildings. 1969, CI/SfB [(F4j)]
160 Think metric. London, 1969, The Ministry CI/SfB (F7)
161 Going metric in the construction industry 2: Dimensional co-ordination. 1968, HMSO CI/SfB (F4j)
162 Simple steps to metric
163 Application of dimensional co-ordination. 1969 CI/SfB (F4j)
164 Going metric—a case study of the Crown Office building at Penrith. 1969 CI/SfB 313 (F7)
165 Education and training in the construction industry for the change to the metric system: a survey CI/SfB (F7)

14 Ministry of Housing and Local Government (*see also p. 190*)
166 Circular 1/68 Metrication of house building. 1968, HMSO CI/SfB 81 (F7)
167 Circular 31/67 (Welsh Office 27/67) Vertical dimensional standards in housing. HMSO CI/SfB 81 (F4j)
168 Metric units with reference to water, sewerage and related subjects: report of a working party. W. H. Norris. 1968, HMSO CI/SfB 17 (F7)
169 Design Bulletin 1 Some aspects of designing for old people: metric edition. 1968, HMSO CI/SfB (E3c)
170 Design Bulletin 2 Grouped flatlets for old people: a sociological study: metric edition. 1968, HMSO CI/SfB 843 (E2u)
171 Design Bulletin 6 Space in the home: metric edition. 1968 HMSO CI/SfB 810 (E2d)
172 Design Bulletin 16 Co-ordination of components in housing: metric dimensional framework. 1968, HMSO CI/SfB 81 (F4j)
173 Public Health Act 1961: Building Regulations 1965: Metric equivalents of dimensions. 1968, HMSO CI/SfB (A3j) (F4j)
174 The Building regulations 1965: metric values. Consultative proposals. 1969 HMSO CI/SfB [(Ajn) (F7)]
175 Circular 48/69 (Welsh Office 43/69) Metrication and the Building Regulations 1965 (guidance for local authorities). 1969, HMSO CI/SfB (A3j) (F7)

15 Department of Education and Science
176 Building Bulletin 42 Co-ordination of components for educational buildings. 1968, HMSO CI/SfB 71 (F4j)
177 Notes on the procedures for the approval of college of education building projects. 1969 CI/SfB 722 *Revised edition*
178 Circular 4/69 The Standards for school premises (middle schools and minor amendments) regulations 1969. CI/SfB 722 (Ajp)
179 Administrative memorandum 14/68 Metrication in the construction industry CI/SfB 7 (Ajp) *With appendixes giving metric analogues for the Standards for School Premises Regulations 1959 and for Building Bulletin 7*

180 Notes on procedure for the approval of further education projects. 1969 CI/SFB 722 *Revised edition*
181 Amendments to Building Work for aided and special agreement schools. 1964 CI/SFB 722

16 Department of Health and Social Security
182 Health Service Design Note 5 Co-ordination of components for health building CI/SFB 4 (F4j)
183 Circular HM(68)25 Metrication of health building. 1968 CI/SFB 4 (F4j) *Free*

17 Ministry of Technology
184 Change to the metric system in the United Kingdom. Report by the standing joint committee on metrication. 1968, HMSO CI/SFB (F7)
185 Computer installations: accommodation and fire precautions. 1968, HMSO CI/SFB 736 (R1)

18 Ministry of Transport
186 Technical Memorandum T8/68 Metrication: highway design. CI/SFB 122 (F7) *Obtainable free from the ministry, St Christopher House, Southwark Street, London SE1*

19 Home Office
187 Fire Service Building Note 1. Metric edition. London 1968, Home Office, CI/SFB 383

20 National Building Agency
188 Metric house shells: two-storey. London, 1968, The Agency CI/SFB 8112 (F4j)
189 Metric housing—the transitional period. London, 1968, The Agency CI/SFB 81 Xy (F4j)
190 Metric housing—what it means. London, 1969, The Agency CI/SFB 81 (F7)
191 The SMM in metric: a comparison with imperial. London, 1969, The Agency CI/SFB (A4s) (F7)

21 Scottish Development Department
192 Circular 27/68 Metrication and dimensional co-ordination in public sector housing, Edinburgh, 1968, HMSO CI/SFB 81 (F4j) *Free*
193 The New Scottish Housing Handbook Bulletin 1 Metric space standards. Edinburgh, 1968, HMSO CI/SFB 810 (E2d)
194 Building Standards (Scotland) Regulations 1963-1967. Metric equivalents of dimensions. Edinburgh, 1968, HMSO CI/SFB (A3j) (F4j)

22 Scottish Education Department
195 Circular 701(VII) Social Work Services Group Circular 4/1968. CI/SFB 7

23 Scottish Home and Health Department
196 Scottish Hospital Memorandum 50/68 Metrication of health building. Edinburgh, 1968, The Department CI/SFB 4 (F7) *Free*

24 Greater London Council
197 London Building Acts 1930-39 and constructional by-laws, metric equivalents of dimensions. (Department of Architecture and Civic design), 1969 CI/SFB (Ajn)
198 Inner London Education Authority. Preambles for bills of quantities: metric supplement. London, 1969, The Council CI/SFB (A4s) *This supplement is issued free to holders of the Preambles which cost £5 5s*

199 Code of practice for means of escape in case of fire. Houses in multiple occupation. Housing Act 1961, sec 16. Publication 88 CI/SFB (Ajn)
200 Means of escape in case of fire. Code of practice for the guidance of applicants. Publication 3868. CI/SFB (Ajn)
201 Technical regulations for places of public entertainment. Publication 185. CI/SFB (Ajn)
202 Rules of management for places of public entertainment. Publication 163 CI/SFB (Ajn)
203 Rules of management for smaller premises—additional rules for smaller premises. Publication 162 CI/SFB (Ajn)

25 National Housebuilders' Registration Council
204 Registered housebuilders' handbook: Part 2 Technical (with metric equivalents). London, 1968, The Council, CI/SFB 812 (Ajp) *Pocket edition*

26 Other bibliographies
205 MINISTRY OF PUBLIC BUILDING AND WORKS, Lambeth Bridge House, London SE1. Contains material up to March 1970 CI/SFB (F7) (Abe) *Free*
206 BLACKBURN COLLEGE OF TECHNOLOGY AND DESIGN Metrication gathers momentum. Contains material up to January 1970. CI/SFB (F7) (Abe)

27 Addendum *(March 1970)*
For a more expanded list see AJ 4.3.70 pp 568, 569

British Standards Institution
PD 6440 Draft for development: accuracy in building: part 1 Imperial units, part 2 Metric units [(F6j)] Price 36s each
PD 6444 Recommendations for the co-ordination of dimensions in building: basic spaces for structure, external envelope and internal subdivision: part 1 Functional groups 1, 2 and 3 [(F4j)] Price 40s
PD 6445 Recommendations for the co-ordination of dimensions in building: tolerances and joints, the derivation of building component manufacturing sizes from co-ordinating sizes [(F4j)] (*See reviews AJ 4.3.70 pp 553-556*).
PD 6432 Recommendations for the co-ordination of dimensions in building: part 2 Arrangement of building components and assemblies within functional groups: Functional group 5 [(F4j)]
PD 6286: November 1969 Metric Standards published and in progress (completed to 31 July 1969) [(F7)]
BS 4 Specification for structural steel sections: part 2 Hot rolled hollow sections [Hh2]
BS 449 The use of structural steel in building: part 2 Metric units [(2–) Gh2 (K)] price 30s
BS 1178 Milled lead sheet and strip for building purposes [Mh8]
BS 1347 Architects', engineers' and surveyors' scales: part 3 Metric scales [(F7)]
BS 3921 Bricks and blocks of fired brickearth, clay or shale: part 2 Metric units [Fg]
BS 4483 Steel fabric for the reinforcement of concrete, metric units [Jh2]
BS 4484 Measuring instruments for constructional works: part 1 Metric graduation and figuring of instruments for linear measurements [(A3s)(F7)]

National Building Agency (*see also p 42*)
Metric housing in brick: vol 1 Design guide: vol 2 Construction guide. 1969, The Agency, [81 (F7)] 25gns.

Ministry of Housing and Local Government
Circular 69/69 Metric house shells. 1969, HMSO [81 (F7)]

32 Conversion factors and tables

Contents

Table I Conversion factors

Quantity	Imperial unit	Metric unit	Metric unit symbol	Conversion factor (imperial to metric)	Notes
GENERAL PURPOSES					
Length	mile	kilometre	km	1 mile = 1·609km	
	chain	metre	m	1 chain = 20·1168m	
	yard	metre	m	1 yard = 0·9144m ⎫	
		millimetre	mm	= 914·4mm ⎬	
	foot	metre	m	1 foot = 0·3048m ⎬	Exact conversion
		millimetre	mm	= 304·8mm ⎪	
	inch	millimetre	mm	1 inch = 25·40mm ⎭	
Area	square mile	square kilometre	km²	1 mile² = 2·590km²	
		hectare	ha	= 259·0ha	
	acre	hectare	ha	1 acre = 0·4047ha	1 hectare = 10 000m²
		square metre	m²	= 4046·9m²	
	square yard	square metre	m²	1 yd² = 0·8361m²	
	square foot	square metre	m²	1 ft² = 0·092 90m²	
		square centimetre	cm²	= 929·03cm²	
	square inch	square millimetre	mm²	1 in² = 645·2mm²	
		square centimetre	cm²	= 6·452cm²	

Table I Conversion factors *continued*

Quantity	Imperial unit	Metric unit	Metric unit symbol	Conversion factor (imperial to metric)	Notes
Volume	cubic yard	cubic metre	m³	1 yd³ = 0·7646m³	
	cubic foot	cubic metre	m³	1 ft³ = 0·028 32m³	
		litre	litre	= 28·32 litre	1000 litre = 1m³
		cubic decimetre	dm³	= 28·32dm³	1 litre = 1dm³
	petrograd standard	cubic metre	m³	1 standard = 4·672m³	
	cubic inch	cubic millimetre	mm³	1 in³ = 16390mm³	
		cubic centimetre	cm³	= 16·39cm³	
		millilitre	ml	= 16·39ml	Possibly write out unit in full to avoid confusion
		litre	litre	= 0·016 39litre	
Capacity	UK gallon	litre	litre	1 gal = 4·546litre	
	quart	litre	litre	1 qt = 1·137litre	
	pint	litre	litre	1 pt = 0·568litre	
	fluid ounce	cubic centimetre	cm³	1 fl oz = 28·413cm³	
Mass	ton	tonne	tonne or t	1 ton = 1·016 tonne	1 tonne = 1000kg
		kilogramme	kg	= 1016·05kg	
	kip (1000 pounds)	kilogramme	kg	1 kip = 453·59kg	
	hundredweight	kilogramme	kg	1 cwt = 50·80kg	
	pound	kilogramme	kg	1 lb = 0·4536kg	
	ounce	gramme	g	1 oz = 28·35g	
Mass per unit length	ton per mile	kilogramme per metre	kg/m	1 ton/mile = 0·6313kg/m	
		tonne per kilometre	tonne/km	= 0·6313tonne/km	
	pound per yard	kilogramme per metre	kg/m	1 lb/yd = 0·4961kg/m	
	pound per foot	kilogramme per metre	kg/m	1 lb/ft = 1·4882kg/m	
	pound per inch	kilogramme per metre	kg/m	1 lb/in =17·858kg/m	
	ounce per inch	kilogramme per metre	kg/m	1 oz/in = 1·1161kg/m	
Length per unit mass	yard per pound	metre per kilogramme	m/kg	1 yd/lb = 2·016m/kg	
Mass per unit area	ton per square mile	kilogramme per square kilometre	kg/km²	1 ton/mile² = 392·3kg/km²	
		grammes per square metre	g/m²	= 0·3923g/m²	
		kilogramme per hectare	kg/ha	= 3·923kg/ha	
	ton per acre	kilogramme per square metre	kg/m²	1 ton/acre = 0·2511kg/m²	
	hundredweight per acre	kilogramme per square metre	kg/m²	1 cwt/acre = 0·012 55kg/m²	
	pound per square foot	kilogramme per square metre	kg/m²	1 lb/ft² = 4·882kg/m²	
	pound per square inch	kilogramme per square metre	kg/m²	1 lb/in² = 703·07kg/m²	
	ounce per square yard	gramme per square metre	g/m²	1 oz/yd² = 33·91g/m²	
	ounce per square foot	gramme per square metre	g/m²	1 oz/ft² = 305·15g/m²	
Rates of coverage	square yard per ton	kilogramme per square metre	kg/m²	ηyd²/ton = $\frac{1}{\eta}$ × 1215kg/m²	Reciprocal conversion
	square yard per gallon	litre per square metre	litre/m²	ηyd²/gal = $\frac{1}{\eta}$ × 5·437litre/m²	Reciprocal conversion eg 2yd²/gal = ½ × 5·437 = 2·719litre/m²
Volume rate of flow	cubic foot per minute	litre per second	litre/s	1 ft³/min = 0·4719litre/s	
		cubic metre per second	m³/s	= 0·000 4719m³/s	
	cubic foot per second (cusec)	cubic metre per second	m³/s	1 ft³/s = 0·028 32m³/s	Im³/s is commonly called the 'cumec'
		cubic decimetre per second	dm³/s	= 28·32dm³/s	
	cubic foot per thousand acres	litre per hectare	litre/ha	1 ft³/1000 acres = 0·069 97litre/ha	
		cubic metre per square kilometre	m³/km²	= 0·006 997m³/km²	
	cubic inch per second	millilitre per second	ml/s	1 in³/s = 16·39ml/s	Flow calculations
	gallon per year	cubic metre per year	m³/year	1 gal/year = 0·004 546m³/year	
	gallon per day	cubic metre per day	m³/day	1 gal/day = 0·004 546m³/day	1 litre/s = 86·40m³/day
	million gallon per day	cubic metre per second	m³/s	1 million gal/day = 0·052 62m³/s	
	gallon per person per day	litre per person day	litre/person day	1 gal/person day = 4·546 litre/person day	
	gallon per square yard per day	cubic metre per square metre day	m³/m² day	1 gal/yd² day = 0·005 437 m³/m² day	If this is converted to a velocity this should be expressed in mm/s (1mm/s = 86·40m³/m² day)
	gallon per cubic yard per day	cubic metre per cubic metre day	m³/m³ day	1 gal/yd³ day = 0·005 946 m³/m³day	
	gallon per hour	litre per hour	litre/h	1 gal/h = 4·5461 litre/h	
	gallon per minute	litre per second	litre/s	1 gal/min = 0·075 77 litre/s	
	gallon per second	litre per second	litre/s	1 gal/s = 4·5461 litre/s	
Fuel consumption	gallon per mile	litre per kilometre	litre/km	1 gal/mile = 2·825 litre/km	
	mile per gallon	kilometre per litre	km/litre	1 mile/gal = 0·354 km/litre	
Density	ton per cubic yard	kilogramme per cubic metre	kg/m³	1 ton/yd³ = 1328·9kg/m³	
		tonne per cubic metre	t/m³	= 1·3289t/m³	
	pound per cubic yard	kilogramme per cubic metre	kg/m³	1 lb/yd³ = 0·5933kg/m³	
	pound per cubic foot	kilogramme per cubic metre	kg/m³	1 lb/ft³ = 16·02kg/m³	
	pound per cubic inch	gramme per cubic centimetre	g/cm³	1 lb/in³ = 27·68g/cm³	
		megagramme per cubic metre	Mg/m³	= 27·68Mg/m³	

Table I Conversion factors *continued*

Quantity	Imperial unit	Metric unit	Metric unit symbol	Conversion factor (imperial to metric)	Notes
Velocity	miles per hour	kilometre per hour	km/h	1 mile/h = 1·609km/h	
	feet per minute	metre per minute	m/min	1 ft/min = 0·3048m/min	
		metre per second	m/s	= 0·0051m/s	
	feet per second	metre per second	m/s	1 ft/s = 0·3048m/s	
	inch per second	millimetre per second	mm/s	1 in/s = 25·4mm/s	
Acceleration	feet per second per second	metre per second per second	m/s²	1 ft/s²= 0·3048m/s²	
HEATING **Temperature**	°F	degree Celsius	°C	t°F = 0·5556(t—32)°C or 5/9(t—32)°C	is the expressed temperature in °F Celsius is the preferred name for Centigrade
Temperature interval	degree Fahrenheit	degree Celsius	degC or °C	1 degF = 0·5556degC	Temperature range or difference
Heat	Btu	joule	J	1 Btu = 1055J	
		kilojoule	kJ	1 Btu = 1·055kJ	
	Therm	megajoule	MJ	1 Therm = 105·5MJ	
Heat flow rate	Btu per hour	watt	W	1 Btu/h = 0·2931W	
		kilowatt	kW	1 Btu/h = 0·000 2931kW	
Density of heat flow rate	Btu per square foot hour	watt per square metre	W/m²	1 Btu/ft²h = 3·155W/m²	
Thermal conductivity	Btu inch per square foot hour degree F	watt per metre degree Celsius	W/m°C	1 Btu in/ft²h degF = 0·1442 W/m°C	'k' value
Coefficient of heat transfer (thermal transmittance)	Btu per square foot hour degree F	watt per square metre degree Celsius	W/m²°C	1 Btu/ft²h degF = 5·678W/m²°C	'U' value. Thermal conductance has the same value
Thermal resistivity	square foot hour degree F per Btu inch	metre degree Celsius per watt	m°C/W	1 ft²h degF/Btu in = 6·933m°C/W	'1/k' value
Thermal or specific heat capacity	Btu per pound degree F	kilojoule per kilogramme degree Celsius	kJ/kg°C	1 Btu/lb degF = 4·187kJ/kg°C	
	Btu per cubic foot degree F	kilojoule per cubic metre degree Celsius	kJ/m²°C	1 Btu/ft² degF = 67·07kJ/m²°C	
Calorific value	Btu per pound	kilojoule per kilogramme	kJ/kg	1 Btu/lb = 2·326kJ/kg	
	Btu per cubic foot	kilojoule per cubic metre	kJ/m²	1 Btu/ft² = 37·26kJ/m²	
	Btu per gallon	joules per litre	J/litre	1 Btu/gal = 232·6J/litre	
Calorific value of gas	Btu per cubic foot	joule per litre	J/litre	1 Btu/ft² = 37·26J/litre	1J/litre = 1kJ/m²
Refrigeration	ton	watt	W	1 ton = 3519W	
Power	horsepower	watt	W	1 hp = 745·7W	
		kilowatt	kW	= 0·7457kW	
LIGHTING **Luminous intensity**	candela	candela	cd		no change
Luminous flux	lumen	lumen	lumen		no change
Illumination	foot-candle	lux	lux	1 ft-candle = 10·76lux	
	lumen per square foot	lux	lux	1 lumen/ft²= 10·76lux	1 lux = 1 lumen/m²
Luminance	candela per square inch	candela per square metre	cd/m²	1 cd/in² = 1550cd/m²	1 cd/m² = 3·142 apostilb
	candela per square foot	candela per square metre	cd/m²	1 cd/ft² = 10·76cd/m²	
STRUCTURAL DESIGN **Force**	pound force	newton	N	1 lbf = 4·448N	
	pound force per foot	newton per metre	N/m	1 lbf/ft = 14·59N/m	
	pound force per inch	newton per millimetre	N/mm	1 lbf/in = 0·1751N/mm	
		kilonewton per metre	kN/m	= 175·1kN/m	
	klp force (1000 lbf)	kilonewton	kN	1 kipf = 4·448kN	
	ton force	kilonewton	kN	1 tonf = 9·964kN	
	ton force per foot	kilonewton per metre	kN/m	1 tonf/ft = 32·69kN/m	

Table Conversion factors *continued*

Quantity	Imperial unit	Metric unit	Metric unit symbol	Conversion factor (Imperial to metric)	Notes
Pressure, stress	pound force per square foot	newton per square metre	N/m²	1 lbf/ft² = 47·88N/m²	1 MN/m² = 1 N/mm²
		kilonewton per square metre	kN/m²	= 0·047 88kN/m²	Approximate pressures may be expressed in
	pound force per square inch	newton per square metre	N/m²	1 lbf/in² = 6895N/m²	metres for hydraulic head, eg for water under
		kilonewton per square metre	kN/m²	= 6·895kN/m²	standard conditions:
		meganewton per square metre	MN/m²	= 0·006 895MN/m²	1 N/m² = 0·1020mm
	ton force per square inch	meganewton per square metre	MN/m²	1 tonf/in² = 15·44MN/m²	The bar (10⁵N/m²) may
		newton per square millimetre	N/mm²	= 15·44N/mm²	be internationally
	ton force per square foot	kilonewton per square metre	kN/m²	1 tonf/ft² = 107·3kN/m²	adopted as a unit of pressure
Bending moment	pound force foot	newton metre	Nm	1 lbf ft = 1·356Nm	
	pound force inch	newton metre	Nm	1 lbf in = 0·1130Nm	
		newton millimetre	Nmm	= 113·0Nmm	
	kip force inch	newton metre	Nm	1 kipf in = 113·0Nm	
	ton force foot	kilonewton metre	kNm	1 tonf ft = 3·037kNm	
	ton force inch	kilonewton metre	kNm	1 tonf in = 0·2531kNm	
Section modulus	inch³	millimetre³	mm³	1 in³ = 16 390mm³	
Second moment of area	inch⁴	millimetre⁴	mm⁴	1 in⁴ = 416 200mm⁴	
Coefficient of compressibility	square foot per ton force	square metre per kilonewton	m²/kN	1 ft²/tonf = 0·009 324 m²/kN	Settlement analysis
Coefficient of consolidation	square foot per year	square millimetre per second	mm²/s	1 ft²/year = 0·002 946mm²/s	Settlement analysis
	square centimetre per second	square millimetre per second	mm²/s	1 cm²/s = 100mm²/s	
Coefficient of permeability	centimetre per second	metre per second	m/s	1 cm/s = 0·01m/s	Flow calculations
	foot per year	metre per second	m/s	1 ft/year = 0·9651 × 10⁻⁸ m/s	
Viscosity (dynamic)	centipoise	newton second per square metre	Ns/m²	1 cP = 10⁻³s/m²	Flow calculations
Viscosity (kinematic)	centistokes	square metre per second	m²/s	1 cSt = m²/s	Centipoise and centistokes may continue in general use

Use this space to note further conversion factors found useful

Table II Inches and fractions of an inch to millimetres ($\frac{1}{32}$ in increments up to $11\frac{31}{32}$ in)

Inches	0	1	2	3	4	5	6	7	8	9	10	11
	Millimetres											
—	—	25·4	50·8	76·2	101·6	127·0	152·4	177·8	203·2	228·6	254·0	279·4
1/32	0·8	26·2	51·6	77·0	102·4	127·8	153·2	178·6	204·0	229·4	254·8	280·2
1/16	1·6	27·0	52·4	77·8	103·2	128·6	154·0	179·4	204·8	230·2	255·6	281·0
3/32	2·4	27·8	53·2	78·6	104·0	129·4	154·8	180·2	205·6	231·0	256·4	281·8
1/8	3·2	28·6	54·0	79·4	104·8	130·2	155·6	181·0	206·4	231·8	257·2	282·6
5/32	4·0	29·4	54·8	80·2	105·6	131·0	156·4	181·8	207·2	232·6	258·0	283·4
3/16	4·8	30·2	55·6	81·0	106·4	131·8	157·2	182·6	208·0	233·4	258·8	284·2
7/32	5·6	31·0	56·4	81·8	107·2	132·6	158·0	183·4	208·8	234·2	259·6	285·0
1/4	6·4	31·8	57·2	82·6	108·0	133·4	158·8	184·2	209·6	235·0	260·4	285·8
9/32	7·1	32·5	57·9	83·3	108·7	134·1	159·5	184·9	210·3	235·7	261·1	286·5
5/16	7·9	33·3	58·7	84·1	109·5	134·9	160·3	185·7	211·1	236·5	261·9	287·3
11/32	8·7	34·1	59·5	84·9	110·3	135·7	161·1	186·5	211·9	237·3	262·7	288·1
3/8	9·5	34·9	60·3	85·7	111·1	136·5	161·9	187·3	212·7	238·1	263·5	288·9
13/32	10·3	35·7	61·1	86·5	111·9	137·3	162·7	188·1	213·5	238·9	264·3	289·7
7/16	11·1	36·5	61·9	87·3	112·7	138·1	163·5	188·9	214·3	239·7	265·1	290·5
15/32	11·9	37·3	62·7	88·1	113·5	138·9	164·3	189·7	215·1	240·5	265·9	291·3
1/2	12·7	38·1	63·5	88·9	114·3	139·7	165·1	190·5	215·9	241·3	266·7	292·1
17/32	13·5	38·9	64·3	89·7	115·1	140·5	165·9	191·3	216·7	242·1	267·5	292·9
9/16	14·3	39·7	65·1	90·5	115·9	141·3	166·7	192·1	217·5	242·9	268·3	293·7
19/32	15·1	40·5	65·9	91·3	116·7	142·1	167·5	192·9	218·3	243·7	269·1	294·5
5/8	15·9	41·3	66·7	92·1	117·5	142·9	168·3	193·7	219·1	244·5	269·9	295·3
21/32	16·7	42·1	67·5	92·9	118·3	143·7	169·1	194·5	219·9	245·3	270·7	296·1
11/16	17·5	42·9	68·3	93·7	119·1	144·5	169·9	195·3	220·7	246·1	271·5	296·9
23/32	18·3	43·7	69·1	94·5	119·9	145·3	170·7	196·1	221·5	246·9	272·3	297·7
3/4	19·1	44·5	69·9	95·3	120·7	146·1	171·5	196·9	222·3	247·7	273·1	298·5
25/32	19·8	45·2	70·6	96·0	121·4	146·8	172·2	197·6	223·0	248·4	273·8	299·2
13/16	20·6	46·0	71·4	96·8	122·2	147·6	173·0	198·4	223·8	249·2	274·6	300·0
27/32	21·4	46·8	72·2	97·6	123·0	148·4	173·8	199·2	224·6	250·0	275·4	300·8
7/8	22·2	47·6	73·0	98·4	123·8	149·2	174·6	200·0	225·4	250·8	276·2	301·6
29/32	23·0	48·4	73·8	99·2	124·6	150·0	175·4	200·8	226·2	251·6	277·0	302·4
15/16	23·8	49·2	74·6	100·0	125·4	150·8	176·2	201·6	227·0	252·4	277·8	303·2
31/32	24·6	50·0	75·4	100·8	126·2	151·6	177·0	202·4	227·8	253·2	278·6	304·0

Table III Feet and inches (up to 20ft) to metres (exact values)

Feet	Inches											
	0	1	2	3	4	5	6	7	8	9	10	11
	Metres											
0	—	0·0254	0·0508	0·0762	0·1016	0·1270	0·1524	0·1778	0·2032	0·2286	0·2540	0·2794
1	0·3048	0·3302	0·3556	0·3810	0·4064	0·4318	0·4572	0·4826	0·5080	0·5334	0·5588	0·5842
2	0·6096	0·6350	0·6604	0·6858	0·7112	0·7366	0·7620	0·7874	0·8128	0·8382	0·8636	0·8890
3	0·9144	0·9398	0·9652	0·9906	1·0160	1·0414	1·0668	1·0922	1·1176	1·1430	1·1684	1·1938
4	1·2192	1·2446	1·2700	1·2954	1·3208	1·3462	1·3716	1·3970	1·4224	1·4478	1·4732	1·4986
5	1·5240	1·5494	1·5748	1·6002	1·6256	1·6510	1·6764	1·7018	1·7272	1·7526	1·7780	1·8034
6	1·8288	1·8542	1·8796	1·9050	1·9304	1·9558	1·9812	2·0066	2·0320	2·0574	2·0828	2·1082
7	2·1336	2·1590	2·1844	2·2098	2·2352	2·2606	2·2860	2·3114	2·3368	2·3622	2·3876	2·4130
8	2·4384	2·4638	2·4892	2·5146	2·5400	2·5654	2·5908	2·6162	2·6416	2·6670	2·6924	2·7178
9	2·7432	2·7686	2·7940	2·8194	2·8448	2·8702	2·8956	2·9210	2·9464	2·9718	2·9972	3·0226
10	3·0480	3·0734	3·0988	3·1242	3·1496	3·1750	3·2004	3·2258	3·2512	3·2766	3·3020	3·3274
11	3·3528	3·3782	3·4036	3·4290	3·4544	3·4798	3·5052	3·5306	3·5560	3·5814	3·6068	3·6322
12	3·6576	3·6830	3·7084	3·7338	3·7592	3·7846	3·8100	3·8354	3·8608	3·8862	3·9116	3·9370
13	3·9624	3·9878	4·0132	4·0386	4·0640	4·0894	4·1148	4·1402	4·1656	4·1910	4·2164	4·2418
14	4·2672	4·2926	4·3180	4·3434	4·3688	4·3942	4·4196	4·4450	4·4704	4·4958	4·5212	4·5466
15	4·5720	4·5974	4·6228	4·6482	4·6736	4·6990	4·7244	4·7498	4·7752	4·8006	4·8260	4·8514
16	4·8768	4·9022	4·9276	4·9530	4·9784	5·0038	5·0292	5·0546	5·0800	5·1054	5·1308	5·1562
17	5·1816	5·2070	5·2324	5·2578	5·2832	5·3086	5·3340	5·3594	5·3848	5·4102	5·4356	5·4610
18	5·4864	5·5118	5·5372	5·5626	5·5880	5·6134	5·6388	5·6642	5·6896	5·7150	5·7404	5·7658
19	5·7912	5·8166	5·8420	5·8674	5·8928	5·9182	5·9436	5·9690	5·9944	6·0198	6·0452	6·0706
20	6·0960	—	—	—	—	—	—	—	—	—	—	—

Table IV Inches and fractions of an inch to millimetres (⅛in increments up to 72in)

in	ft in	mm	in	ft in	mm	in	ft in	mm	in	ft in	mm
⅛		3·2	9⅛		231·8	18⅛	1 6⅛	460·4	27⅛	2 3⅛	689·0
¼		6·4	9¼		235·0	18¼	1 6¼	463·6	27¼	2 3¼	692·2
⅜		9·5	9⅜		238·1	18⅜	1 6⅜	466·7	27⅜	2 3⅜	695·3
½		12·7	9½		241·3	18½	1 6½	469·9	27½	2 3½	698·5
⅝		15·9	9⅝		244·5	18⅝	1 6⅝	473·1	27⅝	2 3⅝	701·7
¾		19·1	9¾		247·7	18¾	1 6¾	476·3	27¾	2 3¾	704·9
⅞		22·2	9⅞		250·8	18⅞	1 6⅞	479·4	27⅞	2 3⅞	708·0
1		25·4	10		254·0	19	1 7	482·6	28	2 4	711·2
1⅛		28·6	10⅛		257·2	19⅛	1 7⅛	485·8	28⅛	2 4⅛	714·4
1¼		31·8	10¼		260·4	19¼	1 7¼	489·0	28¼	2 4¼	717·6
1⅜		34·9	10⅜		263·5	19⅜	1 7⅜	492·1	28⅜	2 4⅜	720·7
1½		38·1	10½		266·7	19½	1 7½	495·3	28½	2 4½	723·9
1⅝		41·3	10⅝		269·9	19⅝	1 7⅝	498·5	28⅝	2 4⅝	727·1
1¾		44·5	10¾		273·1	19¾	1 7¾	501·7	28¾	2 4¾	730·3
1⅞		47·6	10⅞		276·2	19⅞	1 7⅞	504·8	28⅞	2 4⅞	733·4
2		50·8	11		279·4	20	1 8	508·0	29	2 5	736·6
2⅛		54·0	11⅛		282·6	20⅛	1 8⅛	511·2	29⅛	2 5⅛	739·8
2¼		57·2	11¼		285·8	20¼	1 8¼	514·4	29¼	2 5¼	743·0
2⅜		60·3	11⅜		288·9	20⅜	1 8⅜	517·5	29⅜	2 5⅜	746·1
2½		63·5	11½		292·1	20½	1 8½	520·7	29½	2 5½	749·3
2⅝		66·7	11⅝		295·3	20⅝	1 8⅝	523·9	29⅝	2 5⅝	752·5
2¾		69·9	11¾		298·5	20¾	1 8¾	527·1	29¾	2 5¾	755·7
2⅞		73·0	11⅞		301·6	20⅞	1 8⅞	530·2	29⅞	2 5⅞	758·8
3		76·2	12	1 0	304·8	21	1 9	533·4	30	2 6	762·0
3⅛		79·4	12⅛	1 0⅛	308·0	21⅛	1 9⅛	536·6	30⅛	2 6⅛	765·2
3¼		82·6	12¼	1 0¼	311·2	21¼	1 9¼	539·8	30¼	2 6¼	768·4
3⅜		85·7	12⅜	1 0⅜	314·3	21⅜	1 9⅜	542·9	30⅜	2 6⅜	771·5
3½		88·9	12½	1 0½	317·5	21½	1 9½	546·1	30½	2 6½	774·7
3⅝		92·1	12⅝	1 0⅝	320·7	21⅝	1 9⅝	549·3	30⅝	2 6⅝	777·9
3¾		95·3	12¾	1 0¾	323·9	21¾	1 9¾	552·5	30¾	2 6¾	781·1
3⅞		98·4	12⅞	1 0⅞	327·0	21⅞	1 9⅞	555·6	30⅞	2 6⅞	784·2
4		101·6	13	1 1	330·2	22	1 10	558·8	31	2 7	787·4
4⅛		104·8	13⅛	1 1⅛	333·4	22⅛	1 10⅛	562·0	31⅛	2 7⅛	790·6
4¼		108·0	13¼	1 1¼	336·6	22¼	1 10¼	565·2	31¼	2 7¼	793·8
4⅜		111·1	13⅜	1 1⅜	339·7	22⅜	1 10⅜	568·3	31⅜	2 7⅜	796·9
4½		114·3	13½	1 1½	342·9	22½	1 10½	571·5	31½	2 7½	800·1
4⅝		117·5	13⅝	1 1⅝	346·1	22⅝	1 10⅝	574·7	31⅝	2 7⅝	803·3
4¾		120·7	13¾	1 1¾	349·3	22¾	1 10¾	577·9	31¾	2 7¾	806·5
4⅞		123·8	13⅞	1 1⅞	352·4	22⅞	1 10⅞	581·0	31⅞	2 7⅞	809·6
5		127·0	14	1 2	355·6	23	1 11	584·2	32	2 8	812·8
5⅛		130·2	14⅛	1 2⅛	358·8	23⅛	1 11⅛	587·4	32⅛	2 8⅛	816·0
5¼		133·4	14¼	1 2¼	362·0	23¼	1 11¼	590·6	32¼	2 8¼	819·2
5⅜		136·5	14⅜	1 2⅜	365·1	23⅜	1 11⅜	593·7	32⅜	2 8⅜	822·3
5½		139·7	14½	1 2½	368·3	23½	1 11½	596·9	32½	2 8½	825·5
5⅝		142·9	14⅝	1 2⅝	371·5	23⅝	1 11⅝	600·1	32⅝	2 8⅝	828·7
5¾		146·1	14¾	1 2¾	374·7	23¾	1 11¾	603·3	32¾	2 8¾	831·9
5⅞		149·2	41⅞	1 2⅞	377·8	23⅞	1 11⅞	606·4	32⅞	2 8⅞	835·0
6		152·4	15	1 3	381·0	24	2 0	609·6	33	2 9	838·2
6⅛		155·6	15⅛	1 3⅛	384·2	24⅛	2 0⅛	612·8	33⅛	2 9⅛	841·4
6¼		158·8	15¼	1 3¼	387·4	24¼	2 0¼	616·0	33¼	2 9¼	844·6
6⅜		161·9	15⅜	1 3⅜	390·5	24⅜	2 0⅜	619·1	33⅜	2 9⅜	847·7
6½		165·1	15½	1 3½	393·7	24½	2 0½	622·3	33½	2 9½	850·9
6⅝		168·3	15⅝	1 3⅝	396·9	24⅝	2 0⅝	625·5	33⅝	2 9⅝	854·1
6¾		171·5	15¾	1 3¾	400·1	24¾	2 0¾	628·7	33¾	2 9¾	857·3
6⅞		174·6	15⅞	1 3⅞	403·2	24⅞	2 0⅞	631·8	33⅞	2 9⅞	860·4
7		177·8	16	1 4	406·4	25	2 1	635·0	34	2 10	863·6
7⅛		181·0	16⅛	1 4⅛	409·6	25⅛	2 1⅛	638·2	34⅛	2 10⅛	866·8
7¼		184·2	16¼	1 4¼	412·8	25¼	2 1¼	641·4	34¼	2 10¼	870·0
7⅜		187·3	16⅜	1 4⅜	415·9	25⅜	2 1⅜	644·5	34⅜	2 10⅜	873·1
7½		190·5	16½	1 4½	419·1	25½	2 1½	647·7	34½	2 10½	876·3
7⅝		193·7	16⅝	1 4⅝	422·3	25⅝	2 1⅝	650·9	34⅝	2 10⅝	879·5
7¾		196·9	16¾	1 4¾	425·5	25¾	2 1¾	654·1	34¾	2 10¾	882·7
7⅞		200·0	16⅞	1 4⅞	428·6	25⅞	2 1⅞	657·2	34⅞	2 10⅞	885·8
8		203·2	17	1 5	431·8	26	2 2	660·4	35	2 11	889·0
8⅛		206·4	17⅛	1 5⅛	435·0	26⅛	2 2⅛	663·6	35⅛	2 11⅛	892·2
8¼		209·6	17¼	1 5¼	438·2	26¼	2 2¼	666·8	35¼	2 11¼	895·4
8⅜		212·7	17⅜	1 5⅜	441·3	26⅜	2 2⅜	669·9	35⅜	2 11⅜	898·5
8½		215·9	17½	1 5½	444·5	26½	2 2½	673·1	35½	2 11½	901·7
8⅝		219·1	17⅝	1 5⅝	447·7	26⅝	2 2⅝	676·3	35⅝	2 11⅝	904·2
8¾		222·3	17¾	1 5¾	450·0	26¾	2 2¾	679·5	35¾	2 11¾	908·9
8⅞		225·4	17⅞	1 5⅞	454·0	26⅞	2 2⅞	682·6	35⅞	2 11⅞	911·1
9		228·6	18	1 6	457·2	27	2 3	685·8	36	3 0	914·4

Table IV Inches and fractions of an inch to millimetres *continued*

in	ft	in	mm	in	ft	in	mm	in	ft	in	mm	in	ft	in	mm
36⅛	3	0⅛	917·6	45⅛	3	9⅛	1146·2	54⅛	4	6⅛	1374·8	63⅛	5	3⅛	1603·4
36¼	3	0¼	920·8	45¼	3	9¼	1149·4	54¼	4	6¼	1378·0	63¼	5	3¼	1606·6
36⅜	3	0⅜	923·9	45⅜	3	9⅜	1152·5	54⅜	4	6⅜	1381·1	63⅜	5	3⅜	1609·7
36½	3	0½	927·1	45½	3	9½	1155·7	54½	4	6½	1384·3	63½	5	3½	1612·9
36⅝	3	0⅝	930·3	45⅝	3	9⅝	1158·9	54⅝	4	6⅝	1387·5	63⅝	5	3⅝	1616·1
36¾	3	0¾	933·5	45¾	3	9¾	1162·1	54¾	4	6¾	1390·7	63¾	5	3¾	1619·3
36⅞	3	0⅞	936·6	45⅞	3	9⅞	1165·2	54⅞	4	6⅞	1393·8	63⅞	5	3⅞	1622·4
37	3	1	939·8	46	3	10	1168·4	55	4	7	1397·0	64	5	4	1625·6
37⅛	3	1⅛	943·0	46⅛	3	10⅛	1171·6	55⅛	4	7⅛	1400·2	64⅛	5	4⅛	1628·8
37¼	3	1¼	946·2	46¼	3	10¼	1174·8	55¼	4	7¼	1403·4	64¼	5	4¼	1632·0
37⅜	3	1⅜	949·3	46⅜	3	10⅜	1177·9	55⅜	4	7⅜	1406·5	64⅜	5	4⅜	1635·1
37½	3	1½	952·5	46½	3	10½	1181·1	55½	4	7½	1409·7	64½	5	4½	1638·3
37⅝	3	1⅝	955·7	46⅝	3	10⅝	1184·3	55⅝	4	7⅝	1412·9	64⅝	5	4⅝	1641·5
37¾	3	1¾	958·9	46¾	3	10¾	1187·5	55¾	4	7¾	1416·1	64¾	5	4¾	1644·7
37⅞	3	1⅞	962·0	46⅞	3	10⅞	1190·6	55⅞	4	7⅞	1419·2	64⅞	5	4⅞	1647·8
38	3	2	965·2	47	3	11	1193·8	56	4	8	1422·4	65	5	5	1651·0
38⅛	3	2⅛	968·4	47⅛	3	11⅛	1197·0	56⅛	4	8⅛	1425·6	65⅛	5	5⅛	1654·2
38¼	3	2¼	971·6	47¼	3	11¼	1200·2	56¼	4	8¼	1428·8	65¼	5	5¼	1657·4
38⅜	3	2⅜	974·7	47⅜	3	11⅜	1203·3	56⅜	4	8⅜	1431·9	65⅜	5	5⅜	1660·5
38½	3	2½	977·9	47½	3	11½	1206·5	56½	4	8½	1435·1	65½	5	5½	1663·7
38⅝	3	2⅝	981·1	47⅝	3	11⅝	1209·7	56⅝	4	8⅝	1438·3	65⅝	5	5⅝	1666·9
38¾	3	2¾	984·3	47¾	3	11¾	1212·9	56¾	4	8¾	1441·5	65¾	5	5¾	1670·1
38⅞	3	2⅞	987·4	47⅞	3	11⅞	1216·0	56⅞	4	8⅞	1444·6	65⅞	5	5⅞	1673·2
39	3	3	990·6	48	4	0	1219·2	57	4	9	1447·8	66	5	6	1676·4
39⅛	3	3⅛	993·8	48⅛	4	0⅛	1222·4	57⅛	4	9⅛	1451·0	66⅛	5	6⅛	1679·6
39¼	3	3¼	997·0	48¼	4	0¼	1225·6	57¼	4	9¼	1454·2	66¼	5	6¼	1682·8
39⅜	3	3⅜	1000·1	48⅜	4	0⅜	1228·7	57⅜	4	9⅜	1457·3	66⅜	5	6⅜	1685·9
39½	3	3½	1003·3	48½	4	0½	1231·9	57½	4	9½	1460·5	66½	5	6½	1689·1
39⅝	3	3⅝	1006·5	48⅝	4	0⅝	1235·1	57⅝	4	9⅝	1463·7	66⅝	5	6⅝	1692·3
39¾	3	3¾	1009·7	48¾	4	0¾	1238·3	57¾	4	9¾	1466·9	66¾	5	6¾	1695·5
39⅞	3	3⅞	1012·8	48⅞	4	0⅞	1241·4	57⅞	4	9⅞	1470·0	66⅞	5	6⅞	1698·6
40	3	4	1016·0	49	4	1	1244·6	58	4	10	1473·2	67	5	7	1701·8
40⅛	3	4⅛	1019·2	49⅛	4	1⅛	1247·8	58⅛	4	10⅛	1476·4	67⅛	5	7⅛	1705·0
40¼	3	4¼	1022·4	49¼	4	1¼	1251·0	58¼	4	10¼	1479·6	67¼	5	7¼	1708·2
40⅜	3	4⅜	1025·5	49⅜	4	1⅜	1254·1	58⅜	4	10⅜	1482·7	67⅜	5	7⅜	1711·3
40½	3	4½	1028·7	49½	4	1½	1257·3	58½	4	10½	1485·9	67½	5	7½	1714·5
40⅝	3	4⅝	1031·9	49⅝	4	1⅝	1260·5	58⅝	4	10⅝	1489·1	67⅝	5	7⅞	1717·7
40¾	3	4¾	1035·1	49¾	4	1¾	1263·7	58¾	4	10¾	1492·3	67¾	5	7¾	1720·9
40⅞	3	4⅞	1038·2	49⅞	4	1⅞	1266·8	58⅞	4	10⅞	1495·4	67⅞	5	7⅞	1724·0
41	3	5	1041·4	50	4	2	1270·0	59	4	11	1498·6	68	5	8	1727·2
41⅛	3	5⅛	1044·6	50⅛	4	2⅛	1273·2	59⅛	4	11⅛	1501·8	68⅛	5	8⅛	1730·4
41¼	3	5¼	1047·8	50¼	4	2¼	1276·4	59¼	4	11¼	1505·0	68¼	5	8¼	1733·6
41⅜	3	5⅜	1050·9	50⅜	4	2⅜	1279·5	59⅜	4	11⅜	1508·1	68⅜	5	8⅜	1736·7
41½	3	5½	1054·1	50½	4	2½	1282·7	59½	4	11½	1511·3	68½	5	8½	1739·9
41⅝	3	5⅝	1057·3	50⅝	4	2⅝	1285·9	59⅝	4	11⅝	1514·5	68⅝	5	8⅝	1743·1
41¾	3	5¾	1060·5	50¾	4	2¾	1289·1	59¾	4	11¾	1517·7	68¾	5	8¾	1746·3
41⅞	3	5⅞	1063·6	50⅞	4	2⅞	1292·2	59⅞	4	11⅞	1520·8	68⅞	5	8⅞	1749·4
42	3	6	1066·8	51	4	3	1295·4	60	5	0	1524·0	69	5	9	1752·6
42⅛	3	6⅛	1070·0	51⅛	4	3⅛	1298·6	60⅛	5	0⅛	1527·2	69⅛	5	9⅛	1755·8
42¼	3	6¼	1073·2	51¼	4	3¼	1301·8	60¼	5	0¼	1530·4	69¼	5	9¼	1759·0
42⅜	3	6⅜	1076·3	51⅜	4	3⅜	1304·9	60⅜	5	0⅜	1533·5	69⅜	5	9⅜	1762·1
42½	3	6½	1079·5	51½	4	3½	1308·1	60½	5	0½	1536·7	69½	5	9½	1765·3
42⅝	3	6⅝	1082·7	51⅝	4	3⅝	1311·3	60⅝	5	0⅝	1539·9	69⅝	5	9⅝	1768·5
42¾	3	6¾	1085·9	51¾	4	3¾	1314·5	60¾	5	0¾	1543·1	69¾	5	9¾	1771·7
42⅞	3	6⅞	1089·0	51⅞	4	3⅞	1317·6	60⅞	5	0⅞	1546·2	69⅞	5	9⅞	1774·8
43	3	7	1092·2	52	4	4	1320·8	61	5	1	1549·4	70	5	10	1778·0
43⅛	3	7⅛	1095·4	52⅛	4	4⅛	1324·0	61⅛	5	1⅛	1552·6	70⅛	5	10⅛	1781·2
43¼	3	7¼	1098·6	52¼	4	4¼	1327·2	61¼	5	1¼	1555·8	70¼	5	10¼	1784·4
43⅜	3	7⅜	1101·7	52⅜	4	4⅜	1330·3	61⅜	5	1⅜	1558·9	70⅜	5	10⅜	1787·5
43½	3	7½	1104·9	52½	4	4½	1333·5	61½	5	1½	1562·1	70½	5	10½	1790·7
43⅝	3	7⅝	1108·1	52⅝	4	4⅝	1336·7	61⅝	5	1⅝	1565·3	70⅝	5	10⅝	1793·9
43¾	3	7¾	1111·3	52¾	4	4¾	1339·9	61¾	5	1¾	1568·5	70¾	5	10¾	1797·1
43⅞	3	7⅞	1114·4	52⅞	4	4⅞	1343·0	61⅞	5	1⅞	1571·6	70⅞	5	10⅞	1800·2
44	3	8	1117·6	53	4	5	1346·2	62	5	2	1574·8	71	5	11	1803·4
44⅛	3	8⅛	1120·8	53⅛	4	5⅛	1349·4	62⅛	5	2⅛	1578·0	71⅛	5	11⅛	1806·6
44¼	3	8¼	1124·0	53¼	4	5¼	1352·6	62¼	5	2¼	1581·2	71¼	5	11¼	1809·8
44⅜	3	8⅜	1127·1	53⅜	4	5⅜	1355·7	62⅜	5	2⅜	1584·3	71⅜	5	11⅜	1812·9
44½	3	8½	1130·3	53½	4	5½	1358·9	62½	5	2½	1587·5	71½	5	11½	1816·1
44⅝	3	8⅝	1133·5	53⅝	4	5⅝	1362·1	62⅝	5	2⅝	1590·7	71⅝	5	11⅝	1819·3
44¾	3	8¾	1136·7	53¾	4	5¾	1365·3	62¾	5	2¾	1593·9	71¾	5	11¾	1822·5
44⅞	3	8⅞	1139·8	53⅞	4	5⅞	1368·4	62⅞	5	2⅞	1597·0	71⅞	5	11⅞	1825·6
45	3	9	1143·0	54	4	6	1371·6	63	5	3	1600·2	72	6	0	1828·8

Table V Feet and inches (up to 100ft) to metres and millimetres (to nearest millimetre)

Metres and millimetres

Feet	0	1	2	3	4	5	6	7	8	9	10	11
0	—	25	51	76	102	127	152	178	203	229	254	279
1	305	330	356	381	406	432	457	483	508	533	559	584
2	610	635	660	686	711	737	762	787	813	838	864	889
3	914	940	965	991	1·016	1·041	1·067	1·092	1·118	1·143	1·168	1·194
4	1·219	1·245	1·270	1·295	1·321	1·346	1·372	1·397	1·422	1·448	1·473	1·499
5	1·524	1·549	1·575	1·600	1·626	1·651	1·676	1·702	1·727	1·753	1·778	1·803
6	1·829	1·854	1·880	1·905	1·930	1·956	1·981	2·007	2·032	2·057	2·083	2·108
7	2·134	2·159	2·184	2·210	2·235	2·261	2·286	2·311	2·337	2·362	2·388	2·413
8	2·438	2·464	2·489	2·515	2·540	2·565	2·591	2·616	2·642	2·667	2·692	2·718
9	2·743	2·769	2·794	2·819	2·845	2·870	2·896	2·921	2·946	2·972	2·997	3·023
10	3·048	3·073	3·099	3·124	3·150	3·175	3·200	3·226	3·251	3·277	3·302	3·327
11	3·353	3·378	3·404	3·429	3·454	3·480	3·505	3·531	3·556	3·581	3·607	3·632
12	3·658	3·683	3·708	3·734	3·759	3·785	3·810	3·835	3·861	3·886	3·912	3·937
13	3·962	3·988	4·013	4·039	4·064	4·089	4·115	4·140	4·166	4·191	4·216	4·242
14	4·267	4·293	4·318	4·343	4·369	4·394	4·420	4·445	4·470	4·496	4·521	4·547
15	4·572	4·597	4·623	4·648	4·674	4·699	4·724	4·750	4·775	4·801	4·826	4·851
16	4·877	4·902	4·928	4·953	4·978	5·004	5·029	5·055	5·080	5·105	5·131	5·156
17	5·182	5·207	5·232	5·258	5·283	5·309	5·334	5·359	5·385	5·410	5·436	5·461
18	5·486	5·512	5·537	5·563	5·588	5·613	5·639	5·664	5·690	5·715	5·740	5·766
19	5·791	5·817	5·842	5·867	5·893	5·918	5·944	5·969	5·994	6·020	6·045	6·071
20	6·096	6·121	6·147	6·172	6·198	6·223	6·248	6·274	6·299	6·325	6·350	6·375
21	6·401	6·426	6·452	6·477	6·502	6·528	6·553	6·579	6·604	6·629	6·655	6·680
22	6·706	6·731	6·756	6·782	6·807	6·833	6·858	6·883	6·909	6·934	6·960	6·985
23	7·010	7·036	7·061	7·087	7·112	7·137	7·163	7·188	7·214	7·239	7·264	7·290
24	7·315	7·341	7·366	7·391	7·417	7·442	7·468	7·493	7·518	7·544	7·569	7·595
25	7·620	7·645	7·671	7·696	7·722	7·747	7·772	7·798	7·823	7·849	7·874	7·899
26	7·925	7·950	7·976	8·001	8·026	8·052	8·077	8·103	8·128	8·153	8·179	8·204
27	8·230	8·255	8·280	8·306	8·331	8·357	8·382	8·407	8·433	8·458	8·484	8·509
28	8·534	8·560	8·585	8·611	8·636	8·661	8·687	8·712	8·738	8·763	8·788	8·814
29	8·839	8·865	8·890	8·915	8·941	8·966	8·992	9·017	9·042	9·068	9·093	9·119
30	9·144	9·169	9·195	9·220	9·246	9·271	9·296	9·322	9·347	9·373	9·398	9·423
31	9·449	9·474	9·500	9·525	9·550	9·576	9·601	9·627	9·652	9·677	9·703	9·728
32	9·754	9·779	9·804	9·830	9·855	9·881	9·906	9·931	9·957	9·982	10·008	10·033
33	10·058	10·084	10·109	10·135	10·160	10·185	10·211	10·236	10·262	10·287	10·312	10·338
34	10·363	10·389	10·414	10·439	10·465	10·490	10·516	10·541	10·566	10·592	10·617	10·643
35	10·668	10·693	10·719	10·744	10·770	10·795	10·820	10·846	10·871	10·897	10·922	10·947
36	10·973	10·998	11·024	11·049	11·074	11·100	11·125	11·151	11·176	11·201	11·227	11·252
37	11·278	11·303	11·328	11·354	11·379	11·405	11·430	11·455	11·481	11·506	11·532	11·557
38	11·582	11·608	11·633	11·659	11·684	11·709	11·735	11·760	11·786	11·811	11·836	11·862
39	11·887	11·913	11·938	11·963	11·989	12·014	12·040	12·065	12·090	12·116	12·141	12·167
40	12·192	12·217	12·243	12·268	12·294	12·319	12·344	12·370	12·395	12·421	12·446	12·471
41	12·497	12·522	12·548	12·573	12·598	12·624	12·649	12·675	12·700	12·725	12·751	12·776
42	12·802	12·827	12·852	12·878	12·903	12·929	12·954	12·979	13·005	13·030	13·056	13·081
43	13·106	13·132	13·157	13·183	13·208	13·233	13·259	13·284	13·310	13·335	13·360	13·386
44	13·411	13·437	13·462	13·487	13·513	13·538	13·564	13·589	13·614	13·640	13·665	13·691
45	13·716	13·741	13·767	13·792	13·818	13·843	13·868	13·894	13·919	13·945	13·970	13·995
46	14·021	14·046	14·072	14·097	14·122	14·148	14·173	14·199	14·224	14·249	14·275	14·300
47	14·326	14·351	14·376	14·402	14·427	14·453	14·478	14·503	14·529	14·554	14·580	14·605
48	14·630	14·656	14·681	14·707	14·732	14·757	14·783	14·808	14·834	14·859	14·884	14·910
49	14·935	14·961	14·986	15·011	15·037	15·062	15·088	15·113	15·138	15·164	15·189	15·215
50	15·240	15·265	15·291	15·316	15·342	15·367	15·392	15·418	18·443	15·469	15·494	15·519
51	15·545	15·570	15·596	15·621	15·646	15·672	15·697	15·723	15·748	15·773	15·799	15·824
52	15·850	15·875	15·900	15·926	15·951	15·977	16·002	16·027	16·053	16·078	16·104	16·129
53	16·154	16·180	16·205	16·231	16·256	16·281	16·307	16·332	16·358	16·383	16·408	16·434
54	16·459	16·485	16·510	16·535	16·561	16·586	16·612	16·637	16·662	16·688	16·713	16·739
55	16·764	16·789	16·815	16·840	16·866	16·891	16·916	16·942	16·967	16·993	17·018	17·043
56	17·069	17·094	17·120	17·145	17·170	17·196	17·221	17·247	17·272	17·297	17·323	17·348
57	17·374	17·399	17·424	17·450	17·475	17·501	17·526	17·551	17·577	17·602	17·628	17·653
58	17·678	17·704	17·729	17·755	17·780	17·805	17·830	17·856	17·882	17·907	17·932	17·958
59	17·983	18·009	18·034	18·059	18·085	18·110	18·136	18·161	18·186	18·212	18·237	18·263
60	18·288	18·313	18·339	18·364	18·390	18·415	18·440	18·466	18·491	18·517	18·542	18·567
61	18·593	18·618	18·644	18·669	18·694	18·720	18·745	18·771	18·796	18·821	18·847	18·872
62	18·898	18·923	18·948	18·974	18·999	19·025	19·050	19·075	19·101	19·126	19·152	19·177
63	19·202	19·228	19·253	19·279	19·304	19·329	19·355	19·380	19·406	19·431	19·456	19·482
64	19·507	19·533	19·558	19·583	19·609	19·634	19·660	19·685	19·710	19·736	19·761	19·787
65	19·812	19·837	19·863	19·888	19·914	19·939	19·964	19·990	20·015	20·041	20·066	20·091
66	20·117	20·142	20·168	20·193	20·218	20·244	20·269	20·295	20·320	20·345	20·371	20·396
67	20·422	20·447	20·472	20·498	20·523	20·549	20·574	20·599	20·625	20·650	20·676	20·701
68	20·726	20·752	20·777	20·803	20·828	20·853	20·879	20·904	20·930	20·955	20·980	21·006
69	21·031	21·057	21·082	21·107	21·133	21·158	21·184	21·209	21·234	21·260	21·285	21·311

Table V Feet and inches (up to 100 ft.) to metres and millimetres (to nearest millimetre) *continued*

Feet	Inches											
	0	1	2	3	4	5	6	7	8	9	10	11
	Metres and millimetres											
70	21·336	21·361	21·387	21·412	21·438	21·463	21·488	21·514	21·539	21·565	21·590	21·615
71	21·641	21·666	21·692	21·717	21·742	21·768	21·793	21·819	21·844	21·869	21·895	21·920
72	21·946	21·971	21·996	22·022	22·047	22·073	22·098	22·123	22·149	22·174	22·200	22·225
73	22·250	22·276	22·301	22·327	22·352	22·377	22·403	22·428	22·454	22·479	22·504	22·530
74	22·555	22·581	22·606	22·631	22·657	22·682	22·708	22·733	22·758	22·784	22·809	22·835
75	22·860	22·885	22·911	22·936	22·962	22·987	23·012	23·038	23·063	23·089	23·114	23·139
76	23·165	23·190	23·216	23·241	23·266	23·292	23·317	23·343	23·368	23·393	23·419	23·444
77	23·470	23·495	23·520	23·546	23·571	23·597	23·622	23·647	23·673	23·698	23·724	23·749
78	23·774	23·800	23·825	23·851	23·876	23·901	23·927	23·952	23·978	24·003	24·028	24·054
79	24·079	24·105	24·130	24·155	24·181	24·206	24·232	24·257	24·282	24·308	24·333	24·359
80	24·384	24·409	24·435	24·460	24·486	24·511	24·536	24·562	24·587	24·613	24·638	24·663
81	24·689	24·714	24·740	24·765	24·790	24·816	24·841	24·867	24·892	24·917	24·943	24·968
82	24·994	25·019	25·044	25·070	25·095	25·121	25·146	25·171	25·197	25·222	25·248	25·273
83	25·298	25·324	25·349	25·375	25·400	25·425	25·451	25·476	25·502	25·527	25·552	25·578
84	25·603	25·629	25·654	25·679	25·705	25·730	25·756	25·781	25·806	25·832	25·857	25·883
85	25·908	25·933	25·959	25·984	26·010	26·035	26·060	26·086	26·111	26·137	26·162	26·187
86	26·213	26·238	26·264	26·289	26·314	26·340	26·365	26·391	26·416	26·441	26·467	26·492
87	26·518	26·543	26·568	26·594	26·619	26·645	26·670	26·695	26·721	26·746	26·772	26·797
88	26·822	26·848	26·873	26·899	26·924	26·949	26·975	27·000	27·026	27·051	27·076	27·102
89	27·127	27·153	27·178	27·203	27·229	27·254	27·280	27·305	27·330	27·356	27·381	27·407
90	27·432	27·457	27·483	27·508	27·534	27·559	27·584	27·610	27·635	27·661	27·686	27·711
91	27·737	27·762	27·788	27·813	27·838	27·864	27·889	27·915	27·940	27·965	27·991	28·016
92	28·042	28·067	28·092	28·118	28·143	28·169	28·194	28·219	28·245	28·270	28·296	28·321
93	28·346	28·372	28·397	28·423	28·448	28·473	28·499	28·524	28·550	28·575	28·600	28·626
94	28·651	28·677	28·702	28·727	28·753	28·778	28·804	28·829	28·854	28·880	28·905	28·931
95	28·956	28·981	29·007	29·032	29·058	29·083	29·108	29·134	29·159	29·185	29·210	29·235
96	29·261	29·286	29·312	29·337	29·362	29·388	29·413	29·439	29·464	29·489	29·515	29·540
97	29·566	29·591	29·616	29·642	29·667	29·693	29·718	29·743	29·769	29·794	29·820	29·845
98	29·870	29·896	29·921	29·947	29·972	29·997	30·023	30·048	30·074	30·099	30·124	30·150
99	30·175	30·201	30·226	30·251	30·277	30·302	30·328	30·353	30·378	30·404	30·429	30·455
100	30·480	—	—	—	—	—	—	—	—	—	—	—

Table VI Miles (up to 100 miles) to kilometres (to two places of decimals)

Miles	0	1	2	3	4	5	6	7	8	9
	Kilometres									
0	—	1·61	3·22	4·83	6·44	8·05	9·66	11·27	12·87	14·48
10	16·09	17·70	19·31	20·92	22·53	24·14	25·75	27·36	28·97	30·58
20	32·19	33·80	35·41	37·01	38·62	40·23	41·84	43·45	45·06	46·67
30	48·28	49·89	51·50	53·11	54·72	56·33	57·94	59·55	61·16	62·76
40	64·37	65·98	67·59	69·20	70·81	72·42	74·03	75·64	77·25	78·86
50	80·47	82·08	83·69	85·30	86·90	88·51	90·12	91·73	93·34	94·95
60	96·56	98·17	99·78	101·39	103·00	104·61	106·22	107·83	109·44	111·05
70	112·65	114·26	115·87	117·48	119·09	120·70	122·31	123·92	125·53	127·14
80	128·75	130·36	131·97	133·58	135·19	136·79	138·40	140·01	141·62	143·23
90	144·84	146·45	148·06	149·67	151·28	152·89	154·50	156·11	157·72	159·33
100	160·93	—	—	—	—	—	—	—	—	—

Table VII Square inches (up to 100 sq in) to square millimetres (to one place of decimals)

Square inches	0	1	2	3	4	5	6	7	8	9
	Square millimetres (mm²)									
0	—	645·2	1290·3	1935·5	2580·6	3225·8	3871·0	4516·1	5161·3	5806·4
10	6451·6	7096·8	7741·9	8387·1	9032·2	9677·4	10322·6	10967·7	11612·9	12258·0
20	12903·2	13548·4	14193·5	14838·7	15483·8	16129·0	16774·2	17419·3	18064·5	18709·6
30	19354·8	20000·0	20645·1	21290·3	21935·4	22580·6	23225·8	23870·9	24516·1	25161·2
40	25806·4	26451·6	27096·7	27741·9	28387·0	29032·2	29677·4	30322·5	30967·7	31612·8
50	32258·0	32903·2	33548·3	34193·5	34838·6	35483·8	36129·0	36774·1	37419·3	38064·4
60	38709·6	39354·8	39999·9	40645·1	41290·2	41935·4	42580·6	43225·7	43870·9	44516·0
70	45161·2	45806·4	46451·5	47096·7	47741·8	48387·0	49032·2	49677·3	50322·5	50967·6
80	51612·8	52258·0	52903·1	53548·3	54193·4	54838·6	55483·8	56128·9	56774·1	57419·2
90	58064·4	58709·6	59354·7	59999·9	60645·0	61290·2	61935·4	62580·5	63225·7	63870·8
100	64516·0									

Table VIII Square feet (up to 500ft²) to square metres (to two places of decimals)

Square feet	0	1	2	3	4	5	6	7	8	9
	Square metres (m²)									
0	—	0·09	0·19	0·28	0·37	0·46	0·56	0·65	0·74	0·84
10	0·93	1·02	1·11	1·21	1·30	1·39	1·49	1·58	1·67	1·77
20	1·86	1·95	2·04	2·14	2·23	2·32	2·42	2·51	2·60	2·69
30	2·79	2·88	2·97	3·07	3·16	3·25	3·34	3·44	3·53	3·62
40	3·72	3·81	3·90	3·99	4·09	4·18	4·27	4·37	4·46	4·55
50	4·65	4·74	4·83	4·92	5·02	5·11	5·20	5·30	5·39	5·48
60	5·57	5·67	5·76	5·85	5·95	6·04	6·13	6·22	6·32	6·41
70	6·50	6·60	6·69	6·78	6·87	6·97	7·06	7·15	7·25	7·34
80	7·43	7·53	7·62	7·71	7·80	7·90	7·99	8·08	8·18	8·27
90	8·36	8·45	8·55	8·64	8·73	8·83	8·92	9·01	9·10	9·20
100	9·29	9·38	9·48	9·57	9·66	9·75	9·85	9·94	10·03	10·13
110	10·22	10·31	10·41	10·50	10·59	10·68	10·78	10·87	10·96	11·06
120	11·15	11·24	11·33	11·43	11·52	11·61	11·71	11·80	11·89	11·98
130	12·08	12·17	12·26	12·36	12·45	12·54	12·63	12·73	12·82	12·91
140	13·01	13·10	13·19	13·29	13·38	13·47	13·56	13·66	13·75	13·84
150	13·94	14·03	14·12	14·21	14·31	14·40	14·49	14·59	14·68	14·77
160	14·86	14·96	15·05	15·14	15·24	15·33	15·42	15·51	15·61	15·70
170	15·79	15·89	15·98	16·07	16·17	16·26	16·35	16·44	16·54	16·63
180	16·72	16·82	16·91	17·00	17·09	17·19	17·28	17·37	17·47	17·56
190	17·65	17·74	17·84	17·93	18·02	18·12	18·21	18·30	18·39	18·49
200	18·58	18·67	18·77	18·86	18·95	19·05	19·14	19·23	19·32	19·42
210	19·51	19·60	19·70	19·79	19·88	19·97	20·07	20·16	20·25	20·35
220	20·44	20·53	20·62	20·72	20·81	20·90	21·00	21·09	21·18	21·27
230	21·37	21·46	21·55	21·65	21·74	21·83	21·93	22·02	22·11	22·20
240	22·30	22·39	22·48	22·58	22·67	22·76	22·85	22·95	23·04	23·13
250	23·23	23·32	23·41	23·50	23·60	23·69	23·78	23·88	23·97	24·06
260	24·15	24·25	24·34	24·43	24·53	24·62	24·71	24·81	24·90	24·99
270	25·08	25·18	25·27	25·36	25·46	25·55	25·64	25·73	25·83	25·92
280	26·01	26·11	26·20	26·29	26·38	26·48	26·57	26·66	26·76	26·85
290	26·94	27·03	27·13	27·22	27·31	27·41	27·50	27·59	27·69	27·78
300	27·87	27·96	28·06	28·15	28·24	28·34	28·43	28·52	28·61	28·71
310	28·80	28·89	28·99	29·08	29·17	29·26	29·36	29·45	29·54	29·64
320	29·73	29·82	29·91	30·01	30·10	30·19	30·29	30·38	30·47	30·57
330	30·66	30·75	30·84	30·94	31·03	31·12	31·22	31·31	31·40	31·49
340	31·59	31·68	31·77	31·87	31·96	32·05	32·14	32·24	32·33	32·42
350	32·52	32·61	32·70	32·79	32·89	32·98	33·07	33·17	33·26	33·35
360	33·45	33·54	33·63	33·72	33·82	33·91	34·00	34·10	34·19	34·28
370	34·37	34·47	34·56	34·65	34·75	34·84	34·93	35·02	35·12	35·21
380	35·30	35·40	35·49	35·58	35·67	35·77	35·86	35·95	36·05	36·14
390	36·23	36·33	36·42	36·51	36·60	36·70	36·79	36·88	36·98	37·07
400	37·16	37·25	37·35	37·44	37·53	37·63	37·72	37·81	37·90	38·00
410	38·09	38·18	38·28	38·37	38·46	38·55	38·65	38·74	38·83	38·93
420	39·02	39·11	39·21	39·30	39·39	39·48	39·58	39·67	39·76	39·86
430	39·95	40·04	40·13	40·23	40·32	40·41	40·51	40·60	40·69	40·78
440	40·88	40·97	41·06	41·16	41·25	41·34	41·43	41·53	41·62	41·71
450	41·81	41·90	41·99	42·09	42·18	42·27	42·36	42·46	42·55	42·64
460	42·74	42·83	42·92	43·01	43·11	43·20	43·29	43·39	43·48	43·57
470	43·66	43·76	43·85	43·94	44·04	44·13	44·22	44·31	44·41	44·50
480	44·59	44·69	44·78	44·87	44·97	45·06	45·15	45·24	45·34	45·43
490	45·52	45·62	45·71	45·80	45·89	45·99	46·08	46·17	46·27	46·36
500	46·45									

Table IX Cubic feet (up to 100ft³) to cubic metres (to two places of decimals)

Cubic feet	0	1	2	3	4	5	6	7	8	9
	Cubic metres (m³)									
0	—	0·03	0·06	0·08	0·11	0·14	0·17	0·20	0·23	0·25
10	0·28	0·31	0·34	0·37	0·40	0·42	0·45	0·48	0·51	0·54
20	0·57	0·59	0·62	0·65	0·68	0·71	0·73	0·76	0·79	0·82
30	0·85	0·88	0·91	0·93	0·96	0·99	1·02	1·05	1·08	1·10
40	1·13	1·16	1·19	1·22	1·25	1·27	1·30	1·33	1·36	1·39
50	1·42	1·44	1·47	1·50	1·53	1·56	1·59	1·61	1·64	1·67
60	1·70	1·73	1·76	1·78	1·81	1·84	1·87	1·90	1·93	1·95
70	1·98	2·01	2·04	2·07	2·10	2·12	2·15	2·18	2·21	2·24
80	2·27	2·29	2·32	2·35	2·38	2·41	2·44	2·46	2·49	2·52
90	2·55	2·58	2·61	2·63	2·66	2·69	2·72	2·75	2·78	2·80
100	2·83	—	—	—	—	—	—	—	—	—

Table X Pounds (up to 500lb) to kilogrammes (to two places of decimals)

Pounds	0	1	2	3	4	5	6	7	8	9
	Kilogrammes (kg)									
0	—	0·45	0·91	1·36	1·81	2·27	2·72	3·18	3·63	4·08
10	4·54	4·99	5·44	5·90	6·35	6·80	7·26	7·71	8·16	8·62
20	9·07	9·53	9·98	10·43	10·89	11·34	11·79	12·25	12·70	13·15
30	13·61	14·06	14·52	14·97	15·42	15·88	16·33	16·78	17·24	17·69
40	18·14	18·60	19·05	19·50	19·96	20·41	20·87	21·32	21·77	22·23
50	22·68	23·13	23·59	24·04	24·49	24·95	25·40	25·85	26·31	26·76
60	27·22	27·67	28·12	28·58	29·03	29·48	29·94	30·39	30·84	31·30
70	31·75	32·21	32·66	33·11	33·57	34·02	34·47	34·93	35·38	35·83
80	36·29	36·74	37·19	37·65	38·10	38·56	39·01	39·46	39·92	40·37
90	40·82	41·28	41·73	42·18	42·64	43·09	43·54	44·00	44·45	44·91
100	45·36	45·81	46·27	46·72	47·17	47·63	48·08	48·53	48·99	49·44
110	49·90	50·35	50·80	51·26	51·71	52·16	52·62	53·07	53·52	53·98
120	54·43	54·88	55·34	55·79	56·25	56·70	57·15	57·61	58·06	58·51
130	58·97	59·42	59·87	60·33	60·78	61·24	61·69	62·14	62·60	63·05
140	63·50	63·96	64·41	64·86	65·32	65·77	66·22	66·68	67·13	67·59
150	68·04	68·49	68·95	69·40	69·85	70·31	70·76	71·21	71·67	72·12
160	72·57	73·03	73·48	73·94	74·39	74·84	75·30	75·75	76·20	76·66
170	77·11	77·56	78·02	78·47	78·93	79·38	79·83	80·29	80·74	81·19
180	81·65	82·10	82·55	83·01	83·46	83·91	84·37	84·82	85·28	85·73
190	86·18	86·64	87·09	87·54	88·00	88·45	88·90	89·36	89·81	90·26
200	90·72	91·17	91·63	92·08	92·53	92·99	93·44	93·89	94·35	94·80
210	95·25	95·71	96·16	96·62	97·07	97·52	97·98	98·43	98·88	99·34
220	99·79	100·24	100·70	101·15	101·61	102·06	102·51	102·97	103·42	103·87
230	104·33	104·78	105·23	105·69	106·14	106·59	107·05	107·50	107·96	108·41
240	108·86	109·32	109·77	110·22	110·68	111·13	111·58	112·04	112·49	112·95
250	113·40	113·85	114·31	114·76	115·21	115·67	116·12	116·57	117·03	117·48
260	117·93	118·39	118·84	119·30	119·75	120·20	120·66	121·11	121·56	122·02
270	122·47	122·92	123·38	123·83	124·28	124·74	125·19	125·65	126·10	126·55
280	127·01	127·46	127·91	128·37	128·82	129·27	129·73	130·18	130·64	131·09
290	131·54	132·00	132·45	132·90	133·36	133·81	134·26	134·72	135·17	135·62
300	136·08	136·53	136·99	137·44	137·89	138·35	138·80	139·25	139·71	140·16
310	140·61	141·07	141·52	141·97	142·43	142·88	143·34	143·79	144·24	144·70
320	145·15	145·60	146·06	146·51	146·96	147·42	147·87	148·33	148·78	149·23
330	149·69	150·14	150·59	151·05	151·50	151·95	152·41	152·86	153·31	153·77
340	154·22	154·68	155·13	155·58	156·04	156·49	156·94	157·40	157·85	158·30
350	158·76	159·21	159·67	160·12	160·57	161·03	161·48	161·93	162·39	162·84
360	163·29	163·75	164·20	164·65	165·11	165·56	166·02	166·47	166·92	167·38
370	167·83	168·28	168·74	169·10	169·64	170·10	170·55	171·00	171·46	171·91
380	172·37	172·82	173·27	173·73	174·18	174·63	175·09	175·54	175·99	176·45
390	176·90	177·36	177·81	178·26	178·72	179·17	179·62	180·08	180·53	180·98
400	181·44	181·89	183·34	182·80	183·25	183·71	184·16	184·61	185·07	185·52
410	185·97	186·43	186·88	187·33	187·79	188·24	188·69	189·15	189·60	190·06
420	190·51	190·96	191·42	191·87	192·32	192·78	193·23	193·68	194·14	194·59
430	195·05	195·50	195·95	196·41	196·86	197·31	197·77	198·22	198·67	199·13
440	199·58	200·03	200·49	200·94	201·40	201·85	202·30	202·76	203·21	203·66
450	204·12	204·57	205·02	205·48	205·93	206·39	206·84	207·29	207·75	208·20
460	208·65	209·11	209·56	210·01	210·47	210·92	211·37	211·83	212·28	212·74
470	213·19	213·64	214·10	214·55	215·00	215·46	215·91	216·36	216·82	217·27
480	217·72	218·18	218·63	219·09	219·54	219·99	220·45	220·90	221·35	221·81
490	222·26	222·71	223·17	223·62	224·08	224·53	224·98	225·44	225·89	226·34
500	226·80	—	—	—	—	—	—	—	—	—

Table XI Pounds per cubic foot to kilogrammes per cubic metre (to one place of decimals)

Pounds per cubic foot	0	1	2	3	4	5	6	7	8	9
	Kilogrammes per cubic metre (kg/m³)									
0	—	16·0	32·0	48·1	64·1	80·1	96·1	112·1	128·1	144·2
10	160·2	176·2	192·2	208·2	224·3	240·3	256·3	272·3	288·3	304·4
20	320·4	336·4	352·4	368·4	384·4	400·5	416·5	432·5	448·5	464·5
30	480·6	496·6	512·6	528·6	544·6	560·6	576·7	592·7	608·7	624·7
40	640·7	656·8	672·8	688·8	704·8	720·8	736·8	752·9	768·9	784·9
50	800·9	816·9	833·0	849·0	865·0	881·0	897·0	913·1	929·1	945·1
60	961·1	977·1	993·1	1009·2	1025·2	1041·2	1057·2	1073·2	1089·3	1105·3
70	1121·3	1137·3	1153·3	1169·4	1185·4	1201·4	1217·4	1233·4	1249·4	1265·5
80	1281·5	1297·5	1313·5	1329·5	1345·6	1361·6	1377·6	1393·6	1409·6	1425·6
90	1441·7	1457·7	1473·7	1489·7	1505·7	1521·8	1537·8	1553·8	1569·8	1585·8
100	1601·9	—	—	—	—	—	—	—	—	—

Table XII Pounds force per square foot (up to 1000lb) to kilonewtons per square metre (to three places of decimals)

lbf/ft:	0	10	20	30	40	50	60	70	80	90
	Kilonewtons per square metre									
0	—	0·479	0·958	1·436	1·915	2·394	2·873	3·352	3·830	4·309
100	4·788	5·267	5·746	6·224	6·703	7·182	7·661	8·140	8·618	9·097
200	9·576	10·055	10·534	11·013	11·491	11·970	12·449	12·928	13·407	13·885
300	14·364	14·843	15·322	15·801	16·279	16·758	17·237	17·716	18·195	18·673
400	19·152	19·631	20·110	20·589	21·067	21·546	22·025	22·504	22·983	23·461
500	23·940	24·419	24·898	25·377	25·855	26·334	26·813	27·292	27·771	28·249
600	28·728	29·207	29·686	30·165	30·643	31·122	31·601	32·080	32·559	33·037
700	33·516	33·995	34·474	34·953	35·431	35·910	36·389	36·868	37·347	37·825
800	38·304	38·783	39·262	39·741	40·219	40·698	41·177	41·656	42·135	42·613
900	43·092	43·571	44·050	44·529	45·007	45·486	45·965	46·444	46·923	47·402
1000	47·880									

Table XIII UK gallons (up to 100 galls) to litres (to two places of decimals)

UK gallons	0	1	2	3	4	5	6	7	8	9
	Litres									
0	—	4·55	9·09	13·64	18·18	22·73	27·28	31·82	36·37	40·91
10	45·46	50·01	54·55	59·10	63·64	68·19	72·74	77·28	81·83	86·37
20	90·92	95·47	100·01	104·56	109·10	113·65	118·20	122·74	127·29	131·83
30	136·38	140·93	145·47	150·02	154·56	159·11	163·66	168·20	172·75	177·29
40	181·84	186·38	190·93	195·48	200·02	204·57	209·11	213·66	218·21	222·75
50	227·30	231·84	236·39	240·94	245·48	250·03	254·57	259·12	263·67	268·21
60	272.76	277·30	281·85	286·40	290·94	295·49	300·03	304·58	309·13	313.67
70	318·22	322·76	327·31	331·86	336·40	340·95	345·49	350·04	354·59	359·13
80	363·68	368·22	372·77	377·32	381·86	386·41	390·95	395·50	400·04	404·59
90	409·14	413·68	418·23	422·77	427·32	431·87	436·41	440·96	445·50	450·05
100	454·60	—	—	—	—	—	—	—	—	—

Table XIV Acres (up to 1000 acres) to hectares (to two places of decimals)

Acres	0	1	2	3	4	5	6	7	8	9
	Hectares									
	—	0·40	0·81	1·21	1·62	2·02	2·43	2·83	3·24	3·64

Acres	0	10	20	30	40	50	60	70	80	90
	Hectares									
0	—	4·05	8·09	12·14	16·19	20·23	24·28	28·33	32·37	36·42
100	40·47	44·52	48·56	52·61	56·66	60·70	64·75	68·80	72·84	76·89
200	80·94	84·98	89·03	93·08	97·12	101·17	105·22	109·27	113·31	117·36
300	121·41	125·45	129·50	133·55	137·59	141·64	145·69	149·73	153·78	157·83
400	161·87	165·92	169·97	174·02	178·06	182·11	186·16	190·20	194·25	198·30
500	202·34	206·39	210·44	214·48	218·53	222·58	226·62	230·67	234·72	238·77
600	242·81	246·86	250·91	254·95	259·00	263·05	267·09	271·14	275·19	279·23
700	283·28	287·33	291·37	295·42	299·47	303·51	307·56	311·61	315·66	319·70
800	323·75	327·80	331·84	335·89	339·94	343·98	348·03	352·08	356·12	360·17
900	364·22	368·26	372·31	376·36	380·41	384·45	388·50	392·55	396·59	400·64
1000	404·69	—	—	—	—	—	—	—	—	—

Table XV Miles per hour (up to 100mph) to metres per second (to two places of decimals)

miles per hour	0	1	2	3	4	5	6	7	8	9
	Metres per second									
0	—	0·45	0·89	1·34	1·79	2·24	2·68	3·13	3·58	4·02
10	4·47	4·92	5·36	5·81	6·26	6·71	7·15	7·60	8·05	8·49
20	8·94	9·39	9·83	10·28	10·73	11·18	11·62	12·07	12·52	12·96
30	13·41	13·86	14·31	14·75	15·20	15·65	16·09	16·54	16·99	17·43
40	17·88	18·33	18·78	19·22	19·67	20·12	20·56	21·01	21·46	21·91
50	22·35	22·80	23·25	23·69	24·14	24·59	25·03	25·48	25·93	26·38
60	26·82	27·27	27·72	28·16	28·61	29·06	29·50	29·95	30·40	30·85
70	31·29	31·74	32·19	32·63	33·08	33·53	33·98	34·42	34·87	35·32
80	35·76	36·21	36·66	37·10	37·55	38·00	38·45	38·89	39·34	39·79
90	40·23	40·68	41·13	41·57	42·02	42·47	42·92	43·36	43·81	44·26
100	44·70	—	—	—	—	—	—	—	—	—

Table XVI Degrees Fahrenheit (up to 500°F) to degrees Celsius (Centigrade)

Part A: Below freezing point of water (below 32°F or 0°C) Figures are negative values

°F	0°	1°	2°	3°	4°	5°	6°	7°	8°	9°
	°Celsius (below freezing point)									
0°	17·8	17·2	16·7	16·1	15·6	15·0	14·4	13·9	13·3	12·8
10°	12·2	11·7	11·1	10·6	10·0	9·4	8·9	8·3	7·8	7·2
20°	6·7	6·1	5·6	5·0	4·4	3·9	3·3	2·8	2·2	1·7
30°	1·1	0·6	0	—	—	—	—	—	—	—

Part B: Above freezing point of water (above 32°F or 0°C) Figures are positive values

°F	0°	1°	2°	3°	4°	5°	6°	7°	8°	9°
	°Celsius (above freezing point)									
30°	—	—	0	0·6	1·1	1·7	2·2	2·8	3·3	3·9
40°	4·4	5·0	5·6	6·1	6·7	7·2	7·8	8·3	8·9	9·4
50°	10·0	10·6	11·1	11·7	12·2	12·8	13·3	13·9	14·4	15·0
60°	15·6	16·1	16·7	17·2	17·8	18·3	18·9	19·4	20·0	20·6
70°	21·1	21·7	22·2	22·8	23·3	23·9	24·4	25·0	25·6	26·1
80°	26·7	27·2	27·8	28·3	28·9	29·4	30·0	30·6	31·1	31·7
90°	32·2	32·8	33·3	33·9	34·4	35·0	35·6	36·1	36·7	37·2
100°	37·8	38·3	38·9	39·4	40·0	40·6	41·1	41·7	42·2	42·8
110°	43·3	43·9	44·4	45·0	45·6	46·1	46·7	47·2	47·8	48·3
120°	48·9	49·4	50·0	50·6	51·1	51·7	52·2	52·8	53·3	53·9
130°	54·4	55·0	55·6	56·1	56·7	57·2	57·8	58·3	58·9	59·4
140°	60·0	60·6	61·1	61·7	62·2	62·8	63·3	63·9	64·4	65·0
150°	65·6	66·1	66·7	67·2	67·8	68·3	68·9	69·4	70·0	70·6
160°	71·1	71·7	72·2	72·8	73·3	73·9	74·4	75·0	75·6	76·1
170°	76·7	77·2	77·8	78·3	78·9	79·4	80·0	80·6	81·1	81·7
180°	82·2	82·8	83·3	83·9	84·4	85·0	85·6	86·1	86·7	87·2
190°	87·8	88·3	88·9	89·4	90·0	90·6	91·1	91·7	92·2	92·8
200°	93·3	93·9	94·4	95·0	95·6	96·1	96·7	97·2	97·8	98·3
210°	98·9	99·4	100·0	100·6	101·1	101·7	102·2	102·8	103·3	103·9
220°	104·4	105·0	105·6	106·1	106·7	107·2	107·8	108·3	108·9	109·4
230°	110·0	110·6	111·1	111·7	112·2	112·8	113·3	113·9	114·4	115·0
240°	115·6	116·1	116·7	117·2	117·8	118·3	118·9	119·4	120·0	120·6
250°	121·1	121·7	122·2	122·8	123·3	123·9	124·4	125·0	125·6	126·1
260°	126·7	127·2	127·8	128·3	128·9	129·4	130·0	130·6	131·1	131·7
270°	132·2	132·8	133·3	133·9	134·4	135·0	135·6	136·1	136·7	137·2
280°	137·8	138·3	138·9	139·4	140·0	140·6	141·1	141·7	142·2	142·8
290°	143·3	143·9	144·5	145·0	145·6	146·1	146·7	147·2	147·8	148·3
300°	148·9	149·4	150·0	150·6	151·1	151·7	152·2	152·8	153·3	153·9
310°	154·4	155·0	155·6	156·1	156·7	157·2	157·8	158·3	158·9	159·4
320°	160·0	160·6	161·1	161·7	162·2	162·8	163·3	163·9	164·4	165·0
330°	165·6	166·1	166·7	167·2	167·8	168·3	168·9	169·4	170·0	170·6
340°	171·1	171·7	172·2	172·8	173·2	173·9	174·4	175·0	175·6	176·1
350°	176·7	177·2	177·8	178·3	178·9	179·4	180·0	180·6	181·1	181·7
360°	182·2	182·8	183·3	183·9	184·4	185·0	185·6	186·1	186·7	187·2
370°	187·8	188·3	188·9	189·4	190·0	190·6	191·1	191·7	192·2	192·8
380°	193·3	193·9	194·4	195·0	195·6	196·1	196·7	197·2	197·8	198·3
390°	198·9	199·4	200·0	200·6	201·1	201·7	202·2	202·8	203·3	203·9
400°	204·4	205·0	205·6	206·1	206·7	207·2	207·8	208·3	208·9	209·4
410°	210·0	210·6	211·1	211·7	212·2	212·8	213·3	213·9	214·4	215·0
420°	215·6	216·1	216·7	217·2	217·8	218·3	218·9	219·4	220·0	220·6
430°	221·1	221·7	222·2	222·8	223·3	223·3	224·4	225·0	225·6	226·1
440°	226·7	227·2	227·8	228·3	228·9	229·4	230·0	230·6	231·1	231·7
450°	232·2	232·8	233·3	233·9	234·4	235·0	235·6	236·1	236·7	237·2
460°	237·8	238·3	238·9	239·4	240·0	240·6	241·1	241·7	242·2	242·8
470°	243·3	243·9	244·4	245·0	245·6	246·1	246·7	247·2	247·8	248·3
480°	248·9	249·4	250·0	250·6	251·1	251·7	252·2	252·8	253·3	253·9
490°	254·4	255·0	255·6	256·1	256·7	257·2	257·8	258·3	258·9	259·4
500°	260·0	—	—	—	—	—	—	—	—	—

Table XVII New halfpenny conversions for cash transactions

£sd	£p	£sd	£p
1d	½p	7d	3p
2d, 3d	1p	8d	3½p
4d	1½p	9d, 10d	4p
5d	2p	11d	4½p
6d equals 2½p		12d (1/-) equals 5p	

Table XVIII Whole new penny conversions for banking and accounting purposes

£sd	£p	£sd	£p
1d	0	1/2, 1/3	6p
2d, 3d	1p	1/4, 1/5, 1/6	7p
4d, 5d	2p	1/7, 1/8	8p
6d, 7d, 8d	3p	1/9, 1/10	9p
9d, 10d	4p	1/11, 2/-	10p
11d, 1/-, 1/1	5p		

Table XIX £sd to £p in 1d increments

Shillings	Pence											
	0	1	2	3	4	5	6	7	8	9	10	11
	£p equivalents											
0	0·00	0·00	0·01	0·01	0·02	0·02	0·03	0·03	0·03	0·04	0·04	0·05
1	0·05	0·05	0·06	0·06	0·07	0·07	0·07	0·08	0·08	0·09	0·09	0·10
2	0·10	0·10	0·11	0·11	0·12	0·12	0·13	0·13	0·13	0·14	0·14	0·15
3	0·15	0·15	0·16	0·16	0·17	0·17	0·17	0·18	0·18	0·19	0·19	0·20
4	0·20	0·20	0·21	0·21	0·22	0·22	0·23	0·23	0·23	0·24	0·24	0·25
5	0·25	0·25	0·26	0·26	0·27	0·27	0·27	0·28	0·28	0·29	0·29	0·30
6	0·30	0·30	0·31	0·31	0·32	0·32	0·33	0·33	0·33	0·34	0·34	0·35
7	0·35	0·35	0·36	0·36	0·37	0·37	0·37	0·38	0·38	0·39	0·39	0·40
8	0·40	0·40	0·41	0·41	0·42	0·42	0·43	0·43	0·43	0·44	0·44	0·45
9	0·45	0·45	0·46	0·46	0·47	0·47	0·47	0·48	0·48	0·49	0·49	0·50
10	0·50	0·50	0·51	0·51	0·52	0·52	0·53	0·53	0·53	0·54	0·54	0·55
11	0·55	0·55	0·56	0·56	0·57	0·57	0·57	0·58	0·58	0·59	0·59	0·60
12	0·60	0·60	0·61	0·61	0·62	0·62	0·63	0·63	0·63	0·64	0·64	0·65
13	0·65	0·65	0·66	0·66	0·67	0·67	0·67	0·68	0·68	0·69	0·69	0·70
14	0·70	0·70	0·71	0·71	0·72	0·72	0·73	0·73	0·73	0·74	0·74	0·75
15	0·75	0·75	0·76	0·76	0·77	0·77	0·77	0·78	0·78	0·79	0·79	0·80
16	0·80	0·80	0·81	0·81	0·82	0·82	0·83	0·83	0·83	0·84	0·84	0·85
17	0·85	0·85	0·86	0·86	0·87	0·87	0·87	0·88	0·88	0·89	0·89	0·90
18	0·90	0·90	0·91	0·91	0·92	0·92	0·93	0·93	0·93	0·94	0·94	0·95
19	0·95	0·95	0·96	0·96	0·97	0·97	0·97	0·98	0·98	0·99	0·99	1·00
20	1·00	—	—	—	—	—	—	—	—	—	—	—

This table has been compiled to conform to banking and accounting requirements

Table XX £sd/sq ft to £p/m²

Shillings per sq ft	Pence												
	0	½d	1	2	3	4	5	6	7	8	9	10	11
	£p equivalents per m²												
0	—	0·022	0·045	0·090	0·135	0·179	0·224	0·269	0·314	0·359	0·404	0·448	0·493
1	0·538	0·561	0·583	0·628	0·673	0·718	0·762	0·807	0·852	0·897	0·942	0·987	1·032
2	1·076	1·099	1·121	1·166	1·211	1·256	1·301	1·345	1·390	1·435	1·480	1·525	1·570
3	1·615	1·637	1·659	1·704	1·749	1·794	1·839	1·884	1·929	1·973	2·018	2·063	2·108
4	2·153	2·175	2·198	2·242	2·287	2·332	2·337	2·422	2·467	2·512	2·556	2·601	2·646
5	2·691	2·713	2·736	2·781	2·826	2·870	2·915	2·960	3·005	3·050	3·095	3·139	3·184
6	3·229	3·252	3·274	3·319	3·364	3·409	3·453	3·498	3·543	3·588	3·633	3·678	3·723
7	3·767	3·790	3·812	3·857	3·902	3·947	3·992	4·036	4·081	4·126	4·171	4·216	4·261
8	4·306	4·328	4·350	4·395	4·440	4·485	4·530	4·575	4·620	4·664	4·709	4·754	4·799
9	4·844	4·867	4·889	4·933	4·978	5·023	5·068	5·113	5·158	5·203	5·247	5·292	5·337
10	5·382	5·405	5·427	5·472	5·516	5·561	5·606	5·651	5·696	5·741	5·786	5·830	5·875
11	5·920	5·943	5·965	6·010	6·055	6·100	6·144	6·189	6·234	6·279	6·324	6·369	6·413
12	6·458	6·481	6·503	6·548	6·593	6·638	6·683	6·727	6·772	6·817	6·862	6·907	6·952
13	6·997	7·019	7·041	7·086	7·131	7·176	7·221	7·226	7·310	7·355	7·400	7·445	7·490
14	7·535	7·557	7·580	7·624	7·669	7·714	7·759	7·804	7·849	7·894	7·938	7·983	8·028
15	8·073	8·095	8·118	8·163	8·207	8·252	8·297	8·342	8·387	8·432	8·477	8·521	8·566
16	8·611	8·634	8·656	8·701	8·746	8·791	8·835	8·880	8·925	8·970	9·015	9·060	9·104
17	9·149	9·172	9·914	9·239	9·284	9·329	9·374	9·418	9·463	9·508	9·553	9·598	9·643
18	9·687	9·710	9·732	9·777	9·822	9·867	9·912	9·957	10·001	10·046	10·091	10·136	10·181
19	10·226	10·248	10·271	10·315	10·360	10·405	10·450	10·495	10·540	10·584	10·629	10·674	10·719
20	10·764	—	—	—	—	—	—	—	—	—	—	—	—

Table XXI £sd/cub ft to £p/m³

Shillings per cubic foot	Pence												
	0	¼d	1	2	3	4	5	6	7	8	9	10	11
	£p equivalent per m³												
0	—	0·074	0·147	0·294	0·441	0·589	0·736	0·883	1·030	1·177	1·324	1·471	1·619
1	1·766	1·839	1·913	2·060	2·207	2·354	2·501	2·649	2·796	2·943	3·090	3·237	3·384
2	3·531	3·605	3·679	3·826	3·973	4·120	4·267	4·414	4·561	4·709	4·586	5·003	5·150
3	5·297	5·371	5·444	5·591	5·739	5·886	6·033	6·180	6·327	6·474	6·621	6·769	6·916
4	7·063	7·137	7·210	7·357	7·504	7·652	7·799	7·946	8·093	8·240	8·387	8·534	8·682
5	8·829	8·902	8·976	9·123	9·270	9·417	9·564	9·712	9·859	10·006	10·153	10·300	10·447
6	10·594	10·668	10·742	10·889	11·036	11·183	11·330	11·477	11·624	11·772	11·919	12·066	12·213
7	12·360	12·433	12·507	12·654	12·802	12·949	13·096	13·243	13·390	13·537	13·684	13·832	13·979
8	14·126	14·199	14·273	14·420	14·567	14·714	14·862	15·009	15·156	15·303	15·450	15·597	15·744
9	15·892	15·965	16·039	16·186	16·333	16·480	16·627	16·774	16·922	17·069	17·216	17·363	17·510
10	17·657	17·730	17·804	17·952	18·099	18·246	18·393	18·540	18·687	18·834	18·982	19·129	19·276
11	19·423	19·496	19·570	19·717	19·864	20·012	20·159	20·306	20·453	20·600	20·747	20·894	21·042
12	21·189	21·262	21·336	21·483	21·630	21·777	21·925	22·072	22·219	22·366	22·513	22·660	22·807
13	22·955	23·028	23·102	23·249	23·396	23·543	23·690	23·837	23·985	24·132	24·279	24·426	24·573
14	24·720	24·793	24·867	25·015	25·162	25·309	25·456	25·603	25·750	25·897	26·045	26·192	26·339
15	26·486	26·559	26·633	26·780	26·927	27·075	27·222	27·369	27·516	27·663	27·810	27·957	28·105
16	28·252	28·325	28·399	28·546	28·693	28·840	28·987	29·135	29·282	29·429	29·576	29·723	29·870
17	30·017	30·091	30·165	30·312	30·459	30·606	30·753	30·900	31·047	31·195	31·342	31·489	31·636
18	31·783	31·856	31·930	32·077	32·225	32·372	32·519	32·666	32·813	32·960	33·107	33·255	33·402
19	33·549	33·622	33·696	33·843	33·990	34·137	34·285	34·432	34·579	34·726	34·873	35·020	35·167
20	35·315	—	—	—	—	—	—	—	—	—	—	—	—

Table XXII £sd/lb to £p/kg

Shillings per lb	Pence												
	0	1	2	3	4	5	6	7	8	9	10	11	
	£p equivalent per kg												
0	—	0·011	0·022	0·022	0·033	0·044	0·055	0·066	0·077	0·088	0·088	0·099	
1	0·110	0·121	0·132	0·132	0·143	0·154	0·165	0·176	0·187	0·198	0·198	0·209	
2	0·221	0·232	0·243	0·243	0·254	0·265	0·276	0·287	0·298	0·309	0·309	0·320	
3	0·331	0·342	0·353	0·353	0·364	0·375	0·386	0·397	0·408	0·419	0·419	0·430	
4	0·441	0·452	0·463	0·463	0·474	0·485	0·496	0·507	0·518	0·529	0·529	0·540	
5	0·551	0·562	0·573	0·573	0·584	0·595	0·606	0·617	0·628	0·639	0·639	0·650	
6	0·662	0·673	0·684	0·684	0·695	0·706	0·717	0·728	0·739	0·750	0·750	0·761	
7	0·772	0·783	0·794	0·794	0·805	0·816	0·827	0·838	0·849	0·860	0·860	0·871	
8	0·882	0·893	0·904	0·904	0·915	0·926	0·937	0·948	0·959	0·970	0·970	0·981	
9	0·992	1·003	1·014	1·014	1·025	1·036	1·047	1·058	1·069	1·080	1·080	1·091	
10	1·103	1·114	1·125	1·125	1·136	1·147	1·158	1·169	1·180	1·191	1·191	1·202	
11	1·213	1·224	1·235	1·235	1·246	1·257	1·268	1·279	1·290	1·301	1·301	1 312	
12	1·323	1·334	1·345	1·345	1·356	1·367	1·378	1·389	1·400	1·411	1·411	1 422	
13	1·433	1·444	1·455	1·455	1·466	1·477	1·488	1·499	1·510	1·521	1·521	1 532	
14	1·544	1·555	1·566	1·566	1·577	1·588	1·599	1·610	1·621	1·632	1·632	1 643	
15	1·654	1·665	1·676	1·676	1·687	1·698	1·709	1·720	1·731	1·742	1·742	1·753	
16	1·764	1·775	1·786	1·786	1·797	1·808	1·819	1·830	1·841	1·852	1·852	1·863	
17	1·874	1·885	1·896	1·896	1·907	1·918	1·929	1·940	1·951	1·962	1·962	1·973	
18	1·985	1·996	2·007	2·007	2·018	2·029	2·040	2·051	2·062	2·073	2·073	2·084	
19	2·095	2·106	2·117	2·117	2·128	2·139	2·150	2·161	2·172	2·183	2·183	2·194	
20	2·205	—	—	—	—	—	—	—	—	—	—	—	

See also RICS Metric Conversion tables page 186 ref. 44

Additional tables in this handbook

Further conversion tables can be found as below:

HEATING AND AIR CONDITIONING

STRUCTURAL DESIGN

33 Index